Use of Gamma Radiation Techniques in Peaceful Applications

Edited by Basim A. Almayahi

Published in London, United Kingdom

IntechOpen

Supporting open minds since 2005

Use of Gamma Radiation Techniques in Peaceful Applications
http://dx.doi.org/10.5772/intechopen.78481
Edited by Basim A. Almayahi

Contributors
Wanchai De-Eknamkul, Thongchai Koobkokkruad, Praderm Wanichananan, Chalermpol Kirdmanee, Bertil Rolf Ragnar Persson, Hai Lin, Seyed Mahyar Mirmajlessi, Hossein Ahari Mostafavi, Evelin Loit, Neda Najdabbasi, Marika Mänd, Christopher Busby, Hassane Erramli, Jaouad El Asri, Nicolas Thierry Fourches, Magdalena Zielinska, Gabriel Charles, Lubna Mehdi Alasadi, Hasna Albander, Vladimir Krasilnikov, Nikolay Kamyshanchenko, Alexander Parkhomenko, Victor Robuk, Boris Loncar, Luka Rubinjoni, Katarina Karadžić, Kinga Polaczek-Grelik, Aneta Kawa-Iwanicka, Marek Rygielski, Łukasz Michalecki, Alexander Barzilov, Monia Kazemeini, John Vargas, Woosoon Yim, Elizabeth Carvalho Leite Cardoso, Duclerc Fernandes Parra, Sandra Scagliusi, Ademar Benevolo Lugão

First published in London, United Kingdom, 2019 by IntechOpen
IntechOpen is the global imprint of INTECHOPEN LIMITED, registered in England and Wales, registration number: 11086078, The Shard, 25th floor, 32 London Bridge Street
London, SE19SG – United Kingdom
Printed in Croatia

British Library Cataloguing-in-Publication Data
A catalogue record for this book is available from the British Library

Additional hard and PDF copies can be obtained from orders@intechopen.com

Use of Gamma Radiation Techniques in Peaceful Applications
Edited by Basim A. Almayahi
p. cm.
Print ISBN 978-1-83962-259-5
Online ISBN 978-1-83962-260-1
eBook (PDF) ISBN 978-1-83962-261-8

For EU product safety concerns:
IN TECH d.o.o., Prolaz Marije Krucifikse Kozulić 3, 51000 Rijeka, Croatia, info@intechopen.com or visit our website at intechopen.com.

Meet the editor

Basim A. Almayahi, PhD, is a full professor in the Department of Environment, Faculty of Science, University of Kufa. His main areas of interest are applied nuclear physics, radon, alpha, and gamma spectrometry, environmental radioactivity and protection, radionuclides, biomarkers, renewable energy, and synthesis and applications of nanoparticles on bacterial pathogens. For many years he has been working on the development of radon, alpha, and gamma detection dosimeters. He took up the role of Head of the Department of Environment and Director of Scientific Affairs and Cultural Relations. He was the mentor of two masters theses. He is the editor and author of two books and a reviewer in 100 journals. He is editorial board member of 17 boards, international Scientific Program Committee of 44 members, and has membership of different institutions. He has published 75 scientific papers in domestic and international journals and participated in a large number of scientific congresses and conferences.

Contents

Preface

This book brings new research knowledge insights to the properties and behavior of gamma-rays. It encompasses a wide range of options, including gamma applications in nuclear physics, medical physics, industrial processes, environmental sciences, radiation biology, radiation chemistry, radiotherapy, agriculture and forestry, sterilization, and the food industry, as well as a review of both advantages and problems that are present in these applications. The book is primarily intended for scientific workers who have contacts with gamma-rays, such as staff working in nuclear power plants, manufacturing industries, civil engineers, medical equipment manufacturers, oncologists, radiation therapists, dental professionals, universities, and the military, as well as those who intend to enter the world of applications and problems of gamma-rays. Because of the global importance of gamma-rays, the content of this book will be interesting to a wider audience.

Dr. Basim A. Almayahi
Professor,
University of Kufa,
Najaf, Iraq

Section 1

Natural Radioactivity

Chapter 1

Gamma Background Radiations and Measurements with Applications

Lubna A. Al-Asadi

Abstract

This chapter deals with gamma background radiation that is exposed to two major natural sources: normal sources that are earthly gamma rays and astronomical rays. Earthbound gamma rays from radionuclide elements such as thorium, potassium, and uranium. Also, in building and enhancing materials such as medical plants, building purposes, some vegetables and fruits commonly used in markets and soil. The measurements of gamma background radiation differ according to the purpose. So, there are many instruments used such as HPGe detector, NaI (TI) detector, CR-39, a dosimeter, and SSNDT detector. This research will explain in detail the studies and the mechanism for each detector and how it works and the application (practical) studies for each one. The purpose of calculating gamma background radiation is to be in the safe side for human being according to the international union specified with a standard limitation of world average value or with a specific value ranged (standard values) according to the case study in medical, food, or building materials.

Keywords: terrestrial gamma rays, natural gamma ray, background radiation, detectors, annual effective dose, absorbed dose

1. Introduction

How do γ-rays compare with other types of radiation? "Radiation" in common language describes "energy packages" that travel on straight paths. "Electromagnetic radiation" is characterized by variations of electric and magnetic fields in space and time. Another type of "radiation" is "cosmic rays," very energetic particles discovered early in the twentieth century in the upper atmosphere of the Earth and known to pervade interstellar space. These particles are called "cosmic radiation" because with their high energies they propagate at the speed of light and in certain aspects behave like photons of similar energies [1].

Gamma radiation represents the most energetic part of the electromagnetic spectrum (**Figure 1**). Therefore, it is natural that it provides information about the liveliest procedures and wonders in the universe [2].

In general, there are three noteworthy sources: earthly radiation, grandiose radiation, and interior radiation due to the admission of regular radionuclides through inward breath (for the most part radon) and ingestion. Additionally, the most three radionuclide components are thorium (^{232}Th), potassium (^{40}K), and

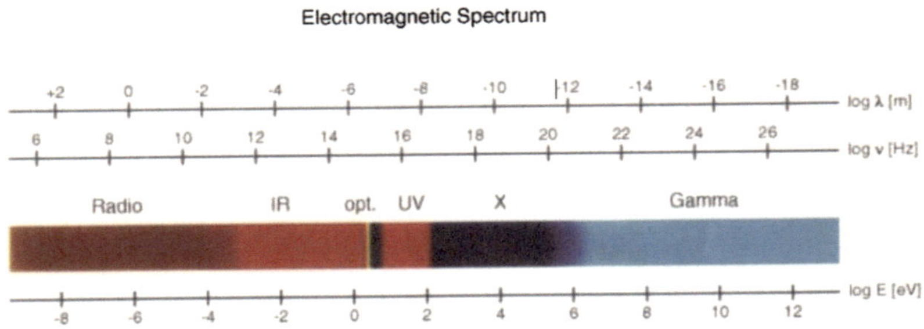

Figure 1.
The electromagnetic spectrum from radio to γ-ray energies. The electromagnetic radiation can be characterized by its photon energy (measured in eV), by its frequency (measured in Hz) or by its wavelength (measured in m) [2].

uranium (^{238}U). Natural radiation sources consolidate the ground, rocks, air, building materials, and drinking water supplies. Big amount of radiation rises up out of sun situated, galactic and extragalactic transmissions and contains decidedly charged particles, muons, neutrons, and gamma radiation.

Inward radiation is in our body because of what we eat and drink and the air we unwind. Common radiation in a general sense relies upon topographical and geological conditions. In this way, the portion ratios of both enormous and earthbound gamma radiation will be found to vary depending upon where the estimations are made [3–5].

As the research shows the relationship between the gamma-ray and living beings and all gamma radiation in the earth, it will explain in detail the mechanisms (according to the user of the detector) to calculate gamma background radiation and the measurements depending on the international standard values.

2. Methods of measuring gamma rays

This section explains the most common detectors used in the measurement of gamma background radiation and equipped with the international practical recent researches, to let the researchers who have worked in this field have the basic knowledge on these devices by knowing the mechanism of how they work and how to calculate gamma background radiation.

2.1 Radiation dosimetry

A radiation dosimeter is a gadget, instrument, or framework that measures or assesses specifically or in a roundabout way, the amounts presentation, KERMA, proportional portion, or when they again their time auxiliaries (rates), or related amounts of ionizing radiation. A dosimeter close by its scrutinize is insinuated as a dosimetry structure.

Estimation of a dosimetric aggregate is the course toward finding the respect of the total likely utilizing dosimetry frameworks. The delayed consequence of estimation is the estimation of a dosimetric aggregate imparted as the aftereffect of a numerical regard and a reasonable unit. To fill in as a radiation dosimeter, the dosimeter must have no short of what one physical property that is a component of the intentional dosimetric sum and that can be used for radiation dosimetry with the real arrangement. With the ultimate objective to be significant, radiation dosimeters must show a couple of appealing characteristics.

For instance, in radiotherapy, correct information of both the consumed portion of water at a predefined point and its spatial appropriation are of significance, also the likelihood of determining the portion to an organ of enthusiasm for the patient [6]. In radiation, preparing dosimetry is utilized to evaluate the vitality kept in a material or consumed by a human from radiation sources.

Diverse dosimetry frameworks are utilized for various purposes in industry and research light offices, which have distinctive prerequisites for portion conclusions. Radiation wellbeing norms and issues including the radiation safety of people against radiation presentation have their very own dosimetry metrology.

Radiation dosimeters and dosimetry frameworks come in numerous shapes and structures and they depend on various physical impacts for capacity and readout of the dosimetric flag. The below table demonstrates the most regularly utilized of dosimetric frameworks, and the qualities and shortcomings of these four dosimeters are outlined. The four most commonly used radiation dosimeters are [6]:

- Ionization chambers
- Radiographic films
- TLDs
- Diodes

Dosimetry type	Advantage	Disadvantage
Ionization chambers	1. Accurate and precise. 2. Recommended for beam calibration. 3. Necessary corrections well understood. 4. Instant readout.	1. Connecting cables required. 2. High voltage supply required. 3. Many corrections required for high energy beam dosimetry.
Radiographic films	1. (2-D) spatial goals. 2. Thin: does not irritate the pillar.	1. Darkroom and preparing offices required. 2. Preparing hard to control variety among movies and clumps. 3. Needs legitimate alignment against ionization chamber estimations. 4. Energy reliance issues can't be utilized for pillar alignment.
TLDs	1. Small in size: point portion estimations conceivable. 2. Numerous TLDs can be uncovered in a solitary presentation. 3. Accessible in different structures such as some similar tissue identical. 4. Not costly.	1. Flag deleted amid readout. 2. Simple to lose perusing. 3. No moment readout. 4. Precise outcomes require care readout and alignment tedious. 5. Not suggested for shaft calibration.
Diodes	1. Little size. 2. High affectability. 3. Moment readout. 4. No outer inclination voltage basic instrumentation.	1. Requires partner joins Variability of alteration with temperature. 2. Change in affectability with the aggregated portion. 3. Uncommon consideration expected to guarantee consistency of reaction. 4. Can't be utilized for pillar adjustment.

Radiation dosimetry is a part of physical science investigating diverse strategies for the quantitative assurance of vitality, which is stored in a given material by ionizing radiation, either through an immediate or aberrant presentation. Dosimetry manages conclusions and computations of amounts (portion) that depict the vitality ingested in a material and to some degree its rate of the statement (portion rate). Dosimetry conclusions that are performed by presenting a dosimeter to a radiation source help in assessing the radiation-prompted impacts, physical, substance, and additionally organic, on a lighted material [7]. **Figure 2** shows a Science Photo Library/Getty Images [8].

Know the most recent research to calculate gamma background radiation is measurements of gamma background radiation in Lorestan, Iran.

The researcher used a G.M. detector (RDS-110) "Inspector Alert model RAP RS1, S.E. international, Inc, USA". The outcome demonstrates the normal yearly powerful portion for gamma base radiation in Lorestan area has been 0.72 mSv, with the scope of 0.3–0.6 mSv which was more than the worldwide esteem (0.48 mSv). A poor coefficient relationship in-between was noticed elevating and retaining portion rates [9].

Another research study used the same device and the same properties. Indoor and outdoor absorbed dose rates were measured in the select region of AL-Qizweenia Najaf in Iraq [10]. Many studies in the gamma field of background radiation were conducted in different cities of Iraq [11–13]. All values of these studies compared with limitation of world average values.

2.2 Radiographic films (NaI scintillation detector)

There was an expansive glitter phenomenon in 1948 which was accounted for. It caused by including a follow amount of thallium (Tl) into a precious stone of sodium iodide (NaI). This identifier, in which sodium iodide (NaI) was utilized for radiation estimations, was based on the fact that recently it had been produced by HORIBA. They had the advancement of the NaI(Tl) plate for gamma cameras began during the 1970s.

In the first place, gems with breadths of 1–3 inches were utilized as indicators for atomic material science tests, natural radiation estimations at atomic power plants, or radioimmunoassay. Furthermore, mosaic-type precious stones were likewise created to relate to vast estimated gems notwithstanding the gems with a distance across of around 5 inches. After the expansive estimated gamma camera was created, bigger and bigger precious stones have been requested each year, and researchers have been making a decent attempt to get ever bigger measured NaI(Tl) gems as of late, and **Figure 3** shows atypical scintillation detector [14].

An elective strategy for assurance of exercises of regular, techno genic, and aftermath radionuclide in natural examples was proposed. The strategy utilized a broadly accessible shine spectrometer and depended on the disintegration of tests' γ-spectra into ghastly segments of discrete radionuclide bunches with the assistance of standard sources. The technique was tested on water, soil, and coal which could be effectively utilized in field (endeavor) conditions (without fluid nitrogen for the indicator cooling) [15].

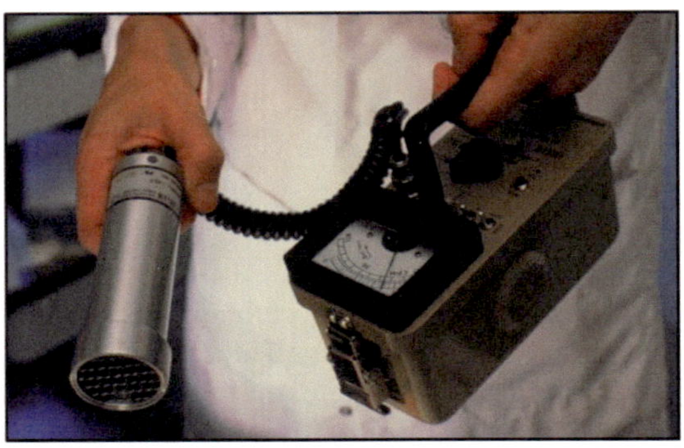

Figure 2.
Science Photo Library/Getty Images [8].

Figure 3.
Atypical scintillation detector [14].

There were many local and international studies used NaI(Tl) detector to evaluate natural or terrestrial of radionuclide or gamma background radiation with a difference of accuracy and efficiency of the device and sometimes equipped with a software program of the trace elements.

In 2013, a nearby report discusses uranium (^{238}U), thorium (^{232}Th), and potassium (^{40}K) with the explicit movement (10) in a few types of vegetables that is accessible at the market in Iraq. Tests had been estimated, and inner risk file, radium identical, and the yearly compelling portion of (^{40}K) in all examples were resolved. The gamma spectrometry methodology with a NaI(Tl) pointer was used for radiometric estimations, outcomes were contrasted and worldwide prescribed qualities and were observed to be inside the global dimension [16].

An investigation was conducted on radionuclides (^{226}Ra, ^{232}Th, and ^{40}K) of natural radioactivity estimations and assessment of radiological hazards in the silt of Oguta Lake, South East Nigeria. NaI(Tl) indicator "(show: Bicron, Pre-intensifier model:2001, Amplifier model:2020, ADC model:8075, HVPS model:3105)" was utilized for the gamma-beam spectrometry estimations. The identifier has a goal of 8% at 0.665 MeV line of ^{137}Cs, which is equipped for recognizing the gamma-beam energies of the radionuclides of enthusiasm for this examination. The investigation could fill in as critical radiometric pattern information whereupon future epidemiological examinations and ecological observing activities could be based [17].

Another examination enduring gamma producers in biscuit samples expired in Iraq, estimated the common radioactivity couple to seemingly perpetual gamma producers in children roll by gamma spectroscopy and appraisal radiation risk records which are the radium comparable action, the delegate of gamma level file, the interior danger file, and yearly powerful portion in kids. The gamma spectrum from each sample was recorded using detector NaI(Tl), and the volume of the crystal is ("3 × 3"), a PC-based multichannel analyzer (4096 channel) and processed using the MAESTRO-32 software. The estimations of eexpress activity, radiation risk records, and a yearly viable portion in all examples in this investigation are discovered lower than the overall middle incentive for all gatherings. Along these lines, these qualities are observed to be protected [18].

Three types of research are in the same field (used NaI detector). First one is in Kütahya, Turkey. The examination of common radioactivity from ^{238}U, ^{232}Th, and ^{40}K in 357 soil tests gathered from territory of Kütahya was completed utilizing a NaI(Tl) gamma-beam spectroscopy. Explicit exercises of ^{238}U, ^{232}Th, and ^{40}K in the dirt examples were assessed. The locator was coupled to a full-featured 16 K channel fused multichannel analyzer "(Canberra DSA-1000)." It is joined with a PC for getting an examination and with reasonable programming "(Genie 2000)."

The identifier was covered in a 0.5 cm thickness lead shield, giving a disguise of the establishment gamma-bar radiation present in the exploration office. The NaI(Tl) gamma-bar spectrometer has high efficiency, and in a similar manner, it might be used for the customary radioactivity. The outcomes acquired in this examination were analyzed inside the cutoff points of values gotten in different urban areas of Turkey, those in different nations [19].

Second research estimated normal radioactivity in chosen tests of therapeutic plants in Iraq, where characteristic dimensions of radiation in some chosen therapeutic plants existing in the Iraqi stores were assessed to decide any action fixation, radium comparable, and inner risk file due to the radionuclide, of ^{238}U, ^{232}Th, and in addition ^{40}K, which happens normally. The movement fixation is recognized by gamma-beam spectroscopy and NaI(Tl). Estimations are done by embracing frameworks of gamma spectrometry from ORTEC, furnished with a high productivity sparkle indicator, a NaI(Tl) locator of (3″ × 3″) precious stone measurement, with goals 9.2% for ^{137}Cs (661.7 keV). A lead shield with a thickness of (10 cm) was put around the finder to decrease the foundation, with a 0.3 cm layer of copper to debilitate X-rays discharged by the lead shield. The spectra were seen disconnected utilizing the ORTEC Maestro-32 information obtaining and examination framework. Regular radionuclides and development of the radium similar to the remedial plant tests were far underneath the world for the ingestion of typically happening radionuclide, as given in UNSCEAR 2000 report [20].

Third research in some vegetables and fruits commonly used in Najaf Governorate, Iraq, determined the natural radioactivity levels. The points of the present work were to gauge the explicit action and yearly compelling portion because of the admission of vegetables and organic products gathered from the nearby market in Najaf governorate. Characteristic radioactivity was estimated in tests utilizing gamma beam spectrometer. Gamma-ray spectroscopy with scintillation detector NaI(TI) from ORTEC had an active area of "3 × 3" inches, the efficiency of 4.6% at the 662 KeV, and energy resolution 7.9%. The qualities found for explicit action and the yearly powerful portion in all examples in this investigation were lower than overall middle qualities for all gatherings as indicated by UNSCEAR (2000) and ICRP (1996) individually; subsequently, these qualities are observed to be sheltered [21].

Also, a study in Turkey for the three radionuclide elements (^{226}Ra, ^{232} Th, and ^{40}K) had been made in some granite samples. The action grouping of primordial radionuclides in rock tests was assessed by utilizing (3 × 3) NaI(Tl) indicator-based gamma-beam spectrometry. This gamma spectrometer has vitality goals 8% for 662 keV and the relative checking productivity about 20%. It was critical that the productivity adjustment of the framework ought to be made before estimation for the right outcome. Estimation of radioactivity for, 14 various types of stone examples have been breaking down for their regular radioactivity content. The end was: the ^{40}K action focus levels in stone examples in these investigations were lower than fixation estimations of various examinations on the planet. A few examples and other stone examples were not observed to be reasonable for utilizing in the human life regions due to every single radiological incentive in these examples are higher than CLV [22].

2.3 TLD detector (thermoluminescence detector)

Roentgen in 1928 was held onto as a unit to describe radiation introduction. Roentgen evaluates the number of electrons made in air, yet not the genuine damage following in a man.

In 1953, a unit known as a rad (100 ergs g^{-1}) was grasped to portray the proportion of essentialness spared in a material. Clearly, it was found that different sorts of radiation affected the tissue in the body all of a sudden. Another unit was made to address this, known as a rem. A rem is proportionate to a rad duplicated by a factor known as a "Quality Factor" (Q) which numerically portrays the relative trademark impact of the express kind of radiation. As these new units of measure were made and executed, new radiation materials and exposure instruments were delivered with the objective that contradictory exposures and doses could be accurately evaluated [23].

Although extremely accurate active radiation detectors are now available, TLDs are small, inexpensive, and if the correct material is chosen, tissue equivalent. They can be used to detect photons, beta particles, and slow neutrons, and with appropriate filters, they can be used to determine the shallow and deep dose. Their biggest advantage is long-term deplorability, possible due to a power source being unnecessary until readout. This allows time-efficient monitoring of typically uninhabited areas. In order to ensure accurate results from long deployments in diverse interior and exterior environments, various aspects of their performance must be examined. This work serves to improve the effectiveness of TLD systems by analyzing several factors which may affect the sensitivity and precision of TLD measurements, as well as determining a practical minimum detectable dose incorporating those factors.

TLDs must be individually calibrated, meaning that the amount of signal response to a known dose must be measured before use. The light response to doses generally between 0.1 mGy and 10 Gy, but varying by material, has a two linear relationship with dose. This makes calibration at only one dose necessary if staying within the linear range [24]. **Figure 4** below shows the thermoluminescence detector (model THERMO SCIENTIFIC 4500 TLD READER) in the laboratory at Ankara University, Institute of nuclear sciences [25].

Some examples of modern researches will be taken. Researches of detector will show it had much application with different branches in physical science. First one in network communication in which the point is to seek after an ordinary free quality review in Czech radiotherapy focuses and to help state supervision. The results appears to be that there are 34 radiotherapy focuses in the Czech Republic. They experience the essential method of the TLD review routinely at regular intervals.

On the off chance that an inside demonstrates a deviation outside the acknowledgment level, it is evaluated more frequently. Step by step, a large portion of the checked shafts conform to the acknowledgment level.

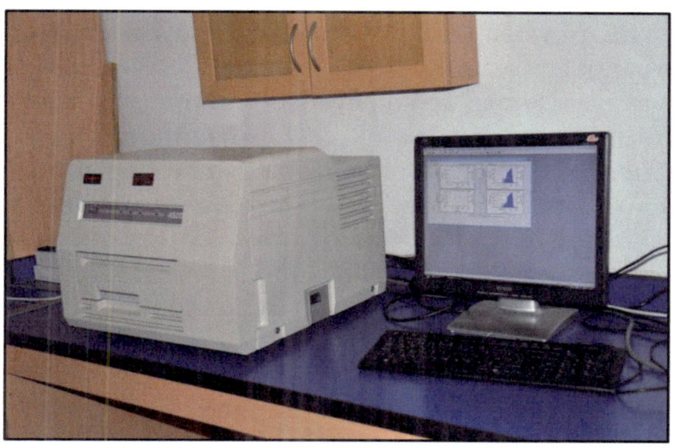

Figure 4.
THERMO SCIENTIFIC 4500 TLD READER [25].

Results were for the most part inside as far as possible for the estimations on-hub, though for off-hub focuses they fell past the limit all the more oftentimes, particularly for set-ups with in-homogeneities, diagonal occurrence, and wedges.

The outcomes demonstrate the significance of the national TLD quality affirmation arrange. It has added to the enhancement of clinical dosimetry in the Czech Republic. What's more, it causes administrative specialist to screen successfully and consistently radiotherapy focuses [26].

An exploration in therapeutic material science including patients and apparition dosimetry in the two cases, thermoluminescence dosimetry (TLD) is the most suitable strategy for estimating the assimilated portion. In this chapter, thermoluminescence wonder and in addition the utilization of TLD in radiodiagnosis and radiotherapy for in vivo or in apparition estimations is talked about. A few aftereffects of estimations made in radiotherapy and radio diagnosis utilizing natively constructed LiF: Mg, Cu, P + PTFE TLD are exhibited [27].

In nuclear atomic material science, an examination in the robust assurance of successful nuclear numbers for electron together with "TLD-100 and TLD-100H thermoluminescent dosimeters," lithium fluoride thermoluminescent dosimeters (TLD) are frequently not completed for clinical dosimetry. The little physical enormity of TLDs makes them applicable, for instance, little field estimation, in vivo dosimetry, and estimation of out-of-field bits of essential structures. The most exhaustively used TLD can't avoid being "TLD-100 (LiF: Mg, Ti)," and for applications requiring higher affectability to low-parcels, TLD-100H (LiF: Mg, Cu, P) is consistently used. The radiological properties of these TLDs are along these lines of significant interest. All of a sudden, in this examination, convincing atomic numbers for radiative, collisional, and mean electron association frames is resolved for "TLD-100 and TLD-100H" dosimeters over the essential expansion of 1 keV–100 MeV. This is applied by using a solid, essentialness subordinate system for calculation instead of normal power-law approximations. The effect of dopant obsessions and unwanted impacts is further analyzed. The two TLDs show relative convincing atomic numbers, generally ranging from 5.77 to 6.51. Differentiations rising up out of the particular dopants are most enunciated in low-imperativeness radiative effects. The TLDs have atomic numbers around (1.48–2.06) events than that of water [28].

Measurement of computed tomography dose profile with pitch variation uses Gafchromic XR-QA2 and thermoluminescence dosimeter (TLD). This examination was meant to point the examples of portion profile on a grown-up and pediatric head filter. They thought about estimation depended on portion profile along the z-hub turn at peripheries and focus apparition with an assortment of the pitch, that is, 0.75, 1, and 1.5 for grown-up and pediatric head convention, keeping whatever is left of the sweep parameters steady.

Estimations were performed on homogeneous, round, and void PMMA ghost with widths of 16 and 10 cm utilizing XR-QA2 Gafchromic film and TLD as dosimeters. The estimation result showed a diminishing in the part about half and 47% for grown-up and pediatric head check with the advancement of pitch. Part profile for adult and pediatric head channel traditions has configuration twist with the most outrageous bit in the inside and inclination of symmetry near the edges, with difference in the dimension length along z-center point bearing incomprehension to the estimation position in the nebulous vision [29].

A research related with nuclear physics was unfolding neutron spectra from the simulated response of thermoluminescence dosimeters, and neutron spectrometry utilizing a solitary circle containing dosimeters has been produced as of late, as a powerful swap for Bonner circle spectrometry. The purpose of the examination is spreading out the neutron essentialness spectra using the GRNN fake neural framework, from the response of thermoluminescence dosimeters, TLDs, arranged inside a polyethylene circle. The spectrometer was reproduced using MCNP5.

TLD-600 and TLD-700 dosimeters were replicated in different positions all over. By then, the GRNN was used for neutron spectrum gauge, using the TLDs' readings. Examination of spectra foreseen by the framework with real spectra shows that the single-circle dosimeter is an incredible instrument in spreading out neutron spectra [30].

Solid state physic took apart with thermoluminescence detector, preliminary studies of thermoluminescence dosimeter "(TLD) CaSO4: Dy Synthesis. thermoluminescence dosimeter (TLD) CaSO4: Dy" was orchestrated by coprecipitation.

The TLD was seen after radiation introduction to Strontium-90. The thermoluminescence drive was scrutinized using a TLD Reader Harshaw 3500. The thermoluminescent response obtained was 59.29 nC. By then, refortifying was driven with the temperature vacillated at 700, 800, and 900°C. The thermoluminescent control got at temperatures of 700–900°C was 66.12, 169.45, and 552.37 nC independently. The affectability of the TLD extended in light of the retoughening temperature rise. Despite viewing the thermoluminescence properties, a relationship was made between the TLD got from this attempt distinctive things with a current TLD in the market. At long last, likewise, the gleam bend attributes of the TLD were watched [31].

The U.S. Naval force utilizes the "Harshaw 8840/8841 dosimetric (DT-702/PD)" framework, which utilizes LiF: Mg, Cu, P thermoluminescent dosimeters (TLDs), created and delivered by Thermo Fisher Scientific (TFS). The dosimeter comprises four LiF: Mg, Cu, P components, mounted in Teflon on an aluminum card and put in a possessor made from plastic. The possessor consists of an interesting channel for each chip made of copper.

The Naval Dosimetry Center (NDC) has created and tried another nondamaging method, which empowers the check and the assessment of installed channels in the possessors. Testing depends on weakening estimations of low-vitality radiation transmitted through each channel in an agent test gathering of possessors to confirm that right channel type and thickness are available. The deliberate reaction proportions are then contrasted and the normal reaction proportions. Moreover, every component's deliberate reaction is contrasted with the mean reaction of the gathering. The test was organized and endeavored to recognize basic singularities, for instance, missing copper or tin channels, twofold copper or twofold tin channels, or diverse discords that may influence TLD response extents. In the midst of the execution of the made strategy, testing revealed a possessor with a twofold copper channel. To finish the assessment, the effect of the qualifications on limit testing was destitute down. The examination uncovered disappointments in capacity testing orders III and IV when these dosimeters were edified to high-significance betas [32].

2.4 Diode detector

The diode indicator is the least difficult and most essential type of abundancy tweak, AM locator, and it distinguishes the envelope of the AM flag as shown in **Figure 5**. The AM diode locator could be worked only on a diode with couple of different segments, and therefore it is a minimal effort circuit hinder inside a general recipient. Because of its expense and comfort, the AM diode envelope identifier has been broadly utilized for a long time in transistor compact radios.

2.4.1 AM diode detection process

In changing the RF hail, the AM diode discoverer gives a yield proportionate to the envelope of one part of the banner, and this implies an envelope locator. In

Figure 5.
Circuit of an envelope locator as utilized in an AM radio collector [33].

context of the errand of the diode marker, it may every so often be implied as an envelope discoverer. The moving toward abundancy changed RF hail includes a waveform of both positive and negative going voltages as shown in **Figure 6**. Any stable transducer would not respond to that.

The diode envelope discoverer changes the waveform leaving only the positive or negative segment of the waveform. The high repeat part of this is filtered through, normally using a capacitor that outlines the low pass channel and suitable "fills" in the high repeat segments, leaving a waveform to which a transducer like two or three earphones or an enhancer could respond to and convert into sound waves.

2.4.2 Focal points and drawbacks of a diode envelope locator

The AM diode envelope discoverer had been viably used to quite a while. The most envelope pointer purposes of intrigue are: ease that means the diode indicator just requires the utilization of a couple of ease parts. This means it is perfect for use in transistor (and valve/vacuum tube) radios utilizing discrete segments, effortlessness means utilizing not a lot of parts, and the diode AM identifier was definitely not hard to complete. It was dependable and did not need any setup, while an envelope identifier inconveniences are:

2.4.2.1 Contortion

As the diode indicator is nondirect, it presents mutilation onto the identified sound flag.

2.4.2.2 Particular blurring

One of the issues a significant part of the time experienced on the short and medium wavebands where the AM transmissions are found is that of express darkening. The diode envelope identifier cannot battle the impacts of this on how some different locators are capable, and therefore, contortion happens when specific blurring happens.

2.4.2.3 Affectability

The diode locator is not as precarious as some remarkable sorts. On the off chance that silicon diodes are utilized, these have a turn-on voltage of around 0.6

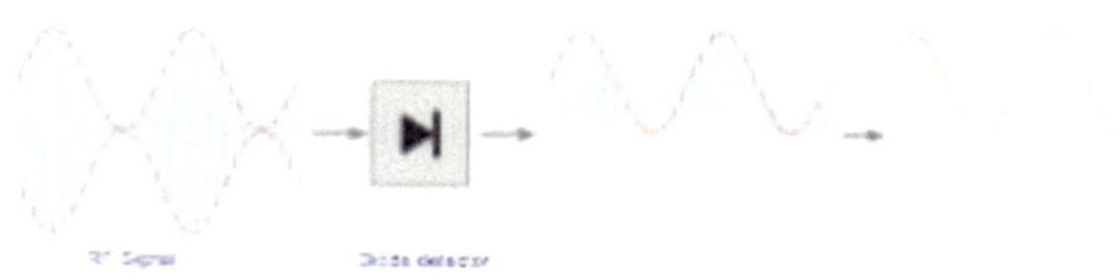

Figure 6.
AM diode envelop detection process [33].

volts, and therefore, germanium or Schottky diodes are utilized which have a lower turn-on voltage of around 0.2–0.3 volts. Without a doubt, even with the utilization of the Schottky diode, the diode envelope identifier still experiences a pore estimation of affectability.

The AM diode envelope indicator has been accessible for a long time. Despite the fact that abundancy balance is utilized less nowadays, and different types of AM finder can be effortlessly consolidated into coordinated circuits, the basic diode identifier still has a few points of interest [33].

After a brief explanation about diode detector, some recent applications of diode detector will be discussed.

In remedial material science, in light of the way is that the skin diode is made on a thin epitaxial layer and bundled utilizing the "drop-in" advancement. It was portrayed comparably as rate hugeness isolate, segment linearity, and section rate reliance and benchmarked against the Attix ionization chamber. The reaction of the skin diode in the enhancement zone of the rate importance divide touch of a 6 MV clinical photon bar was explored. The radiation hardness of the skin diode up to an amassed bit of 80 kGy using photons from a Co-60 gamma source was surveyed [34].

Another application in the same field is evaluation of the dosimetric properties of a diode detector for small field proton radiosurgery; the little fields and sharp angles regularly experienced in proton radiosurgery require high spatial goal dosimetric estimations, particularly underneath 1–2 cm measurements. The radiochromic film gives high goals; however, it requires postprocessing and unique taking care of. Promising choices are diode identifiers with little delicate volumes (SV) that are able to do high goals and continuous portion obtaining.

In this investigation, the analyst assessed the PTW PR60020 proton dosimetry diode utilizing radiation fields and shaft energies pertinent to radiosurgery applications [35].

Schottky diode is well know; therefore, we have given examples. The first one is calculating the diode junction resistance variations with RF power of a series Schottky diode detector. Based on the Ritz-Galerkin method, this research provides a simple formula that can be used to calculate the differential input impedance and frequency response of a diode detector. Calculated results are presented for several circuit configurations that are confirmed by ADS [36].

The second example is high-resolution Schottky CdTe diode detector, with a Schottky intersection created on the Te face of a top-notch CdTe semiconductor by dissipating indium, and they have possessed the capacity to accomplish a CdTe diode including high vitality goals. The identifiers demonstrate the best execution when they utilize a moderately thin locator of 1 mm. The high vitality goals of the CdTe diode are exceptionally alluring for hard X-beam and gamma-beam recognition. Particularly, an extensive CdTe diode with measurements bigger than 20 × 20 mm can possibly supplant shine locators because of its high ceasing force and vitality goals of 3 keV at 100 keV. Numerous ideas dependent on high goal CdTe diodes are currently being examined, and model indicators are being created [37].

3. HPGe detector

Semiconductor identifiers are basically strong state analogs of gas-filled ionization chambers. Since the strong indicator materials utilized in semiconductor identifiers are 2000 to multiple times more than gases, they have the much better halting force and are considerably more productive finders for X-rays and γ-rays.

Semiconductor indicators regularly are poor electrical transmitters; when they were ionized by an ionizing radiation event, the electrical charge delivered could be gathered by an outer connected voltage, for what its worth with gas-filled locators. This guideline could not be connected utilizing a leading material for the locator (e.g., a square of metal) on the grounds that such a material would direct a lot of current even without ionizing occasions. Protectors (e.g., glass) are not appropriate identifier materials either on the grounds that they don't lead even within the sight of ionizing radiation. Subsequently, just semiconductor materials can work as "strong ionization chambers."

The most by and large used semiconductor locator materials are silicon (Si) and germanium (Ge). Even more starting late, cadmium telluride (CdTe) or cadmium zinc telluride (CZT) has been utilized as the major material in insignificant atomic medicine checking and imaging gadgets. One ionization is made per 3–5 eV of radiation importance consumed. By examination, this propelling power for gases (air) is around 34 eV for each ionization. Subsequently, a semiconductor locator not exclusively is more fruitful shield of radiation, in any case, passes on an electrical standard that is around various events more prominent (per unit of radiation centrality ingested) than a gas-filled pioneer. The flag is sufficiently enormous to allow recognizing verification and checking of individual radiation occasions. Also, the proportion of the electrical banner is relating to the proportion of radiation imperativeness acclimatized. Thusly, semiconductor discoverers can be used for essentialness explicit radiation counting [38].

The most advantage of HPGe high purity germanium detectors (HPGe) is the best energy resolution among all detector types. In principle, they work like reverse biased diodes; energy deposition by nuclear radiation causes the flow of a current, which is processed by front-end electronics [39]. **Figure 7** shows a geometrical dimension of the investigated detector (mm).

In 2018, many international studies are about gamma background radiation and using HPGe detector. One of these studies is treating illnesses for many years all over the world. The aim of this study is to determine the radioactivity levels in some anti-carcinogenic medicinal plants that are often used to treat illnesses in Turkey.

The analysis of ^{226}Ra, ^{232}Th, ^{40}K, and ^{137}Cs activity concentration of medicinal plants was performed using a high-resolution gamma-ray spectrometer with HpGe

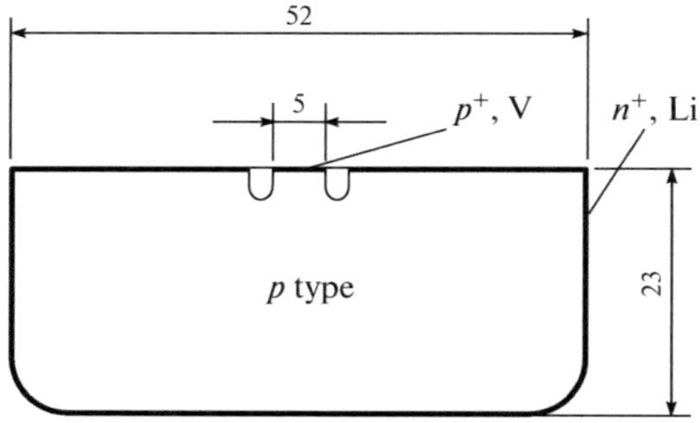

Figure 7.
A geometrical dimension of the investigated detector (mm) [39].

detector. Total committed effective dose value due to ingestion was determined as 55.04 μSv y^{-1} for these medicinal plants and has no risk to public health [40].

A study to analyze natural radioactivity level contents in Nigeria and China and a typical radioactive substance in tiles manufactured in Nigeria and tiles imported from China were evaluated using gamma-shaft spectroscopy. High purity germanium identifier was used to check the combinations of a couple of radioisotopes present in 17 trial of various tiles from Nigeria and China. The mean estimation of annual viable gamma dosages and the lifetime dangers to procure in this examination is not as much as that of the worldwide reference estimation of 370 Bq/kg for the two sorts of tiles [41].

Another place in Nigeria studied building material purpose, which evaluated the activity concentration of natural radionuclides (^{226}Ra, ^{232}Th, and ^{40}K) for fifteen (15) different brands of tile samples, is used for building purposes in Nigeria. The tile samples were analyzed using high purity germanium gamma detector. The deliberate centralizations of these radioactive materials were related with different past outcomes got from comparable tile materials utilized in different nations and observed to be in great concurrence with the global standard, and be that as it may, the tiles are suggested for adornment purposes in Nigeria [42].

A research in Amman calculated hazard indices and annual effective dose due to terrestrial radioactivity in the urban areas of south Amman. The extricated qualities are, by and large, tantamount to the relating ones acquired from different locales in Jordan and different nations and they all fall inside the normal overall reaches. Consequently, the foundation in these regions is actually equivalent to the esteem run of the mill of the upper piece of the Earth's outside layer. Hence, all the samples investigated can be considered as safe materials for use in building constructions [43].

The radioactivity groupings of ^{226}Ra, ^{232}Th, and ^{40}K in 24 tests of normal and made building materials ordinarily utilized in Bangladesh were estimated utilizing HPGe gamma-beam spectrometer. Outcomes were contrasted and the world normal and furthermore with the detailed information. The radium equal action, the assimilated portion rate, yearly successful portion, outer and inside peril records, gamma file, alpha file, yearly gonadal portion equal, and overabundance lifetime malignancy chance were additionally assessed to assess the potential radiation dangers related with these building materials.

All models under investigation were seen to be inside the recommended prosperity limit and don't speak to any essential radiation risks. This examination can be used as a sort of viewpoint for more expansive examinations of a comparable subject in future [44].

The radiological risk from building stone interfaces in Jordanian houses was determined depending on gamma-ray spectrometric techniques. Building stone samples collected from seven types mostly used in Jordanian houses have been analyzed for the naturally occurring radioactive radionuclides. Moreover, different radiological hazardous parameters (absorbed dose, annual effective dose equivalent, AGDE, ELCR, and AUI) were calculated. The results were lower than those of published world average values. Also, the obtained values were comparable with the presented data of other building materials used in Jordan [45].

Two researches in 2017, first one is Evaluation of Natural Radioactivity and its Radiation Hazards in Some Building and Decorative Materials in Iraq, 29 examples of various kinds of building materials, for example, blocks, bond, earthenware, rock, marble, paint, mortar, sand, and soil were examined by a gamma spectrometer dependent on HPGe locator. The outcomes demonstrated that all the building materials in Iraq are sheltered with the exception of the materials that utilized as improving materials that must be directed [46].

The second one is in north focal of Nigeria; measurement of the radiation portion dispersion is critical in surveying the wellbeing hazard a populace and fill in as a kind of perspective in archiving changes to ecological radioactivity in soil because of man-made exercises. The mean assimilated portion rate, yearly successful portion, and gamma radiation record assessed were 66.2 $nGyh^{-1}$, 81.2 μSvy^{-1}, and 1.05 individually which are higher than as far as possible for ordinary foundation radiation. In this manner, it is argued that individuals living in these areas might be presented to higher radiation [47].

There are numerous examinations in Turkey about radioactivity levels and malignancy hazard. Turkey, particularly its northern piece, was one of the nations which were defiled by the Chernobyl mishap. In the Northeastern area of Turkey, there was a city named Rize, which was intensely affected by the Chernobyl atomic mishap.

The action convergences of characteristic (^{226}Ra, ^{232}Th, and ^{40}K) and artificial (^{137}Cs) were estimations in soil tests gathered from 132 distinct focuses in Rize territory of Turkey utilizing gamma spectrometry with HPGe identifier. The open-air assimilated portion rates (D) because of earthbound gamma rays for soil have been determined due to agrarian zone and lived in the encompassing. It is essential to decide foundation radiation level with the end goal to assess the wellbeing dangers.

Yearly viable gamma dosages and the presence dangers of disease were higher than the world's normal. Besides contrast with the World's normal, the existence time danger of malignancy is multiplied for a large portion of the territories [48].

In Egypt, south Sinai granite is generally utilized in the development of homes as a building material. It contains the common radionuclides, ^{238}U and ^{232}Th, and their descendants together with ^{40}K. This guarantees the significance of the evaluation of radiation levels and the related radiological dangers to which the populace may be uncovered. Gamma shaft spectra of standard radioactivity from the ^{238}U and ^{232}Th game-plan and from ^{40}K of eight (tending to 40 gathered models) shake tests collected from Saint Katherine district, South Sinai, Egypt, had been assessed utilizing a gamma-column spectrometer with HPGe identifier. From the exploratory and computational work on normal radioactivity of Egyptian shake tests, we can complete the going with:

First, the region from where they collected the granite samples, South Sinai Governorate, Egypt contains ^{238}U, ^{232}Th, and ^{40}K radionuclides with obsessions higher, comparable and lower than beyond what many would consider possible. Second, the radium tantamount activity is not actually beyond what many would consider possible. Thirdly, the danger lists, the dimension lists, and the actors use lists that are not exactly the world set criteria. Fourth, the Clark esteem is equivalent to around five which implies that district from where they took the rock tests isn't financial for uranium mining and extraction [49].

Southwestern of Nigeria (^{226}Ra, ^{232}Th, and ^{40}K) and artificial (^{137}Cs) were measured using (HPGe) detector (Canberra Industries Inc.). The outdoor absorbed dose rates in the air at about 1.0 m height were estimated from the activity concentrations and dose rate conversion factors for the radionuclides. The annual outdoor effective dose equivalent rates were also estimated for urban and rural areas of the state using the calculated absorbed dose rates in air. The results showed that area named (Igbeti and Eruwa) soils contain the highest level of natural radioactivity, while Egbeda soil contains the lowest level. The study showed that healthy burden caused by natural background radiation from soils on inhabitant area of study is generally low and carried insignificant radiation hazard except for two locations (Igbeti and Eruwa) [50].

A second study in Malaysia 2013, the radiation survey of the ambient environment was conducted using two gamma detectors, and the measurement results were used in the computation of the mean external radiation dose rate, mean-weighted dose rate, and annual effective dose, and also, the mean lifetime dose and lifetime cancer risk for each person living in the area with an average lifetime (70 years). Two strategies have been utilized to assess outer exposures in this examination. The first was a provoke estimation of outside gamma segment rates.

The observation was taken at 497 zones in the Kluang District at 1 m above the ground utilizing two NaI-based gamma locators. The second system depended upon surveyed action revolves around soil tests and the gamma divide (Dc) from the gatherings of ^{238}U, ^{232}Th, and ^{40}K. The advancement groupings of ^{226}Ra, ^{232}Th, and ^{40}K in soil were acquired by utilizing the HPGe gamma spectrometer. The advancement fixations decided for 57 soil tests. The rationale for affirmation of the gamma segment rates from the fixations was gotten a handle on from United Nations Scientific Committee on the Effects of Atomic Radiation (UNSCEAR) 2000 [51].

The fundamental goal of this examination is to decide the action centralizations of primordial radionuclides in soil tests gathered from various of studying locale areas in India such as Ramanagara, Tumkur, and Karnataka districts, using gamma bar spectrometry and besides the radiological dangers records identified with based on soil tests. The radiological hazard files of the common radioactivity have been determined and contrasted and the globally affirmed qualities. The convergence of these radionuclides with various sizes and depth of the dirt examples was contemplated. They were found to pursue [52]:

The ^{232}Th and ^{40}K action focus was seen to be marginally higher than the world normal values reported by UNSCEAR (2000). The normal ^{226}Ra fixation in soil tests of the contemplated zone was observed to be like the world normal.

The assessed normal action grouping of ^{226}Ra, ^{232}Th, and ^{40}K in soil tests of those contemplated zones cbserved to be higher than the Indian normal.

Results acquired had demonstrated radiological dangers, for example, gamma list, radium equal movement, external peril record, and indoor risk list are well inside the world normal esteem. At last, it is presumed that the radiation discharged from the radionuclides present in the dirt of the examination zone doesn't represent any radiological wellbeing risk to the general population of the zone.

The normal indoor and open-air ingested portion rate for the dirt examples of investigation zone were marginally higher than the world normal estimations of 75 nGy h^{-1} for indoor what's more, 59 nGy h^{-1} for open air.

Results show that ordinary indoor, outdoors amazing part, and a total yearly convincing bit in view of trademark radioactivity of soil tests is lower than the typical national and world endorsed estimation of 1 mSv y^{-1}.

Movement assembly of primordial radionuclides increases with an addition in grain gauge.

Information procured in this examination will fill in as a benchmark for looking over the radiation presentation of the inhabitants.

4. Conclusions

This section manages gamma foundation radiation which uncovered with two noteworthy normal sources; earthly gamma rays and astronomical rays.

Recent studies in building materials, medical plants, building purposes, some vegetables and fruits commonly used in markets, and different soil samples in various countries in the world. The measurements of gamma background radiation differ according to the purpose. Each type of detectors is explained which are: radiographic films (scintillation detector), TLD detector (thermoluminescence detector), diode detector, and HPGe detector. The mechanism for each detector is provided with recently applied researches for the past 10 years, focusing on used gamma background radiations measurements.

Acknowledgements

The author expresses a deep sense of gratitude to Kufa University, Faculty of Science, for providing the scientific support to do this work. The author is grateful to Dr. Basim Almayahi, University of Kufa (basimnajaf@yahoo.com), for assisting throughout conducting the present chapter.

Conflict of interest

The author declares that she has no conflicts of interest.

Author details

Lubna A. Al-Asadi
Faculty of Science, Najaf, Iraq

*Address all correspondence to: lubna.alasadi@uokufa.edu.iq

References

[1] Diehl R. Gamma-ray production and absorption processes. In: Schönfelder V, editor. The Universe in Gamma Rays. Berlin Heidelberg: Verlag, Springer; 2001. p. 9. DOI: 10.1007/978-3-662-04593-0.ch2

[2] Schoenfelder V, Diehl R. Gamma-ray production and absorption processes. In: Schönfelder V, editor. The Universe in Gamma Rays. Berlin Heidelberg: Verlag, Springer; 2001. p. 9. DOI: 10.1007/978-3-662-04593-0.ch1

[3] El Samad O, Baydoun R, Nsouli B, Darwish T. Determination of natural and artificial radioactivity in soil at North Lebanon province. Journal of Environmental Radioactivity. 2013;**125**:36-39. DOI: 10.1016/j.jenvrad.2013.02.010

[4] Nageswara Rao MV, Bhati SS, Rama Sheshu P, Reddy AR. Natural radioactivity in soil and radiation levels of Rajasthan. Radiation Protection Dosimetry. 1996;**63**:207-216. DOI: 10.1093/oxfordjournals.rpd.a031531

[5] UNSCEAR. Sources Effects and Risk of Ionizing Radiation. New York: United Nations Scientific Committee on the Effects of Atomic Radiation; 2000. Available from: http://www.unscear.org/docs/publications/2000/UNSCEAR_2000_Report_Vol.I.pdf [Accessed: 2000]

[6] Izewska J, Rajan G. Radiation dosimeters. In: Podgorsak EB, editor. Radiation Oncology Physics: A Handbook for Teachers and Students. Vienna, Austria: International Atomic Energy Association; 2005. pp. 97-98. DOI: 10.1016/j.ijrobp.2006.08.048

[7] Adlienė D, Adlytė R. Dosimetry principles, dose measurements, and radiation protection. In: Sun Y, editor. Applications of Ionizing Radiation in Materials Processing. Vol. 1. Institute of Nuclear Chemistry and Technology, Romania. 2017. p. 55. DOI: 10.1016/0014-3057(72)90048-1

[8] Science Photo Library/Getty Images. 2016. Available from: https://cosmosmagazine.com/technology/how-does-geiger-counter-work

[9] Gholami M, Mirzaei S, Jomehzadeh A. Gamma background radiation measurements in Lorestan province, Iran. Journal of Radiation Research. 2011;**9**(2):89-93

[10] Abojassim AA et al. Survey of absorbed dose rates in the air of buildings agriculture and sciences at the university of Kufa at Al-Najaf Governorate, Iraq. Journal of Chemical and Pharmaceutical Research. 2016;**8**(4):1388-1392. DOI: 10.1093/rpd/ncq579

[11] Khader RB. Measure the background radiation in some parts of Nineveh. Rafidain Journal of Science. 2010;**21**(2):92-104

[12] Al-Mayahi B. Exposure rate measurements of the natural background radiation in the colleges of science & agriculture-Kufa university. Journal of Babylon University. 2008;**15**(3):1-4

[13] Makki NF et al. Natural radioactivity measurements in different regions in Najaf city, Iraq. International Journal of Computer Trends & Technology. 2014;**9**(2):286-289. DOI: 10.14445/22312803/IJCTT-V9P154

[14] Imagawa K et al. NaI scintillation detector. Technical Journal "Readout". 2005;**9**:98-101

[15] Muminov IT et al. Application of NaI(Tl) detector for measurement of natural radionuclides and 137Cs in environmental samples: A new approach

by decomposition of measured spectrum. Journal of Environmental Radioactivity. 2005;**84**(3):321-331. DOI: 10.1016/j.jenvrad.2005.02.017

[16] Al-Hamidawi AA et al. Determination of natural radiation contamination for some types of legumes available in the Iraqi markets. Pelagia Research Library. 2013;**4**(5):245-250

[17] Isinkaye MO et al. Natural radioactivity measurements and evaluation of radiological hazards in the sediment of Oguta Lake, South East Nigeria. Journal of Radiation Research and Applied Science. 2015;**8**:259-469. DOI: 10.1016/j.jrras.2015.05.001

[18] Abid Abojassim A et al. Long lived gamma emitters in biscuit samples consumed in Iraq. International Journal of Environmental Sciences. 2015;**5**(5):956-963. DOI: 10.6088/ijes.2014050100090

[19] Hafizoglu N et al. Assessment of radiological hazard parameters due to natural radioactivity in soils from granite-rich regions in Kütahya province, Turkey. Isotopes in Environmental and Health Studies. 2017;**53**(2):212-221. DOI: 10.1080/10256016.2016.1207640

[20] Hady HN, Abojassim AA. Measurement of natural radioactivity in selected samples of medical plants in Iraq. International Journal of Physical Sciences. 2016;**11**(14):178-182. DOI: 10.5897/IJPS2016.4507

[21] Ali Abid A et al. Natural radioactivity levels in some vegetables and fruits commonly used in Najaf Governorate, Iraq. Journal of Bioenergy and Food Science. 2016;**3**(3):114-123. DOI: 10.18067/jbfs.v3i3.108

[22] Şahin Bal S. The determination of concentrations of radioisotopes in some granite samples in Turkey and their radiation hazards. Radiation Effects and Defects in Solids. 2018;**173**:1-14. DOI: 10.1080/10420150.2018.1462358

[23] Boyd WL. Using Thermoluminescent Dosimeters to Measure the Dose from High and Low Energy X-Ray Sources. Las Vegas: University of Nevada; 2009

[24] Harvey JA. Performance of Thermoluminescent Dosimeters Under As-Deployed Conditions [thesis]. Nuclear Engineering and Radiological Sciences: University of Michigan; 2011

[25] THERMO SCIENTIFIC 4500 TLD READER. 2013. Available from: http://ins.en.ankara.edu.tr/files/2013/01/tld4500.jpg

[26] Kroutilķková D et al. Thermoluminescent dosimeters (TLD) quality assurance network in the Czech Republic. Radiotherapy and Oncology. 2003;**66**(2):235-244. DOI: 10.1016/S0167-8140(02)00328-6

[27] Juan AN. Thermoluminescence-dosimetry (TLD) and its application in medical physics. In: AIP Conference Proceedings; 17-19 March 2004; Guanajuato. Mexico: NY: American Institute of Physics; 2004. pp. 20-27

[28] Taylor ML. Robust determination of effective atomic numbers for electron interactions with TLD-100 and TLD-100H thermoluminescent dosimeters. Nuclear Instruments and Methods in Physics Research Section B: Beam Interactions with Materials and Atoms. 2011;**269**(8):770-773. DOI: 10.1016/j.nimb.2011.02.010

[29] Purwaningsih S et al. Measurement of computed tomography dose profile with pitch variation using Gafchromic XR-QA2 and thermoluminescence dosimeter (TLD). In: 13th South-East Asian Congress of Medical Physics 2015 (SEACOMP); 10-12 December 2015;

Yogyakarta. Indonesia: IOP Science; 2016. pp. 1-5

[30] Lotfalizadeh F et al. Unfolding neutron spectra from the simulated response of thermoluminescence dosimeters inside a polyethylene sphere using GRNN neural network. Journal of Instrumentation. 2017;**12**(7):T07007. DOI: 10.1088/1748-0221/12/07/T07007

[31] Nuraeni N et al. Preliminary studies of thermoluminescence dosimeter (TLD) CaSO4: Dy synthesis. In: International Conference on Energy Sciences (ICES 2016) 25-27 July 2016. IOP Science: Bandung, Indonesia; 2017. pp. 1-5

[32] Romanyukha A et al. Acceptance testing of thermoluminescent dosimeter holders. Health Physics. 2018;**114**(5):543-548. DOI: 10.1097/HP.0000000000000845

[33] https://www.electronics-notes.com/articles/radio/modulation/am-diode-detector-demodulator.php

[34] Vicorosk N et al. Development of a silicon diode detector for skin dosimetry in radiotherapy. Medical Physics. 2017;**44**(10):5420-5412. DOI: 10.1002/mp.12469

[35] McAuley et al. Evaluation of the dosimetric properties of a diode detector for small field proton radiosurgery. Journal of Applied Clinical Medical Physics. 2015;**16**(6):51-64. DOI: 10.1120/jacmp.v16i6.5391

[36] Le Polozec X. A simple formula to calculate the diode junction resistance variations with RF power of a series schottky diode detector. In: X. Le Polozec is with the Engagement Practice IP & Transport. Massy, France: Ericsson; 2016. DOI: 10.13140/RG.2.2.27882.31681

[37] Takahashi T et al. 2002. High-resolution Schottky CdTe diode detector. IEEE Transactions on Nuclear Science 2001; IEEE Transactions on Nuclear Science. 2002;**49**(3):1297-1303. DOI: 10.1109/TNS.2002.1039655

[38] Cherry SR, Sorenson JA, Phelps ME. Radiation detectors. In: Cherry SR, Sorenson JA, Phelps ME, editors. Physics in Nuclear Medicine. 4th ed. Elsevier; 2012. pp. 87-106. DOI: 10.1016/B978-1-4160-5198-5.00007-1

[39] Brudanin VB et al. Large-volume HPGe detectors for rare events with a low deposited energy. Instruments and Experimental Techniques. 2011;**5**(4):470-472. DOI: 10.1134/S0020441211030109

[40] Canbazoglu C et al. A preliminary study on radioactivity concentrations and dose assessment of some anticarcinogenic medicinal plants used in Turkey. Fresenius Environmental Bulletin. 2018;**27**(2):793-798. DOI: 10.1080/10256016.2011.556723

[41] Joel ES et al. Comparative analysis of natural radioactivity content in tiles made in Nigeria and imported tiles from China. Scientific Reports. 2018;**8**:1842. DOI: 10.1038/s41598-018-20309-0

[42] Joel ES et al. Assessment of natural radioactivity in various commercial tiles used for building purposes in Nigeria. MethodsX. 2018;**5**:8-19. DOI: 10.1016/j.mex.2017.12.002

[43] Al-Hamarneh IF. Hazard indices and annual effective dose due to terrestrial radioactivity in the urban areas in the south of Jordan. Journal of Radioanalytical and Nuclear Chemistry. 2018;**316**(1):139-151. DOI: 10.1007/s10967-018-5723-6(01.23456789)

[44] Azmary Khatun M. Natural radioactivity measurement and assessment of radiological hazards in some building materials used in Bangladesh. Journal of Environmental

Protection. 2018;**9**:1034-1048. DOI: 10.4236/jep.2018.910064

[45] Saleh H et al. Radiological risk measurements due to natural radioactivity of building stones used in Jordanian houses. Jordan Journal of Physics. 2018;**11**(3):193-200

[46] Abdellah WM et al. Evaluation of natural radioactivity and its radiation hazards in some building and decorative materials in Iraq. Arab Journal of Nuclear Sciences and Applications. 2017;**51**(1):39-45. DOI: 10.1007/s10967-016-4995-y

[47] Usikalu MR et al. Radiation hazard in soil from Ajaokuta North-central Nigeria. International Journal of Radiation Research. 2017;**15**(2):221-224. DOI: 10.18869/acadpub.ijrr.15.2.219

[48] Dizman S et al. Determination of radioactivity levels of soil samples and the excess of lifetime cancer risk in Rize province, Turkey. International Journal of Radiation Research. 2016;**14**(3):237-244. DOI: 10.18869/acadpub.ijrr.14.3.237

[49] Darwish DAE et al. The assessment of natural radioactivity and its associated radiological hazards and dose parameters in granite samples from South Sinai, Egypt. Journal of Radiation Research and Applied Science. 2015;**8**(1):17-25. DOI: 10.1016/j.jrras.2014.10.003

[50] Ajayi OS, Ibikunle SB. Radioactivity of surface soils from Oyo state, South Western Nigeria. International Journal of Radiation Research. 2013;**11**(4):271-278. DOI: 10.1097/01.HP.0000309766.12663.1d

[51] Saleh MA et al. Assessment of health hazard due to natural radioactivity in Kluang District, Johor, Malaysia. Isotopes in Environmental and Health Studies. 2013;**50**(1):103-113. DOI: 10.1080/10256016.2013.821469

[52] Srilatha MC et al. Measurement of natural radioactivity and radiation hazard assessment in the soil samples of Ramanagara and Tumkur districts, Karnataka, India. Journal of Radioanalytical and Nuclear Chemistry. 2014;**303**(1):993-1003. DOI: 10.1007/s10967-014-3584-1

Chapter 2

Basic Modes of Radioactive Decay

Hasna Albandar

Abstract

This basic modes of radioactive decay review "Gamma Rays" reviews some topics related to radiation, its classification and importance. In general, gamma rays interfere with our life, so we need to comprehend radiation as fact around us all the time and all the time. We live in a naturally radioactive world, but to what extent do physicians, nurses, and medical technicians, who may have to deal with urgent cases of a radiation, know about it? This chapter will address what radiation is and what is its role. This chapter will guide us toward the knowledge of ionizing radiation and its certain forms such as alpha particles, beta particles, gamma rays, and X-rays. as well as it will review on radioactive decay (nuclear decay) as well as help us learn about radioactivity and radiation, in addition to the types of decays, which are divided into beta decay, gamma decay, electron capture, positron decay, and alpha decay. This chapter will focus on radioactive decay, the activity and units of radioactive activity, and half-life of it. The last part of this chapter discusses attenuation as the reduction in the intensity of gamma ray or X-ray beam. The most important subtitles that are scattered from attenuation are HVL mean free path, the linear attenuation coefficient, pair production, and photoelectric scattering.

Keywords: activity, alpha particle, beta particle, decay (radioactive), electron capture, electromagnetic radiation, electron, emissions, energy, electron volt (eV), gamma rays, gray (Gy), half-life (radiological), ionizing radiation, isotopes, molecule, nonionizing radiation, nucleus (of an atom), nuclide, photon, positron, positron emission, proton, radiation, radionuclide, X-ray

1. What is radiation?

Radiation is a form of energy that is released as electromagnetic waves or particles, moves through space, and may be able to penetrate or interact with different materials.

Radiation-caused changes in materials depend on the origin, type of radiation, and the deposited energy [1].

1.1 Classification of radiations

Radiation is classified into ionizing and nonionizing radiation. Ionizing radiation is divided into direct ionizing and indirect ionizing (as shown in **Figure 1**).

1.1.1 Ionizing radiation

Ionization radiation refers to the capability of ionizing material either directly or indirectly because of their higher energy such as X-rays, γ-rays, energetic neutrons, electrons, protons, and heavier particles (as shown in **Figure 2**).

IntechOpen

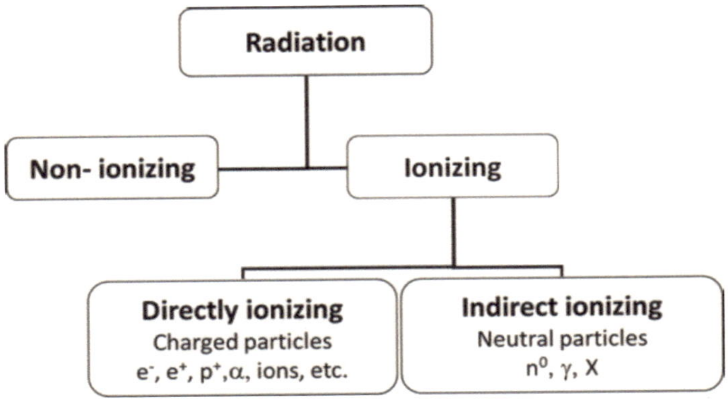

Figure 1.
Classification of radiations.

- If the energy of electromagnetic waves is high, the frequency will be high with short wavelength such as those of gamma rays or heavy particles (beta and alpha).
- Enough high energy to pull electron from orbit [2].

1.1.2 Nonionizing radiation

Nonionizing radiation refers to the inability of ionizing materials because of their lower energy, Such as ultraviolet radiation, visible light, infrared photons, microwaves and radio waves (as shown in **Figure 2**).

- If the energy of electromagnetic waves is low, the frequency will be low with long wavelength such as those of radio waves and microwaves.
- Not enough energy to pull electron from orbit, but the electron can exit [3].

1.2 Classification of ionizing radiation

Ionizing radiation is classified into two types:

i. Directly ionizing radiation

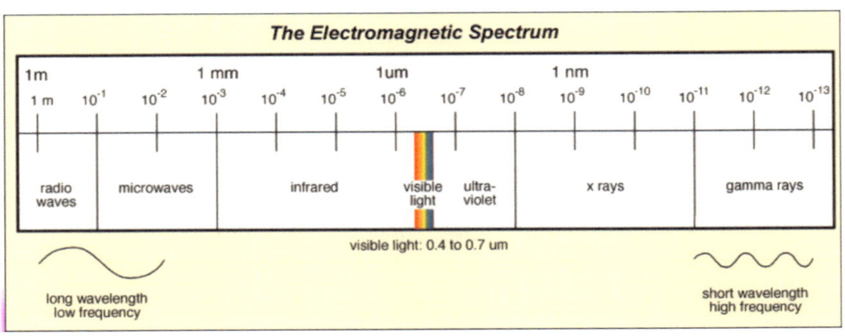

Figure 2.
The schematic representation of the different regions of the electromagnetic spectrum.

- This consists of charged particles, such as electrons, protons, α particles, and heavy ions.
    - Energy can be deposited by directly ionizing radiation in the medium through direct coulomb interaction between the directly ionizing charged particles and orbital electrons of atoms.

ii. Indirectly ionizing radiation

- This consists of neutral particles, photons (X-ray and γ-rays), and neutrons.
    - Energy can be deposited by indirectly ionizing radiation (photons or neutrons) in the medium through two steps:
        1. First step: a charged molecule is transmitted in the medium (photons discharge electrons or positrons; neutrons discharge protons or heavier particles).
        2. The second step: the produced charged particles store vitality to the medium through direct Coulomb interaction with the orbital electrons of the atoms.

1.2.1 Ionizing radiation takes a few forms

Alpha, beta, neutron particles, gamma rays and X-rays are each caused by unstable atoms, either through the overabundance of vitality or mass (or both of them).

To reach a steady state, they must discharge that additional vitality or mass within the frame of radiation.

Alpha particles (α particle): positive charged particles (+2), which are released in the radioactive decay of some nuclei. An alpha is a particle which is emitted from the nucleus of an atom, which consists of (+2) protons and (2) neutrons with mass number (4) (Helium atom). It is strong ionizing with low penetration power and short range.

Beta particles (β+ or β-): They are particles with electric charge ((+) or (−)) emitted from the nucleus during radioactive decay. They take the form of either an electron or a positron (a particle with the size and mass of an electron, but with a positive charge).

- Electrons or positrons have small mass and variable energy. Electrons are formed when a neutron transforms into a proton and an electron.

Gamma rays: are different from alpha or beta rays, because they do not contain any particles, as they are used in electromagnetic radiation. Instead, they consist of a photon of energy, which is released from an unstable nucleus of an atom.

Isomeric transition: It occurs when the excited atomic nucleus changes from a higher to a lower state of the energy by emitting gamma ray.

Internal conversion electron: This process occurs when the gamma rays are not released sometimes, so they provide their exceed energy to the electron in the atomic orbit; this process usually happens to the nearest nucleus (as shown in **Figure 3**).

X-rays: They are similar to gamma radiation; the only one primary difference is that they originate from the electron shell. This is generally caused by energy

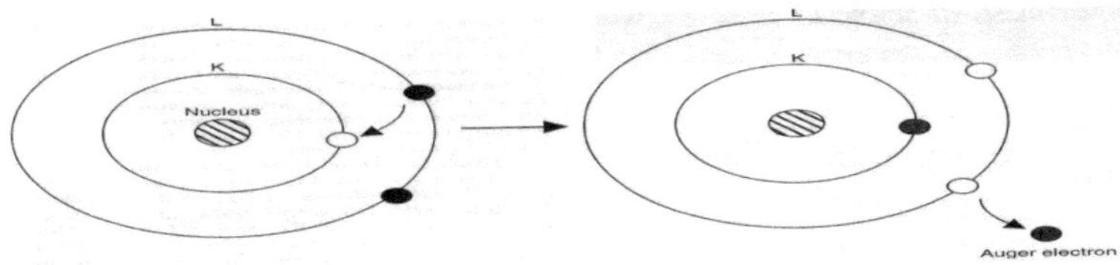

Figure 3.
Emission of an auger electron as an alternative to X-ray emission. No X-ray is emitted.

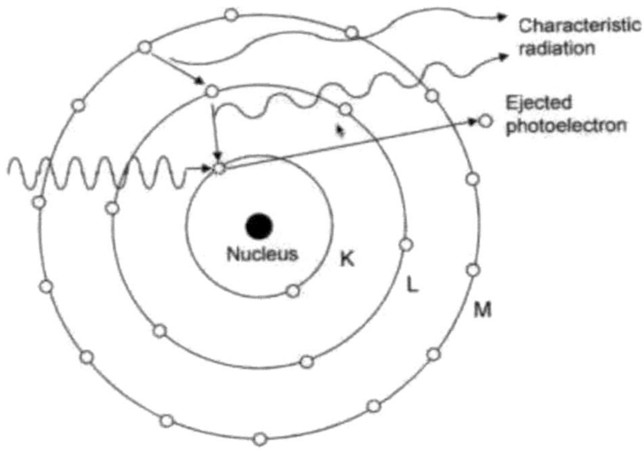

Figure 4.
How the emission of the characteristics of X-ray occurs when the orbital electrons move from an outer shell so as to fill in the inner shell vacancy.

changes in an electron, such as moving from a higher energy level to a lower one. This causes the excess energy to be released. X-rays are called **characteristic X-ray**. It (X-ray) has longer wavelength and possess (usually) lower energy than gamma radiation, as well. The emission of high-energy waves came from the electron of an atom (as shown in **Figure 4**).

A **neutron particle** is an uncharged element particle with a mass that is slightly greater than that of the proton, and it is found in the nucleus. It is usually released because of spontaneous or induced nuclear fission.

2. Radioactive decay (nuclear decay)

Radioactive decay is a process in which an unstable nucleus transforms into a more stable one by releasing particles or photons. In addition, it results in the conversion of mass into energy.

In some decay modes, electron mass is converted into energy as well. The total mass-energy conversion amount is called the transition energy. Most of this energy is imparted as kinetic energy to released particles or is converted to photons with a small portion as kinetic energy.

In a few decay modes, electron mass is changed into vitality as well. The full mass-energy transformation sum is called the transition energy. Most of this energy is imparted as active energy to discharged particles or is changed over to photons with a little portion as kinetic energy [4].

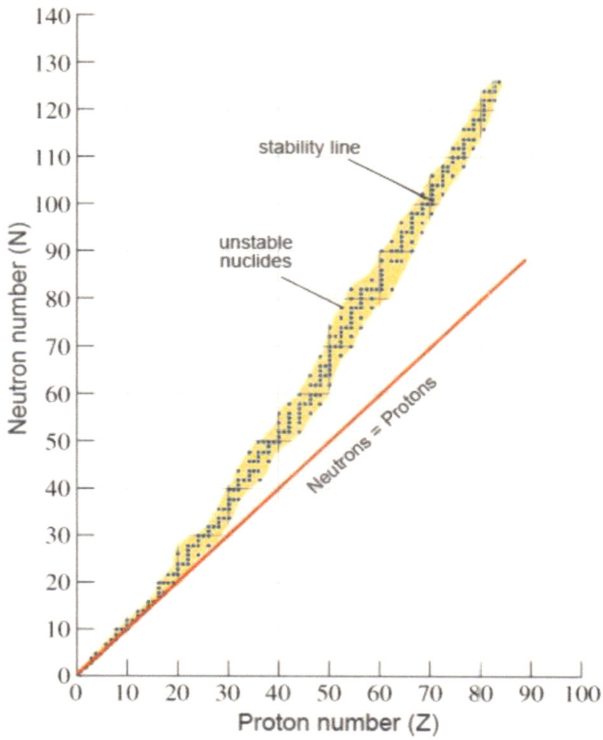

Figure 5.
For proton numbers (Z) up to 20, N = Z could be a straight line. For all nuclei with Z > 20, stable nuclei have less protons than neutrons; the line bends upwards. Unsteady nuclei over the soundness bend are called neutron-rich [1].

The neutron - Proton ratio shows that if the number of protons increase, the number of neutrons must increase even more for stability. This process is shown in **Figure 5**.

2.1 Radioactive decay

Radioactive decay is the process by which the unstable nucleus tries to change into a more stable form. As, it is the process in which the transformation will take place depending on the composition of the nucleus.

2.2 Types of decays

1. Beta decay
2. Gamma decay
3. Electron capture
4. The positron decay
5. Alpha decay

2.2.1 Beta decay

Beta decay or (**β^- decay**) is a process in which the neutron in the nucleus is essentially transformed into a proton and electron:

$$n \rightarrow p + \beta^- + \upsilon + \text{energy}$$

Beta decay is also the decay of one of the neutrons to a proton via the weak interaction:

$$_{Z}^{A}X \overset{\beta^-{}_{Z+1}^{A}}{\rightarrow} Y^{Z+1}$$

The electron is called **β^-** particle (υ), meanwhile the neutrino is a particle that has no mass or electrical charge. It does not virtually undergo interactions with matter and therefore is essentially undetectable.

The energy released in β^-decay is shared between β^-particle and neutrino (υ). This sharing of energy is more or less random from one decay to the next. As shown in **Figure 6**, the plot displays the distribution of β^- particle energy. It is also noticed that beta particles are not monoenergetic for a particular radionuclide, but they are released at varying energy levels over a continuous range (spectrum). The average energy of beta emission can be estimated as one-third of the maximum energy of emission: $E_{avg} = 1/3E_{max}$ (as shown in **Figure 6**) [1].

2.2.2 Gamma decay

It is a mechanism for an excited nucleus to release energy. Emanation could be a sort of radioactivity in which a few unsteady nuclear nuclei disseminate excess energy by an unconstrained electromagnetic radiation.

Within the most common form of gamma decay, which is called gamma emission, gamma rays (photons or bundles of electromagnetic vitality, of highly short wavelength) are radiated.

Gamma rays are electromagnetic radiation (high-energy photons) with an extreme frequency and a high energy. They are created by the decay of nuclei as they travel from a high-energy state to a lower state; this process is called "gamma decay." Most of atomic responses are accompanied by gamma emission.

Gamma decay also includes two other electromagnetic processes: internal conversion and pair production.

Internal conversion (IC) is a process in which the excess energy of the nucleus is directly transferred to one of its own orbital electrons which is ejected instead of the ray. In this case, the ejected electron is called a conversion electron (as shown in **Figure 7**).

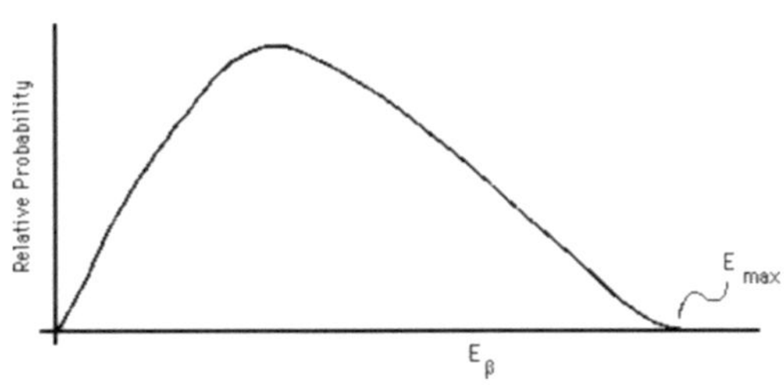

Figure 6.
The distribution or spectrum for β^- particle.

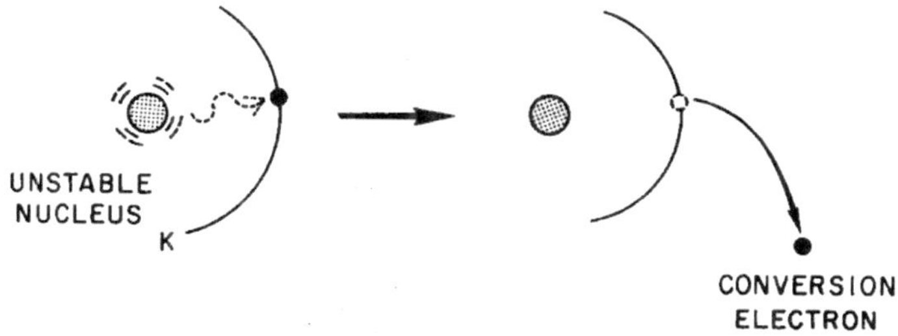

Figure 7.
Schematic representation of internal conversion involving a K shell electron. Unstable nucleus transfers its energy to an orbital electron to release a converted electron.

In internal pair production, the excess energy is converted within the electromagnetic field of a nucleus into an electron and a positron that are released together. Internal conversion always accompanies the predominant process of gamma emission.

Internal pair production needs the excess energy of the unstable nucleus to be at least equivalent to the combined masses of an electron and a positron (as shown in **Figure 8**).

2.2.2.1 Isomeric transition conversion

The daughter of radioactive parent may be formed in a long-lived metastable (isomeric state) as opposed to an excited state. The decay of the metastable (isomeric state) by the emission of a γ-ray is called isomeric transition.

Isomeric transition: a nuclear process in which a nucleus has abundant energy following the emanation of an alpha molecule or a beta molecule and in turn discharges energy without a change in its number of protons or neutrons. Isomeric moves can occur through the emission of a gamma ray or through the process called "internal conversion."

In many nuclides, isomeric transitions produce gamma photons and internal conversion electrons. When an electron is removed from the atom by internal conversion, a vacancy is created. All transitions are usually followed by either gamma or internal conversion electron emission.

The energized atomic state taking after the emission of a beta particle may be nearly steady, and the nucleus may be able to remain in this state for minutes, hours, or even days, sometimes recently discharging a gamma ray.

The isomer (no change of the number of proton or neutron) works as a separate radioactive material, which is decaying exponentially with the emission of a gamma ray only [5].

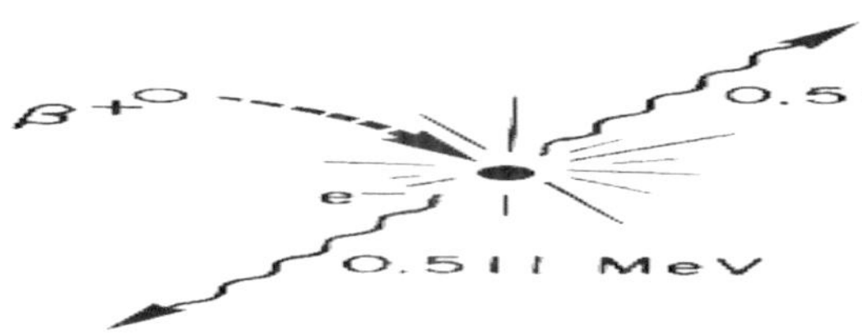

Figure 8.
Schematic representation of mutual annihilation reaction between a positron (β+) and an ordinary electron. A pair of 0.511 MeV annihilation photons is released "back to back" at 180° to each other.

2.2.3 Electron capture

Electron capture decay: it is an inverted β^-decay, whereas an orbital electron is captured by the nucleus and combines with a proton to form a neutron:

$$p^+ + e^- \rightarrow n + \upsilon + \text{energy} \tag{3}$$

In other words, we can say that the electron capture is a process, in which a parent nucleus captures one of its orbital electrons and releases a neutrino. This neutrino is emitted from the nucleus and carries away some of the transitions energy. The remaining energy appears in the form of characteristic X-rays and Auger electrons, which are emitted by daughter product, whereas the resulting orbital electron vacancy is filled (as shown in **Figure 9**).

2.2.4 The positron decay β^+

In case of radioactive decay by positron emission, a proton in the nucleus is transformed into a neutron and a positively charged electron (positron β^+) then a proton ejected from the nucleus. A positron is an antiparticle of an ordinary electron:

$$p^+ \rightarrow n + \beta^+ + \upsilon + \text{energy} \tag{4}$$

After ejection from the nucleus, it loses its kinetic energy in collision with atoms of the surrounding matter and comes to rest; this usually happens within a few millimeters from the site of its origin in body tissue [6].

2.2.5 Alpha decay

Radionuclide that decayed by a particle emission or by nuclear fission has relatively little importance for direct usage as tracers in nuclear medicine.

Both of these decay modes occur primarily among very heavy elements that are of a little interest as physiological tracers [7].

The particles, which are released with kinetic energy, are usually found between 4 and 8 MeV.

Decay by alpha particle emission results in transmission of elements, but it is not isobaric.

Activity: It is the total number of nuclei that are decaying per second. It is the probability that any individual atom will undergo decay during the same period:

$$A = \lambda N \tag{5}$$

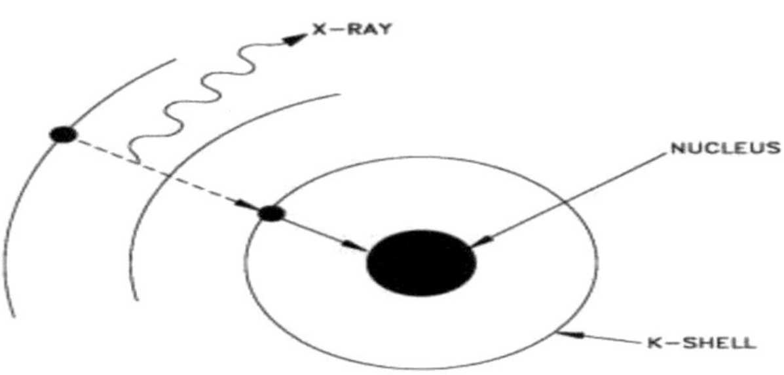

Figure 9.
The nucleus captures one of its orbital electrons and X-ray.

DOI: http://dx.doi.org/10.5772/intechopen.85502

where A = activity; N = the number of decay nuclei in the sample; λ = decay constant.

The decay factor ($e^{-\lambda t}$) is an exponential function of time (t). Exponential decay is characterized by disappearance of a constant function of activity or number of atoms prevented per unit time interval:

$$A = A_0\, e^{-\lambda t} \quad (6)$$

where A is the activity of radionuclide at a given time t; A_0 is the activity of radionuclide at time t = 0; decay constant (λ).

The decay constant (λ) is the probability that a nucleus will decay per second, so its unit is (s^{-1}). Activity can be determined by direct measurement.

2.2.6 Units of radioactivity

1. Conversion unit

   -Curie, where 1 Ci = 3.7 × 10^{10} disintegration per second

2. SI unit

   -Becquerel, where 1 Bq = 1 disintegration per second.

   1 Ci = 3.7 × 10^{10} dps = 37 GBq.

   1 mCi = 3.7 × 10^{7} dps = 37 MBq.

   1 μCi = 3.7 × 10^{4} dps = 37 kBq.

Half-life: It is the amount of time taken for the given quantity so as to be decreased to half of its initial value. As shown in **Figure 10**, the term is most commonly used in relation to atoms undergoing radioactive decay, but it can be used to describe other types of decay, whether exponential or not. One of the most well-known applications of half-life is:

$$T_{1/2} = Ln2/\lambda \quad (7)$$

where $T_{1/2}$ is the half-life of radionuclides; ln2 = 0.693 is the base of natural logarithms; λ is decay constant of radionuclides.

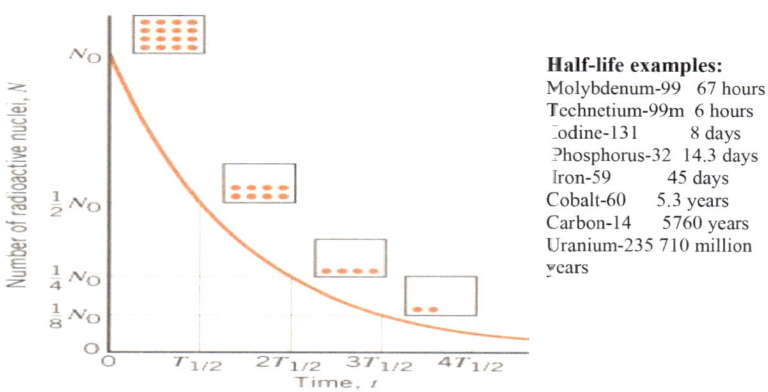

Figure 10.
The time required for it to decay the number of radioactive nuclei to 50% of the (N_O).

3. Attenuation

Attenuation is the reduction in the intensity of gamma ray or X-ray beam, as it traverses matter either by the absorption of photons or by deflection (scattering) of photons from the beam.

Attenuation results from the interaction between penetrating radiation and matter, as it is not a simple process. These interactions include the photoelectric effect, scattering, and pair production [8].

3.1 HVL

Half-value layer (HVL): It is defined as the thickness of material required to reduce intensity of gamma ray or X-ray beam to one-half of its initial value (as shown in **Figure 11**).

3.2 Mean free path

The range of a single photon in matter that cannot be predicted. The distance traveled some time recently interaction can be calculated from direct attenuation coefficient or the HVL of the beam.

Mean free path (MFP) of photon beam is:

$$MFP = \frac{1}{\mu} = \frac{1}{0.693/\mathrm{HVL}} = 1.44\mathrm{HVL} \quad (8)$$

3.3 Linear attenuation coefficient

The linear attenuation coefficient (μ) can be characterized as the division of a beam of X-rays or gamma rays that's retained or scattered per unit thickness of the absorber.

This esteem accounts for the volume of number of atoms in a cubic cm of material and the probability of a photon of being scattered or absorbed from the nucleus or an electron of one of these atoms.

Linear attenuation coefficient is the sum of individual linear attenuation coefficients for each type of interaction:

$$\mu = \mu_{\mathrm{coherent}} + \mu_{\mathrm{photo}} + \mu_{\mathrm{Compton}} + \mu_{\mathrm{pair}} \quad (9)$$

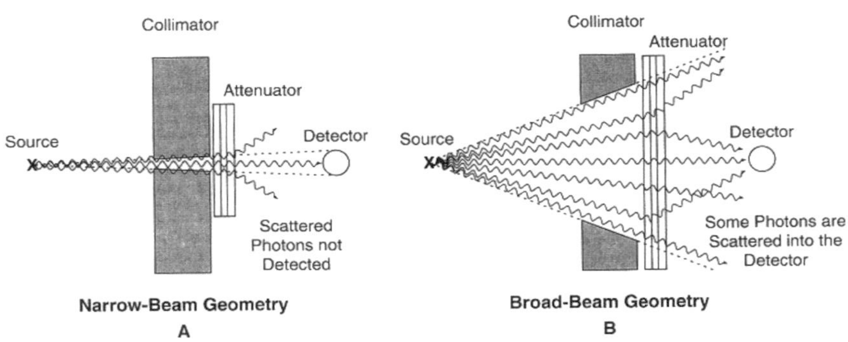

Figure 11.
Monoenergetic photons under narrow-beam geometry conditions. The probability of attenuation remains the same for each additional HVL thickness placed in the beam.

In diagnostic energy range, m decreases with increasing energy except at absorption edges (e.g., K-edge) [9].

3.3.1 Linear attenuation coefficient

- The process of fraction of photons removed from a monoenergetic beam of X-ray or gamma ray per unit thickness of material is called linear attenuation coefficient (μ), and it is typically expressed in cm^{-1}.
- The number of photons removed from the beam traversing a very small thickness μ_x:

$$n = \mu N \Delta x \quad (10)$$

where n = the number removed from beam; N = the number of photons incident on the material; X = the thickness of material.

- For a given thickness of material, the probability of interaction depends on the number of atoms which the X-rays or gamma rays encounter per unit distance.
- The density (μ) of material affects this number.
- -Linear attenuation coefficient is proportional to the density of the material:

$$\mu_{\text{water}} > \mu_{\text{ice}} > \mu_{\text{water vapour}} \quad (11)$$

3.3.2 Mass attenuation coefficient

- For a given thickness, the probability of interaction relies on the number of atoms per volume.
- Dependency can be overcome by normalizing linear attenuation coefficient for thickness of material:

$$\text{mass attenuation coefficient } (\mu/\rho) = \frac{\text{linear attenuation coefficient } (\mu)}{\text{density of material } (\rho)} \quad (12)$$

1. Mass attenuation coefficient ordinarily can be seen in cm^2/g units.
2. Mass attenuation coefficient is autonomous of density.
3. For a given photon energy:

$$\mu_{\text{water}}/\rho_{\text{water}} = \mu_{\text{ice}}/\rho_{\text{ice}} = \mu_{\text{water vapour}}/\rho_{\text{water vapour}} \quad (13)$$

4. In radiology, we usually differentiate between regions of an image that correspond to irradiation of adjacent volumes of tissue.
5. In density, the mass contained within a given volume plays an important role.

$$N = N_o e^{-\left(\frac{\mu}{\rho}\right)\rho x} \quad (14)$$

The photon interactions are dependent on the atomic properties of a material rather than its density; the attenuation coefficients for isolated processes are often given as mass attenuation coefficients (divided by ρ).

3.3.2.1 Attenuation from coherent scattering

Coherent scattering is vital for low kilo voltage photons as it increases with atomic number.

3.3.2.2 Attenuation from photoelectric effect

The mass photoelectric attenuation coefficient is commensurate to the cube of the atomic number (Z^3) and inversely proportional to the cube of the beam energy (E^3).

3.3.2.3 Attenuation from incoherent scattering

The mass incoherent scattering attenuation coefficient is comparative to most values of Z, but it diminishes gradually with the expanding of beam energy. It is most dependent on the electron density [10].

3.3.2.4 Attenuation from pair production

Pair production happens only with higher beam energies (over 1.02 MeV). The mass attenuation coefficient for pair production is linearly related to the atomic number.

Increasing beam energy also raises the attenuation from pair production in a logarithmic style [11].

The attenuation of gamma radiation can be achieved using a wide range of materials. Understanding the basic principles involved in the physical interactions of gamma radiation with matter that lead to gamma attenuation can help in the choice of shielding for a given application. Utilizing this understanding and considering the physical, chemical, and fiscal constraints of a project will lead to better application of resources to develop the most appropriate type of shielding [1].

3.3.3 The methodology

Experimental and analytical methods are methods that are used to describe actions to be taken so as to investigate this subject in detail and the rationale for the application of specific procedures or techniques used to identify, select, process, and analyze information applied for understanding, thereby allowing the reader to critically evaluate a study's overall validity and reliability. Both of these methods are prime methods of inquiry in science. The key features are controlled over variables' careful measurement and establishing cause and effect relationships. An advantage of both is that the experimental and analytical methods should be objective.

4. Conclusion

Radiation has always been present around us. Life has evolved in a world containing significant levels of ionizing radiation. We are also exposed to fabricated radiation from sources such as medical treatments and activities involving radioactive materials. Since the early twentieth century, radiation's impacts have been considered in profundity, in both the research facility and among human populaces. Because dangers of radiation on the well-being are known, it must be carefully utilized and entirely controlled. A balance must be struck between radiation's societal benefits and the dangers that radiation postures to individuals, wellbeing, and the environment. It can be confirmed that ionizing radiation has long been vital in medicine and industry.

Modern medicine would be impossible without ionizing radiation. X-ray imaging, computed tomography scans, diagnostic and therapeutic nuclear medicine, the gamma knife, and linear accelerators are a few of the technologies that have revolutionized medical diagnosis and treatment. Radiation's benefits for human wellbeing can be measured in thousands of lives spared and indeed more prominent numbers of people whose quality of life has been made strides each year by these innovations. Indeed in spite of the fact that the utilization of ionizing radiation in medicine offers gigantic benefits, in any case, it moreover postures potential dangers to patients, restorative faculty, and the public. The diagnostic and helpful devices that remedy moreover can cause intestinal wounds and chronic illness such as cancer.

In expansion to the gamma rays, the attenuation of gamma radiation can be accomplished by employing a wide range of materials. Understanding the fundamental standards included within the physical interactions of gamma radiation with matter that lead to gamma radiation can offer assistance within the choice of protecting for a given application. Utilizing this understanding and considering the physical and chemical limits of a project will lead to a better application of resources to develop the most suitable type of shielding.

Author details

Hasna Albandar
Prince Sultan Medical City, Riyadh, Saudi Arabia

*Address all correspondence to: halbander@psmmc.med.sa

References

[1] Cherry SR. Physics in Nuclear Medicine. Saunders. university of California, Davis, California: An Imprint of Elsevier Inc.; 2012

[2] Nobuyuki H, Yuki F. Classification of Radiation Effects for Dose Limitation Purposes. Safety Research Center, Central Research Institute of Electric Power; 2014

[3] Bailey DL, Todd-Pokropek A, van Aswegen A. Nuclear Medicine Physics. Vienna: International Atomic Energy Agency; 2014

[4] Adlienė D. Basic Radiation Physics and Sources of Radiation. Lithuania: Kaunas University of Technology, Physics Department, Studentų g, Kaunas; 2012

[5] Schneider C. Radioactive Lifetime and Energy of the Low-Energy Isomeric Level. Moscow State University, Leninskie gory; 2018

[6] Martin JE. Physics for Radiation Protection. Singapore: USA Printing and Binding Markono Print Media Pte Ltd; 2013

[7] Thormod H. Radioactivity and X-Rays Applications and Health Effects. University of Oslo; 2009

[8] Diehl R. X Radiation & Gamma Radiation and Neutrons. Max Plank Institute Fur, Garching, Germany: Triangle Park, NC; 2003

[9] Ghose AM. Radiation Physics. India: Department of Atomic Energy; 1974

[10] Poškus A. Attenuation of Gamma Rays. Laboratory of Atomic and Nuclear Physics; 2018

[11] Adlienė D. Basic Radiation Physics and Sources of Radiation. Lithuania: Kaunas University of Technology, Physics Department, Studentų g., LT-51368 Kaunas; 2012

Chapter 3

Electron Oscillation-Based Mono-Color Gamma-Ray Source

Hai Lin, ChengPu Liu and Chen Wang

Abstract

Production of artificial gamma-ray source usually is a conception belonging to the category of experimental nuclear physics. Nuclear physicists achieve this goal through utilizing/manipulating nucleons, such as proton and neutron. Low-energy electrons are often taken as "by-products" when preparing these nucleons by ionizing atoms, molecules and solids, and high-energy electrons or β rays are taken as "wastage" generated in nuclear reaction. Utilization of those "by-products" has not won sufficient attention from the nuclear physics community. In this chapter, we point out a potential, valuable utilization of those "by-products." Based on a universal principle of achieving powerful mono-color radiation source, we propose how to set up an efficient powerful electron-based gamma-ray source through available solid-state components/elements. Larger charge-to-mass ratio of an electron warrants the advantage of electron-based gamma-ray source over its nucleon-based counterpart. Our technique offers a more efficient way of manipulating nuclear matter through its characteristic EM stimulus. It can warrant sufficient dose/brightness/intensity and hence an efficient manipulation of nuclear matter. Especially, the manipulation of a nucleus is not at the cost of destroying many nuclei to generate a desired tool, that is, gamma ray with sufficient intensity, for achieving this goal. This fundamentally warrants a practical manipulation of more nuclei at desirable number.

Keywords: gamma-ray source, electron oscillation, DC fields

1. Introduction

Powerful mono-color gamma-ray source is a very appealing, but also seem-to-be-dream, topic in modern physics. This is because the gamma ray, an electromagnetic (EM) wave with sub-*pm*-level wavelength λ, is generated from quantum transition of nuclear matter in nuclear reaction. At present, using synthesized radioactive heavy elements, or radioisotope, is the main route for achieving a gamma-ray source [1–4]. Finite proton number in a radioactive heavy element is the bottleneck affecting the intensity and power of the gamma-ray output. Higher proton number is favorable not only to higher intensity and power but also faster decay or shorter life of the radioactive heavy element. For practical purposes, such a gamma-ray source should be of a stable output over a sufficiently long time duration. Piling large amount of radioactive heavy elements might be a solution to this requirement but its accompanied environment-protection cost might be overly high. Moreover, because a nucleon has smaller charge-to-mass ratio than an electron, the accelerator cost and the reactor cost at the synthesis stage are also of considerable amount even though it is only aimed at low-energy nuclear physics applications rather than high-energy physics applications. To some extent, obtaining a powerful mono-color gamma-ray source corresponds to an artful skill of manipulating nuclear matter.

Therefore, new working principle of achieving radiation source with narrower output spectrum is of significant application value. Based on Takeuchi's theory [5], we proposed a universal principle of achieving mono-color radiation source at arbitrary wavelength [6, 7]. According to this principle, available parameter values can ensure a powerful mono-color gamma-ray source.

The core of this working principle can be summarized as "tailoring" Takeuchi orbit. Takeuchi's theory reveals that the orbit of a classical charged particle, such as electron, in a DC field configuration $E_s \times B_s$, where E_s and B_s are constant, can be elliptical or parabolic according to values of E_s and B_s and that of initial particle's velocity entering into this configuration [5, 8]. The time cycle of an elliptical orbit can be in principle an arbitrary value by choosing suitable values of these parameters. Thus, for a far-field observer on the normal direction of this 2-D orbit, electrons moving along the orbit will behave like a radiation source whose central frequency is the inverse of the time cycle of the orbit. But a realistic factor affecting its practicality is the geometric size of such an orbit. Overly large geometric size will hurt the practicality of such a radiation source. At present, for available values of E_s and B_s, about $MVolt/meter$-level and $Tesla$-level, the size can be down to m-level for s-level time cycle or Hz-level frequency.

For warranting the practicality of such a radiation source, we propose a scheme for making it compact by "tailoring" Takeuchi orbit through targeted designed DC field configuration [6]. In this configuration, B_s is made space-varying along the direction normal to the unperturbed path of an electron bunch by not letting two Helmholtz coils be co-axial on purpose [6]. By choosing suitable values of related parameters, such as the relative distance between the bunch path and the $B_s = 0$ contour, E_s-values and the slope $\beta = d_x B_s$, where B_s is along the y-direction, its magnitude $|B_s|$ is a function of the coordinate x, and the unperturbed path is along the z-axis.

2. Theory and method

2.1 Theoretical basis

For the convenience of readers, we paste related materials published elsewhere [7]. For a simple configuration containing merely static electric field (along x-direction) and static magnetic field (along z-direction), the behavior of an incident electron can be described by dimensionless 3-D relativistic Newton equations (RNEs)

$$d_s[\Gamma d_s Z] = 0, \tag{1}$$

$$d_s[\Gamma d_s Y] = W_B d_s X \tag{2}$$

$$d_s[\Gamma d_s X] = -W_B[\eta + d_s Y] \tag{3}$$

where

$$\frac{1}{\Gamma} = \sqrt{1 - (d_s X)^2 - (d_s Y)^2 - (d_s Z)^2}. \tag{4}$$

Moreover, E_s and B_s are constant-valued electric and magnetic fields and meet $E_s = \eta c B_s$; $\lambda = c/\omega$ and ω are reference wavelength and frequency, respectively; and $s = \omega t$, $Z = \frac{z}{\lambda}$, $Y = \frac{y}{\lambda}$, $X = \frac{x}{\lambda}$, $W_B = \frac{\omega_B}{\omega}$, where $\omega_B = \frac{eB_s}{m_e}$ is the cyclotron frequency.

Eqs. (1)–(3) lead to

$$d_s Z \equiv 0 \tag{5}$$

$$\Gamma d_s Y - W_B X = const = C_y; \tag{6}$$

$$\Gamma d_s X + W_B[\eta s + Y] = const = C_x, \tag{7}$$

where the values of these constants, *const*, are determined from the initial conditions $(X, Y, Z, d_s X, d_s Y, d_s Z)|_{s=0} = \left(0, 0, 0, \frac{C_x}{\sqrt{1+C_x^2+C_y^2}}, \frac{C_y}{\sqrt{1+C_x^2+C_y^2}}, 0\right)$.

Eqs. (5)–(7) can yield an equation for $d_s X$ and $d_s Y$

$$(d_s Y)^2 = \left[C_y + W_B X\right]^2 * \left[1 - (d_s X)^2 - (d_s Y)^2\right] \tag{8}$$

$$(d_s X)^2 = \left[C_x - W_B * (\eta s + Y)\right]^2 * \left[1 - (d_s X)^2 - (d_s Y)^2\right] \tag{9}$$

whose solution reads

$$(d_s X)^2 = \frac{\left[C_x - W_B * (\eta s + Y)\right]^2}{\left[1 + \left[C_y + W_B X\right]^2 + \left[C_x - W_B * (\eta s + Y)\right]^2\right]} \tag{10}$$

$$(d_s Y)^2 = \frac{\left[C_y + W_B X\right]^2}{\left[1 + \left[C_y + W_B X\right]^2 + \left[C_x - W_B * (\eta s + Y)\right]^2\right]}. \tag{11}$$

It is easy to verify that the solutions (10, 11) will lead to $\Gamma = \sqrt{1 + \left[C_y + W_B X\right]^2 + \left[C_x - W_B * (\eta s + Y)\right]^2}$ and, with the help of Eqs. (6) and (7), $d_s\Gamma = -W_B\eta * d_s X$ (i.e., $m_e c^2 d_t \Gamma = eEd_t X$). Noting Γ can be formally expressed as $\Gamma = \sqrt{1 + C_y^2 + C_x^2} - W_B\eta * X$, which agrees with Takeuchi's theory [15], we can find that the electronic trajectory can be expressed as

$$\left[\sqrt{1 + C_y^2 + C_x^2} - W_B\eta * X\right]^2 = 1 + \left[C_y + W_B X\right]^2 + \left[C_x - W_B * (\eta s + Y)\right]^2, \tag{12}$$

or

$$(1-\eta^2)\left[X + \frac{(\eta + v_{y0})}{1-\eta^2}\frac{\Gamma_0}{W_B}\right]^2 + \left[(Y + \eta s) - v_{x0}\frac{\Gamma_0}{W_B}\right]^2 = \frac{\left[(\eta + v_{y0})^2 + (1-\eta^2)v_{x0}^2\right]}{1-\eta^2}\left(\frac{\Gamma_0}{W_B}\right)^2, \tag{13}$$

where $\Gamma_0 = \sqrt{1 + C_y^2 + C_x^2}$, $v_{x0} = \frac{C_x}{\Gamma_0}$ and $v_{y0} = \frac{C_y}{\Gamma_0}$.

There will be an elliptical trajectory for $\eta < 1$ and a hyperbolic one for $\eta > 1$ [15, 16]. The time cycle for an electron traveling through an elliptical trajectory can be exactly calculated by re-writing Eq. (10) as [15]

$$\pm ds = \frac{\frac{1}{W_B}\Gamma_0 - \eta * X}{\sqrt{aX^2 + bX + c}}dX = \frac{\eta}{\sqrt{-a}}\frac{X_N - X}{\sqrt{\frac{b^2-4ac}{4a^2} - \left(X + \frac{b}{2a}\right)^2}}dX, \tag{14}$$

where $a = (\eta^2 - 1)$, $b = -2\left[\eta\Gamma_0 + C_y\right]\frac{1}{W_B}$, $c = C_x^2\left(\frac{1}{W_B}\right)^2$ and $X_N = \frac{1}{\eta}\frac{1}{W_B}\Gamma_0$. The equation can be written as a more general form

$$\pm ds = \sigma\frac{M - u}{\sqrt{1 - u^2}}du \tag{15}$$

where $u = \frac{X+\frac{b}{2a}}{\sqrt{\frac{b^2-4ac}{4a^2}}} = \frac{X-\frac{X_R+X_L}{2}}{X_R-X_L}$, $X_L = \min\left(\frac{-b-\sqrt{b^2-4ac}}{2a}, \frac{-b+\sqrt{b^2-4ac}}{2a}\right)$ and $X_R = \max\left(\frac{-b-\sqrt{b^2-4ac}}{2a}, \frac{-b+\sqrt{b^2-4ac}}{2a}\right)$. In addition, $\sigma = \frac{\eta}{\sqrt{-a}}\sqrt{\frac{b^2-4ac}{4a^2}}$ and $M = \frac{X_N+\frac{b}{2a}}{\sqrt{\frac{b^2-4ac}{4a^2}}} = \frac{X_N-\frac{X_R+X_L}{2}}{X_R-X_L}$. It is easy to verify that for $\eta^2 - 1 < 0$, there is $M = \frac{1+\eta v_{y0}}{\eta\sqrt{\left(\eta+v_{y0}\right)^2+(1-\eta^2)v_{x0}^2}} > 1$.

Initially, $(X, Y)|_{s=0} = (0, 0)$ and hence $u_{st} = u|_{s=0} = \frac{0+\frac{b}{2a}}{\sqrt{\frac{b^2-4ac}{4a^2}}} = \frac{-\left(\eta+v_{y0}\right)}{\sqrt{\left(\eta+v_{y0}\right)^2+(1-\eta^2)v_{x0}^2}}$.

From strict solution

$$\pm s(u) = \sigma * \left\{M * \arcsin(u) + \sqrt{1 - u^2}\right\} + const, \tag{16}$$

we can find the time for an electron traveling through an elliptical trajectory to meet $s_{cycle} = \omega T_c = 2 * [\sigma M\pi]$ and hence a time cycle $T_c = \frac{(1+\eta v_{y0})\Gamma_0}{\left(\sqrt{1-\eta^2}\right)^3}\frac{2\pi}{\omega_B}$. That is, the oscillation along the elliptical trajectory will have a circular frequency ω_B. Moreover, it is interesting to note that $\left(v_{x0}, v_{y0}\right) = (0, -\eta)$ will lead to $\frac{\left[\left(\eta+v_{y0}\right)^2+\left(1-\eta^2\right)v_{x0}^2\right]}{1-\eta^2} = 0$ and hence a straight-line trajectory $(X(s), Y(s)) = (0, -\eta s)$.

The motion on an elliptical trajectory is very inhomogeneous. The time for finishing the $\eta X > 0$ half might be very short while that for the $\eta X < 0$ half might be very long. We term the two halves as fast-half and slow-half, respectively. If η is fixed over whole space, a fast-half is always linked with a slow-half and hence makes the time cycle for finishing the whole trajectory being at considerable level.

For convenience, our discussion is based on the parameterized ellipse. For the case $\left(v_{x0} = 0, v_{y0} = -\eta - \Delta\right)$, (where Δ is small-valued and positive), the starting position $X = 0$ is the left extreme of the ellipse and hence corresponds to $u = -1$. The time required for an acute-angled rotation from $u = -1$ to $u = -1 + \xi$, (where ξ is small-valued and positive), will be $\sigma M * \left[\arcsin(-1 + \xi) - \frac{\pi}{2}\right] + \sigma\sqrt{2\xi - \xi^2}$, which is $= 0$ if $\xi = 0$.

It is interesting to note that if there is $B = 0$ at the region $u > -1 + \xi$, the electron will enter from $\left(E, B = \frac{E}{\eta c}\right)$-region into $(E, B = 0)$-region with an initial velocity whose x-component is $v_{x1}\tilde{}d_s u|_{u=-1+\xi} = \frac{1}{\sigma}\frac{\sqrt{1-(-1+\xi)^2}}{M-(-1+\xi)} > 0$ and y-component v_{y1} is $\neq 0$. Then, the electron will enter into the $(E, B = 0)$-region at a distance because $v_{x1} > 0$. After a time $T_{tr} = \frac{2v_{x1}}{E\sqrt{1-v_{x1}^2-v_{y1}^2}}$, the electron will return into the $\left(E, B = \frac{E}{\eta c}\right)$-region and the returning velocity will have a x-component $-v_{x1}$. During this stage, the electron will move $v_{y1} * T_{tr}$ along the y direction. Then, the motion in the $\left(E, B = \frac{E}{\eta c}\right)$-region can be described by an acute-angled rotation along the ellipse $u = -1 + \xi \to u = -1$. Thus, a complete closed cycle along the x direction is finished even though the motion along the y direction is not closed. Repeating this closed cycle will lead to an oscillation along the x direction.

Clearly, the time cycle of such an oscillation, or that of a "tailored" Takeuchi orbit, is

$$T_x = T_{tr} + 2\sigma M * \left[\arcsin(-1 + \xi) - \frac{\pi}{2}\right] + 2\sigma\sqrt{2\xi - \xi^2}. \qquad (17)$$

Under fixed values of Δ, E and B, the smaller ξ is, the smaller T_x is. There will be $T_x = 0$ at $\xi = 0$. In principle, arbitrary value of $T_x < T_c$ can be achieved by choosing suitable value of ξ. That is, arbitrarily high center frequency ($> \omega_B$) oscillation can be achieved by choosing a suitable value of ξ. Although the time history of $x(t)$ might cause its Fourier spectrum to have some spread, the center frequency will be $\frac{1}{T_x}$.

This result implies a simple and universal method of setting up quasi-mono-color light source at any desirable center wavelength: by applying vertically static electric field $E = E_x$ and static magnetic field $B = B_z$ and on purpose letting a $B = 0$ region exist and the ratio $\frac{E}{cB} < 1$, then injecting electron along the y-axis with a velocity slightly above $|\frac{E}{cB}|$, and close to the boundary line between the $B = 0$ region and the $B \neq 0$ region. As shown in **Figure 1** of Ref. [7], adjusting the distance $D = \xi * \sqrt{\frac{b^2 - 4ac}{4a^2}}$ can lead to a quasi-monocolor oscillation source with any desired center frequency up to gamma-ray level.

Of course, such a step-like magnetic field profile is overly idealized. Therefore, we propose using a more realistic magnetic slope to achieve such a tailored Takeuchi orbit [6].

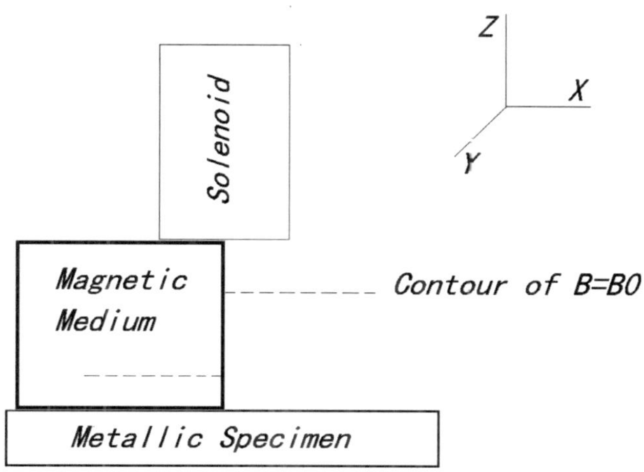

Figure 1.
Sketch of the device. The axis of a finite-sized solenoid is along z direction, the space between the solenoid and the conducting wire/specimen is filled with two kinds of magnetic mediums, which are represented by μ_{high} and μ_{low}. The μ_{low} medium can be the vacuum. A pair of electrodes yields a DC electric field along the x direction on the specimen. The dashed line represents the contour plane of B at a given value B_0. Different μ values cause the contour plane to have different z coordinates in two mediums.

2.2 Details on electron source

The above discussions have revealed theoretically the feasibility of an electron oscillation-based gamma-ray source. It is obvious that the electron oscillation-based radiation source is more advantageous than its proton oscillation-based counterpart because of larger oscillation magnitude, as well as power, available in the former. Utilization of electrons receives less attention than that of protons in experimental nuclear physics. It is really a pity if taking electrons as by-products of preparing protons. Reasonably utilizing those "by-products" is worthy of consideration.

Electron source can be designed to be compact and easily prepared. Among familiar electron sources, thermion-emission cathode is limited by its efficiency, and photocathode needs to be driven by high-intensity laser. The simplest method of achieving a high-efficiency electron source can share the same idea as that embodied in above sections, that is, using Hall effect by a magnetic slope in the above-mentioned discussion. Details are presented as follows.

Hall effect of a metal by a static (DC) magnetic field B_s is a familiar phenomenon caused by magnetic field. But until now, it is merely taken as a method of probing physical property of solid-state materials and hence its applications are often limited to weak magnetic field cases, which are usually at 10*G*-level. Higher strength of B_s needs stronger strength of current which might be beyond what a conducting wire can sustain and also is far beyond what most magnetic materials can produce [9].

Beside the strength of $|B_s|$, the space shape of B_s can also affect its interaction with matter. Such a space shape is described by the contour surface of B_s. For most components, their generated fields are usually of smooth contour surfaces. For example, B_s generated by a solenoid has contour planes normal to the axis of the solenoid, or the electromagnetic (EM) energy density profile $|B_s|^2$ is smooth or has a very small gradient $\nabla|B_s|^2$.

If the strength $|B_s|$ is not easy to be enhanced, adjusting the space shape of B_s is a worthy trial to optimize interaction. A high-gradient $|B_s|^2$ profile is not difficult to be produced. For example, one can on purpose make a pair of Helmholtz coils, which is a well-known device for screening external magnetic field, so that they are not co-axial. **Figure 1** displays a simple scheme for producing a high-gradient $|B_s|^2$ profile. As shown in **Figure 1**, the intrusion of a high-μ (magnetic permeability) medium distorts contours of $|B_s|^2$ to be bent. In other words, in each plane normal to the axis of the solenoid, a huge gradient of $|B_s|^2$ along the direction normal to the side surface of the medium appears. If a metal wire/specimen is arranged in such a high-$\nabla|B_s|^2$ region and an DC electric field E_s along the direction of $\nabla|B_s|^2$ is applied, the Hall effect in such a situation where the DC magnetic field is very space-inhomogeneous is worthy of being studied.

When studying applications such as probing and imagining local magnetic moment and magnetic microscopic structure [10–20], many authors have made in-depth investigation on the Hall effect of semiconductors in highly inhomogeneous magnetic field (HIMF). Because the purpose of these applications is detection or probing, the electric field or bias DC field is designed to avoid the breakdown of the semiconductor and hence its strength is usually not too strong. That is, in applications for detection purpose, Hall current is not required to be large enough.

It is worth noting the potential value of the extension of the same idea to a different case. The purpose of such an extension is aimed at a controllable "break-down" of the metal. Therefore, higher DC field strength is chosen. Now that Hall effect implies that electrons have the potential to run along a direction normal to the applied electric field, it is natural for us to consider the feasibility of side escape of electrons from a conducting wire through Hall effect. This drives us to actively establish a HIMF and apply it to metal under a higher-strength DC electric field.

E_s can achieve $10^5 V/m$-level by letting the inter-plate distance of a pair of plane-plate electrodes as $10^{-3\sim-2}m$ and applied voltage as $10^{2\sim3}V$. Parameter values at such a level is not difficult to be realized technically. Because of the condition $\nabla \cdot B = 0$ and the fact that the solenoid is finite-sized, B_s has two components $B_x\, \vec{e_x}$ and $B_z\, \vec{e_z}$, where $B_x = \sum_i g x_i x^{2i+1}$ and $B_z = \sum_i g z_i x^{2i+1}$ correspond to a vector potential $\vec{A} = A_x \vec{e_x} + A_z \vec{e_z} = -yB_z \vec{e_x} + yB_x\, \vec{e_z}$.

As shown in **Figure 1**, the solenoid is arranged on the demarcation line of two magnetic mediums. The end section of the solenoid is taken as the $z = 0$ plane, and the metal is arranged at $z < -|z_d|$ region. The $(x < 0, |z_u| > z > -|z_d|)$ region is filled with $\mu = \mu_{high}$ medium and the $(x > 0, |z_u| > z > -|z_d|)$ region with $\mu = \mu_{low}$ medium. According to the theory of electromagnetism, the contour plane $B_z = B_0$ in the $(x < 0, -|z_d| > z)$ region and that in the $(x > 0, -|z_d| > z)$ region will have different z coordinates. This implies a discontinuity in B exists near the demarcation line. That is, different values of the dropping rate $\partial_z |B|$ in two mediums cause the specimen in the $(-|z_d| > z)$ region to still feel a gradient $\partial_x |B|^2$. To ensure a sufficiently large gradient, the interface of two mediums is required to be smooth enough and hence should be polished/ground sufficiently. At present, mirror finish grinding can ensure surface roughness to be of $Ra <= 0.01\mu m$. This fundamentally warrants sufficiently large gradient $\partial_x \mu$, as well as sufficiently large $\partial_x B^2$ up to $\frac{T}{nm}$-level, to be feasible.

Because the DC magnetic field can effectively penetrate into metal interior if its direction is normal to the surface of a metal (in normal state), it can affect bulk electron states of the metal. In contrast, the AC magnetic field, or a light beam, is limited to the skin layer of the metal [21, 22].

For Al, its electron-phonon collision relaxation time τ is at $10^{1\sim2} fs$-level [21, 22], its Fermi velocity v_F is at $10^6 m/s$-level and its Fermi energy E_F is about $5.5eV$ [21, 22]. If a cm^3-level Al cubic specimen is placed between a pair of electrode plates with $220V$ voltage, the DC electric field E_s it feels will be at $10^4 V/m$-level. If we merely take into account the work done by E_s, the maximum velocity increment along the E_s direction, $\max\Delta v_x$, can reach $eE_s\tau/m_e = 1.6/9.1 * 10^{-19+4-15+1+31} \approx 20m/s$, which corresponds to $\frac{1}{2} m_e v_x^2 \sim 200 * 9.1 * 10^{-31} J \approx 10^{-9} eV$.

Emission is a many-body process because the sheath field, or space charge effect, left by emitted electrons in turn affects emission [23–26]. This phenomenon can be reflected by following quantum theory (21, 23–26),

$$i\hbar\partial_t \psi_k = \frac{1}{2m_e}\left[i\hbar\nabla + eB_0 y e_x\right]^2 \psi_k + U_i \psi_k + V(x,y)\psi_k + eE_0 x \psi_k + V_{ph}(x,y,t;T)\psi_k. \tag{18}$$

$$\nabla^2 V = \frac{e}{4\pi\varepsilon_0} n_0 \left(1 - \frac{\int\int |\psi_k|^2 f(k,T) dk_x dk_y}{n_0}\right) \tag{19}$$

where

$$V_{ph} = \int\int \left[\vec{u_q} \cdot \nabla_r U_i\right] g(k,T) dq_x dq_y, \tag{20}$$

$$\vec{u_q} = \sin\left(q_x x + q_y y + q_z z - \nu_q t\right)\vec{e_p}, \tag{21}$$

n_0 is the average background ionic density, $\psi_k(x,y,t) = \exp(S_k)\psi_k^0$, ψ_k^0 is the unperturbed wavefunction and f is the Fermi-Dirac distribution function. V_{ph} is the vibrating lattice potential, u is the field of ionic displacement, U_i is the lattice potential at zero-temperature, g is the Bose-Einstein distribution function, ν_q is the phonon dispersion relation and T is the temperature. More comprehensive model should contain a motion equation of the displacement field u, which is derived from the Lagrangian density of the electron-phonon system. This will be done in future work. Here, we approximate u as prescribed. Such an approximation is acceptable because the temporal variation of u is merely obvious over a large time scale $\frac{2\pi}{\nu_q}$ which is usually $> 100fs$.

The equation of S_k reads

$$\begin{aligned} i\hbar\partial_t S_k = & -\frac{\hbar^2}{2m_e}\left[\begin{array}{c}(\partial_{xx}+\partial_{yy}+\partial_{zz})S_k + (\partial_x S_k)^2 + (\partial_y S_k)^2 + (\partial_z S_k)^2 \\ +2ik_x\partial_x S_k + 2ik_y\partial_y S_k + 2ik_z\partial_z S_k\end{array}\right] \\ & -\frac{\hbar B_0}{m_e}ey(k_z f_x - k_x f_z) + \frac{i\hbar B_0}{m_e}ey(f_x\partial_z - f_z\partial_x)S_k \\ & +\frac{e^2B_0^2(f_x^2+f_z^2)}{2m_e}y^2 + eE_0x + V(x,y) + V_{ph}, \end{aligned} \tag{22}$$

where the space inhomogeneity of B_s is reflected by f

$$f = (f_x, 0, f_z) = (B_x/B_0, 0, B_z/B_0). \tag{23}$$

Note that $f = (0,0,1)$ corresponds to a space-homogeneous $B_s = B_0$ along z direction.

Actively applying highly space-inhomogeneous external field, especially DC magnetic field, might be an effective way of enhancing the effect of the external field on the electrons. According to Hamiltonian formula or Eq. (22), there is always an operator $A \cdot \hat{p} \sim Br\nabla$. Space-inhomogeneous B will cause more space-inhomogeneous wavefunction than space-uniform B. Because the energy of an electron is also dependent on the space derivative of the modulus of its wavefunction, more space-inhomogeneous wavefunction often implies larger energy.

To warrant the technique route to be competitive in economics and efficiency among all candidates for a same goal, we avoid more intermediate conversion steps in EM energy utilization, and favor direct usage of EM energy in power frequency (PF), the most primitive EM energy form for all physics laboratories.

3. Conclusion

The application value of such an electron oscillation-based gamma-ray source is obvious. It offers a more efficient way of manipulating nuclear matter through its characteristic EM stimulus, that is, gamma ray. At present, the goal of manipulating nuclear matter is mainly achieved through: (1) using Bremsstrahlung by proton output from accelerators—this implies the application of an EM stimulus of a broad spectrum to the nucleus, and hence the efficiency of this route is poor because most photons are of low frequency relative to nuclear matter; (2) using EM radiations from heavier radioactive elements—the dose, or the brightness, or the intensity of gamma ray generated in this route is limited and hence the manipulation is also less efficient; (3) injecting protons into target nucleus. In contrast, the electron oscillation-based mono-color gamma-ray source proposed in this work can warrant sufficient dose/brightness/intensity and hence an efficient manipulation of nuclear matter. Especially, the manipulation of a nucleus is not at the cost of destroying many nuclei to generate a desired tool, that is gamma ray with sufficient intensity, for achieving this goal. This fundamentally warrants a practical manipulation of more nuclei at desirable number.

Author details

Hai Lin*, ChengPu Liu and Chen Wang
State Key Laboratory of High Field Laser Physics, Shanghai Institute of Optics and Fine Mechanics, Shanghai, China

*Address all correspondence to: linhai@siom.ac.cn

References

[1] Hamilton JH. Radioactivity in Nuclear Spectroscopy. Vol. 1. New York: Gordon and Breach; 1972

[2] Lederer CM, Shirley VS. Table of Isotopes. 7th ed. New York: John Wiley & Sons Inc; 1978

[3] Sparks P. Handbook of Radioisotpes. New York: NY Reaserch Press; 2015

[4] Chaudri MA. IEEE Transactions on Nuclear Science. 1975;**26**:2281

[5] Takeuchi S. Relativistic $E \times B$ acceleration. Physical Review E. 2002; **66**:037402

[6] H Lin. Monocolor radiation source based on low-energy electron beam and DC fields with high gradient of electromagnetic energy density. Journal of Lasers, Optics and Photonics. 2017;**4**. DOI: 10.4172/2469-410X.1000161

[7] H Lin. A simpled and universal setup of quasimonocolor gamma ray source, arXiv. 1503.00815

[8] Lin H. Miniaturization of solid-state accelerator by compact low-energy-loss electron reflecting mirror. Europhysics Letters. 2015;**109**:54004

[9] Strnat KJ. Handbook of Magnetic Materials. Vol. 41988. Elsevier; p. 137

[10] Bending SJ, Local magnetic probes of superconductors. Advances in Physics. 1999;48:449

[11] Bending SJ, Oral A. Hall effect in a highly inhomogeneous magnetic field distribution. Journal of Applied Physics. 1997;**81**:3721

[12] Thiaville A et al. Measurement of the stray field emanating from magnetic force microscope tips by Hall effect microsensors. Journal of Applied Physics. 1997;**82**:3182

[13] Peeters FM, Li XQ. Hall magnetometer in the ballistic regime. Applied Physics Letters. 1998;**72**:572

[14] Nabaei V et al. Optimization of Hall bar response to localized magnetic and electric fields. Journal of Applied Physics. 2013;**113**:064504

[15] Veracke K et al. Size dependence of microscopic Hall sensor detection limits. Review of Scientific Instruments. 2009; **80**:074701

[16] Engel-Herbert R, Schaadt DM, Hesjedal T. Analytical and numerical calculations of the magnetic force microscopy response: A comparison. Journal of Applied Physics. 2006;**99**: 113905

[17] Liu S et al. Effect of probe geometry on the Hall response in an inhomogeneous magnetic field: A numerical study. Journal of Applied Physics. 1998;**83**:6161

[18] Koon DW, Knickerbocker CJ. What do you measure when you measure the Hall effect? Review of Scientific Instruments. 1993;**64**:510

[19] Comelissens YG, Peeters FM. Response function of a Hall magnetosensor in the diffusive regime. Journal of Applied Physics. 2002;**92**: 2006

[20] Folks L et al. Near-surface nanoscale InAs Hall cross sensitivity to localized magnetic and electric fields. Journal of Physics C. 2009;**21**:255802

[21] Ashcroft NW, Mermin ND. Solid State Physics. Orlando: Harcourt. Inc; 1976. p. 5, 36, 37

[22] Kittel C. Introduction to Solid State Physics. New York: John Wiley Sons; 2005

[23] Lang ND, Kohn W. Theory of Metal Surfaces: Charge Density and Surface Energy. Physical Review B. 1970;**1**:4555

[24] Jensen KL. General formulation of thermal, field, and photoinduced electron emission. Journal of Applied Physics. 2007;**102**:024911

[25] Jensen KL. Exchange-correlation, dipole, and image charge potentials for electron sources: Temperature and field variation of the barrier height. Journal of Applied Physics. 1999;**85**:2667

[26] Rokhlenko A, Jensen KL, Lebowitz JL. Journal of Applied Physics. 2010;**107**: 014904

Chapter 4

Gamma-Ray Emitting Radionuclides in People Living in Northern Sub Arctic Regions

Bertil R.R. Persson

Abstract

In 1960 Caesium-137 activity from atmospheric nuclear weapons test was discovered in measurements at the whole body gamma-ray counting laboratory in Lund. This event initiated measurements of the Swedish Sami population, and in products from reindeer that bite lichens in the Swedish mountains. A semi-portable whole-body counter designed with a detection limit for ^{137}Cs at high radiation background areas which was good enough for measuring people with high body concentration of ^{137}Cs. The ^{137}Cs activity concentration in Sami people increased during 1963–1965 from 300 to 600 Bq/kg body weight. Some individual males had values above 1000 Bq/kg. The catastrophic nuclear accident on the 25–26 April 1986 at the Chernobyl Nuclear Power Plant caused a massive release of fission- and neutron-activation products to the atmosphere. Already the following day the atmospheric plume of released radioactivity reached Sweden and was deposited over the central part of Sweden also in Sami populated reindeer raising districts. During 1991 and 1992 whole-body content of ^{137}Cs was measured in the Sami population of northern Sweden and similar levels were found as during 1963–1965. These levels are about twice those estimated in people living in the Chernobyl contaminated area.

Keywords: Caesium-137, nuclear-weapons test, sodium-iodine (Tl), detector, whole-body counting, Sami, Chernobyl, principal component analysis, PCA

1. Introduction

All people contain natural potassium of which a small fraction (0.012%) consists of the radioactive gamma emitting isotope ^{40}K with the half-life of 1,258,300,000 years. A 70 kg adult human contains about 140 g of potassium in the whole body, of which 252,929,911,920,000,000,000 atoms are ^{40}K. Every second a number of 4453 ^{40}K atoms decay in the human body, whereby emitting gamma radiation with the energy 1.46 MeV. Half of the gamma radiation is absorbed in the body while about 200 gamma ray photons per second are leaving the body. By recording these gamma rays in whole-body gamma-ray counting laboratories, the potassium content in the human body can be estimated and used in medical research.

In 1960 when people were measured for potassium at the whole body gamma-ray counting laboratory in Lund, the presence of another gamma-ray emitting radionuclide, Caesium-137, appeared as well. This ^{137}Cs-activity originates from nuclear weapons test performed in the atmosphere. Whole-body measurements of the Swedish Sami population, who consume products from reindeer that bite lichens in the Swedish mountains showed substantial content of ^{137}Cs.

This chapter tells the story of how everything began and finally contributed to the termination of the extensive nuclear weapons testing in the atmosphere.

2. The low-level gamma ray laboratory

The research took place at the low-level gamma-ray laboratory located in the culverts of the children's hospital in Lund. There, a room was fitted with thick walls of iron plates to screen a sensitive Sodium iodide detector from ambient background radiation.

In the room, a crystal of sodium iodide, doped with thallium, was recording gamma radiation from radioactive substances in the human body. When the gamma rays hit the crystal, light flashes appear, which are captured by a photosensitive photo-anode in a photomultiplier tube which, in turn, generate electrical impulses whose magnitude are due to the energy of the gamma-rays. One of the inventors of the NaI (Tl) scintillation crystal was professor Sven Johansson at the Department of Nuclear Physics at Lund University. Recently the first small crystals he produced was found when tidying up at the Physics department in Lund.

Figure 1 shows the principle of how the sodium-iodine (Tl) detector record gamma-rays.

The original idea for the gamma-ray laboratory was to use the equipment to record the most common natural radioactive substance in the human body, potassium-40, for medical research [2]. As a quality control, the collaborators at the

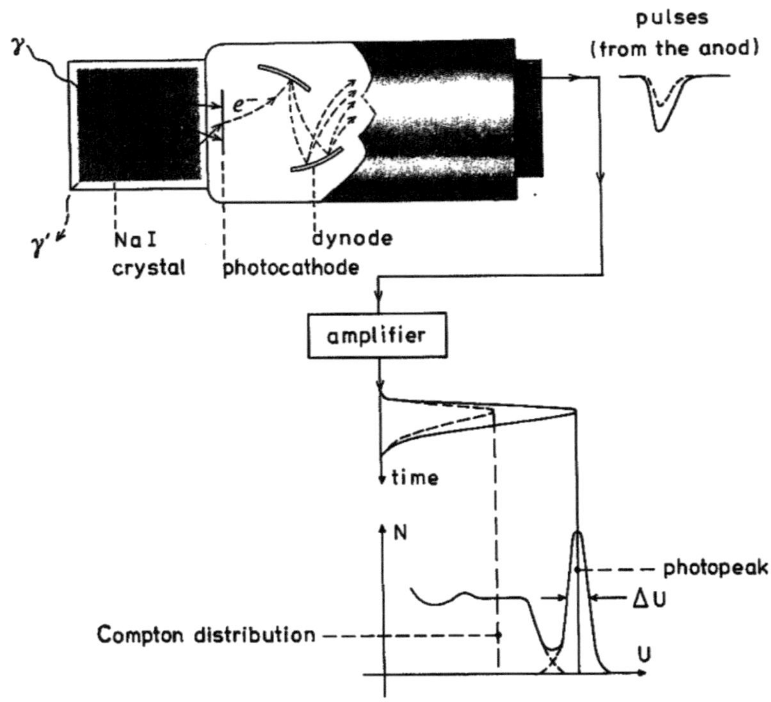

Figure 1.
The principle of how a sodium-iodine (Tl) detector record gamma-rays [1].

gamma-ray laboratory regularly measuring their potassium content. One collaborator was from Norway, and in 1960, when he returned from his vacation in Norway, a broad peak indicating gamma radiation from ^{137}Cs appeared in the whole-body spectrum. He told that he had been eating cheese produced from sheep's baiting in the Norwegian mountains. The cheese contained high levels of radioactive caesium-137 released to the atmosphere by the United States nuclear weapons testing in the Pacific.

A similar relationship would exist in the reindeer raising Sami people who live in Northern Sweden. The transfer of atmospheric fallout through the food-chain Lichen-reindeer-man shown in **Figure 2**. Another collaborator Calle Carlsson (whom later on became a professor in Radiation Physics at Linköping University) was sent to Northern Sweden to collect samples. The piece of reindeer meat he brought from Northern Sweden to Lund was found to contain just over 1000 Bq of ^{137}Cs per kg. That was quite a high level compared to the ^{137}Cs levels in ordinary beef from Southern Sweden which contained about 4 Bq per kg. It became apparent that the Sami people should have high levels of caesium-137 due to the food chain lichen-reindeer-man as shown in **Figure 2**.

At the beginning of May 1961 encouraged by the results of Calle Carlsson's exploration, Kurt Lidén invited two Sami people from Lapland (67°N) to Lund. They measured at the gamma-ray laboratory, and it discovered that they had a caesium-137 content which was much higher than earlier was found in the Norwegian collaborator.

A research program was arranged to study the Caesium-137 activity in the Swedish Sami population, and in products from reindeer that bite lichens in the Swedish mountains. I participated in the annual expeditions to the Sami villages from Funäsdalen up to Karesuando at the Northern Swedish border to Finland. A mobile whole body counter laboratory was built on a military bus, lent by the National Defence Research department in Stockholm (FOA). In the middle of the bus was placed a lead-protected whole body gauge with a large NaI(Tl) crystal which was 5 inch in diameter and 4 inches thick.

Yngve Naversten was responsible for the design of the semi-portable whole-body counter [4]. The 42 cm wide chair arrangement was similar to the stationary whole-body counter at Lund and shielded was lead bricks. Since the ^{137}Cs body

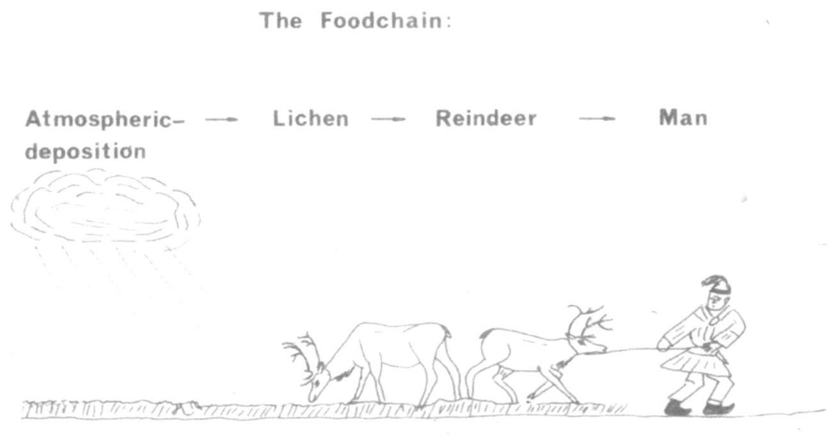

Figure 2.
A sketch illustrating the food chain lichen-reindeer-man drawn by the author's colleague and friend MD Torsten Landberg (d. 2015) as a dedication to my thesis 1970 [3].

burdens in the Sami population was in the order of 4 kBq the background shielding was not critical.

The sensitivity of the NaI(Tl) detector for ^{137}Cs and potassium (^{40}K) in a subject of 70 kg body weight, is 0.048 counts per minute per Becquerel (Bq) ^{137}Cs, and 0.18 counts per minute per gram potassium.

The detection limit for ^{137}Cs in people at the high radiation background level at Jokkmokk, corresponded to about 14 Bq ^{137}Cs/kg, which is good enough for their people, with average body concentration of about 55 Bq ^{137}Cs/kg. In Lund, however, the detection limit corresponded to about 10 Bq/kg which is too low for people in the control group with body concentration of about 12 Bq/kg. They measured in the iron shielded room with detection limit corresponding to about 3 Bq/kg.

3. Atmospheric nuclear weapons testing

The origin of ^{137}Cs in the environment was the atmospheric nuclear weapons tests started by the development of the atomic bombs, which in 1945 felt over Hiroshima and Nagasaki. After World War 2 the developing and testing of nuclear weapons continued by both the USA and the Soviet Union. A summary of the annual Nuclear weapon explosion yield by the USA and the USSR up to 1990 given in **Figure 3** in units of an equivalent amount (kilotons, kt) of the conventional explosive tri-nitro-toluene (TNT) [5, 6].

A Limited Test Ban Treaty, pledging to refrain from testing nuclear weapons in the atmosphere, underwater, or in outer space, was signed by the three (UK, US, Soviet Union) of the four nations developing nuclear weapons in 1963. The fourth nation France continued atmospheric nuclear weapon testing until 1974. Later China started to develop their nuclear weapons and continued with atmospheric testing until 1980.

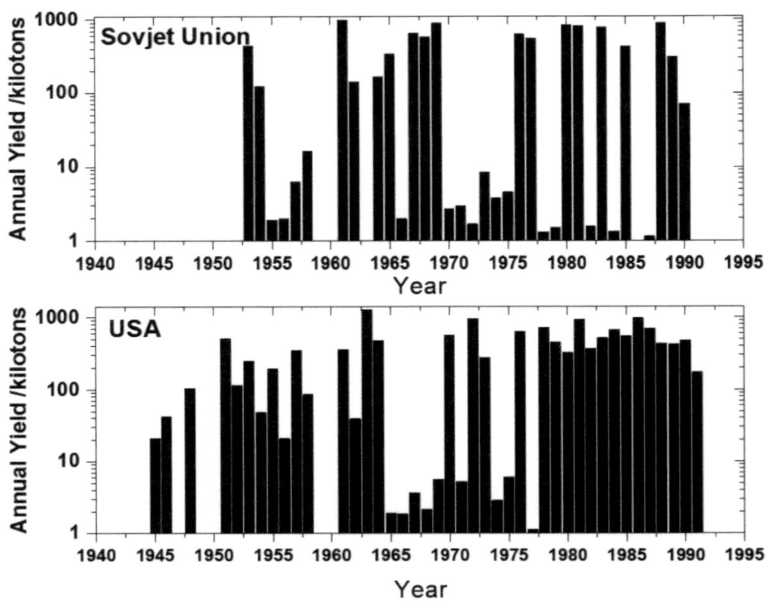

Figure 3.
Annual nuclear weapon explosion yield by the USA and the USSR up to 1990, expressed in an equivalent amount (kilotons, kt) of the conventional explosive tri-nitro-toluene (TNT) [5, 6].

Calle Carlsson declared in his death row over Kurt Lidén:

> *"I am convinced that the high radioactivity of reindeer raising people in The Northern Sub Arctic regions accelerated and actually forced the Test Prohibition Agreement pledging to refrain from testing nuclear weapons in the atmosphere"* [7].

After the Treaty in 1963 to refrain from testing nuclear weapons in the atmosphere, United States continued to perform underground nuclear weapons testing until 1992 (its last nuclear test), and so did the Soviet Union until 1990.

In the explosion of a nuclear weapon, its load of uranium or plutonium splits into two lighter nuclei and release an enormous amount of energy, a process called fission. The most long-lived gamma-ray emitting lighter element released is ^{137}Cs with a half-life of about 30.2 years. During the period of atmospheric testing, ^{137}Cs was spread in the atmosphere and deposited as nuclear fallout and transferred to plants, animals, and man all over the world.

4. Reference groups for whole-body measurements

In January 1960 a reference group for whole-body measurements which comprised of people of different age and both sexes was established in Lund. Since 1964 everyone in this group could not regularly participate in the whole-body measurements, and two utterly new reference groups started—one with 33 school-children aged between 14 and 15 years, from a school in Lund. The other group included 14 women and 20 men who worked nearby in a local factory. Three times a year, these groups experienced whole-body measurements of ^{137}Cs originated from the fallout of atmospheric nuclear weapons tests and natural ^{40}K for estimation of the total potassium content in the body. Since both the element Caesium (Cs) and potassium (K) belong to alkali metals, the first group in the periodic system, their physiological behaviour was believed to be highly correlated. Thus the ^{137}Cs-body content, is sometimes indicated as the ^{137}Cs/K ratio in Bq/gK. **Figure 4** displays the time variation of the ^{137}Cs/K ratio Bq/gK in the control groups at Lund, with the total yields of atmospheric nuclear weapons testing derived from **Figure 3** at the bottom.

In 1964, a comprehensive survey of school children aged 14–15 conducted at five different places, in different parts of Sweden. The selection criteria were as follow.

- That the students should have lived in the resort for at least 1 year.
- That they did not belong to any Same families.

The results given in **Table 1** are ^{137}Cs levels in March 1964 of non-Sami schoolchildren at the age of 14–15 years.

The activity concentration of ^{137}Cs in the Nässjö group was slightly higher than the Lund group, which may be due to higher concentrations of ^{137}Cs in locally produced foods. The high levels in Jokkmokk and Funäsdalen are probably due to consumption of reindeer products. While the unexpectedly low values in Lycksele may be that reindeer products are easier available in Jokkmokk and Funäsdalen than in Lycksele.

A partial least square modelling of the activity concentration of ^{137}Cs given in **Table 1**, with position(°N,°E), age (a), body-weight (kg_{BW}) and potassium concentration (gK/kg_{BW}) as dependent variables was performed for females (FM) and

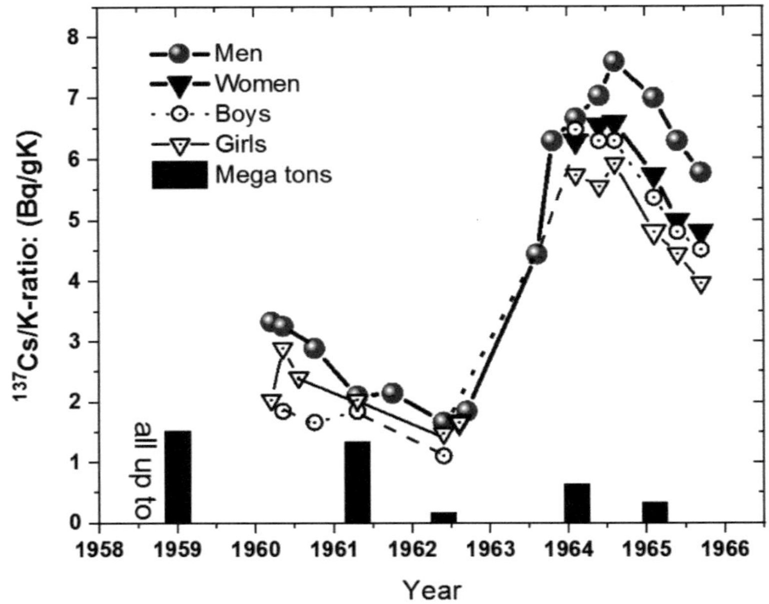

Figure 4.
Time variation of the $^{137}Cs/K$ ratio Bq/gK in the control groups at Lund. The columns at the bottom represent the yields of atmospheric nuclear weapons testing.

Location	Weight	Weight	^{137}Cs	^{137}Cs	K	K
Place	**M kg**	**FM kg**	**M Bq/kg**	**FM Bq/kg**	**M gK/kg**	**FM gK/kg**
Jokkmokk AJ	44	43	60.13	50.34	1.68	1.75
Lycksele ALy	44.5	42	21.53	17.44	1.59	1.61
Funäsdalen AF	41	45.5	45.39	53.26	1.69	1.53
Nässjö AN	45	43.5	15.62	13.78	1.74	1.68
Lund AN	46.5	48	12.89	10.18	2.01	1.84

Table 1.
The ^{137}Cs-concentration, Bq per kg body weight (Bq/kg), the potassium concentration in g per kg body weight gK/kg measured 1964 in male (M) and female (FM) school children (age about 14) at the various locations in Sweden and who did not belong to any Sami families.

males (M) respectively. The equations for prediction of ^{137}Cs activity concentration in Bq per kg body weight are given below for females and males respectively.

$$^{137}\mathrm{Cs}(\mathrm{Bq/kg_{BW}FM}) = -81.6 + 1.366 \times (^{\circ}\mathrm{N}) + 0.379 \times (^{\circ}\mathrm{E}) + 7.13 \times (\mathrm{a}) - 0.569 \times (\mathrm{kg_{BW}}) - 32.88 \times (\mathrm{gK/kg_{BW}}) \quad (1)$$

$R^2 = 0.32$,

$$^{137}\mathrm{Cs}(\mathrm{Bq/kg_{BW}M}) = -169.8 + 3.796 \times (^{\circ}\mathrm{N}) + 1.584 \times (^{\circ}\mathrm{E}) + 5.843 \times (\mathrm{a}) - 7.623 \times (\mathrm{kg_{BW}}) - 113.0 \times (\mathrm{gK/kg_{BW}}) \quad (2)$$

$R^2 = 0.88$.

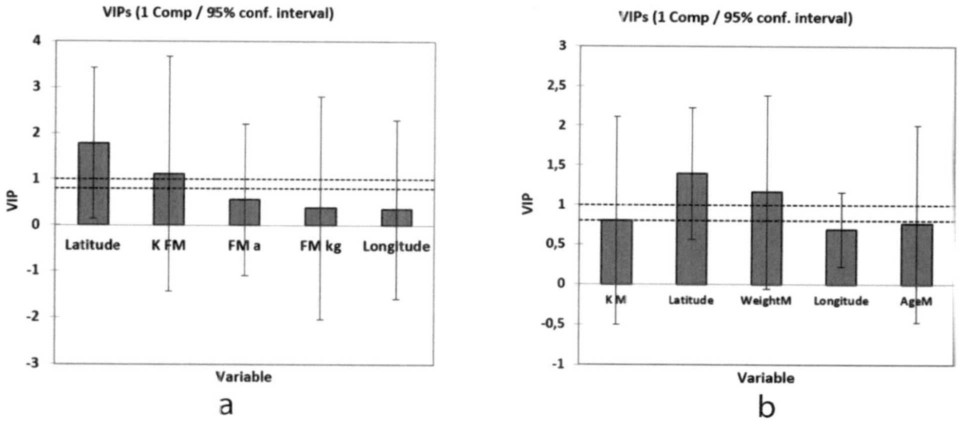

Figure 5.
Variable importance in the projection of the ^{137}Cs prediction equations for (a) female and (b) male.

Variable importance in the projection (VIP) of the ^{137}Cs prediction equations (Eqs. (1) and (2)) are given in **Figure 5a** for female and **Figure 5b** for male respectively. The most important variables for female are latitude and potassium concentration, while for male, the body weight is of great importance as well.

5. Whole-body measurement reference groups from the rural population in northern Sweden

Other whole-body measurement reference groups established at different places in Sweden with farmers, forest workers and other heavy working people with their family members. The aim was to compare the body activity of ^{137}Cs in the Sami people with other groups of people,

As seen from **Table 2** the activity concentration of ^{137}Cs in females are about half of the corresponding values in males.

A principal component analysis (PCA) was performed with the variables ^{137}Cs-concentration (Bq/kg$_{BW}$), potassium concentration (gK/kg$_{BW}$), and position (°N, °E). The results shown in **Figure 6** are those labelled A from **Table 1**, and that labelled B are from **Table 2**.

As seen in **Figure 6** the reference groups are well separated, with Funäsdalen (62.50°N, 12.50°E) and Lycksele (64.6°N,18.7°E) as a subgroup in A and Övre Soppero (68.09°N, 21.70°E), Jokkmokk (66.60°N, 19.80°E) and Arvidsjaur (65.66°N, 19.47°E) as a subgroup in B, and the two groups in Jokkmokk separated.

Location	Male	Female	Male	Female
	^{137}Cs Bq/kg$_{BW}$	^{137}Cs Bq/kg$_{BW}$	gK/kg$_{BW}$	gK/kg$_{BW}$
Övre Soppero BOS	273.8	166.5	2.3	1.8
Jokkmokk BJ	188.7	103.2	2.4	1.8
Arvidsjaur BA	92.1	57.0	2.2	1.8
Funäsdalen BF	80.7	34.4	2.1	1.9

Table 2.
The ^{137}Cs-concentration, Bq per kg body weight (Bq/kg), the potassium concentration in gK per kg body weight gK/kg$_{BW}$ in the reference groups from the rural population measured 1965 at various places in northern Sweden.

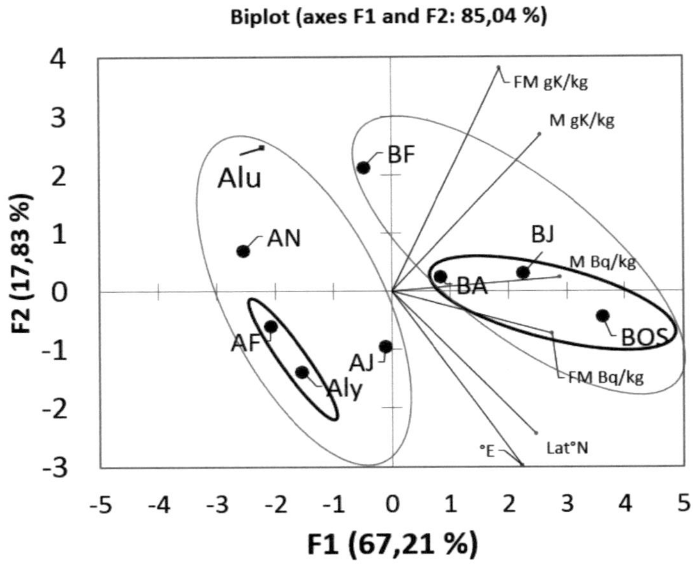

Figure 6.
Biplot of the results of principal component analysis (PCA) of the variables, latitude, longitude, ^{137}Cs-concentration, and potassium concentration ***Table 1*** *Labelled A and in* ***Table 2*** *labelled B.*

A partial least squares modelling of the activity concentration of ^{137}Cs given in **Table 2**, with position (°N,°E), and potassium concentration (gK/kg_{BW}) as dependent variables was performed for FM and M respectively. The equations for prediction of ^{137}Cs activity concentration in Bq per kg body weight are given below for females and males respectively.

Male $R^2 \approx 0.6$

$$^{137}\mathrm{Cs}\ (\mathrm{Bq/gK}) = -474.743 + 5.109 \times (^{\circ}\mathrm{N}) + 2.550 \times (^{\circ}\mathrm{E}) + 70.549 \times (\mathrm{gK/kg_{BW}}) \quad (3)$$

$$^{137}\mathrm{Cs}(\mathrm{Bq/kg_{BW}}) = -1189.764 + 12.436 \times (^{\circ}\mathrm{N}) + 6.253 \times (^{\circ}\mathrm{E}) + 182.643 \times (\mathrm{gK/kg_{BW}}) \quad (4)$$

Female $R^2 \approx 0.98$

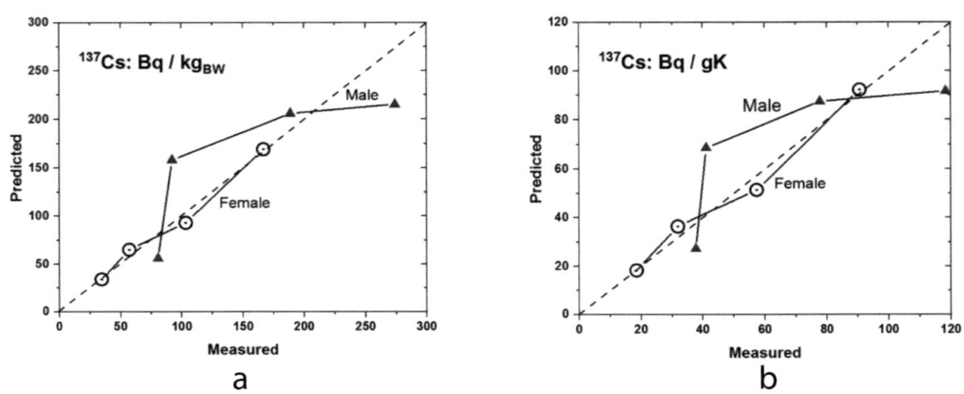

Figure 7.
(a) Predicted versus the measured ^{137}Cs body concentration in Bq/kg body weight. (b) Predicted versus the measured value of the ratio of ^{137}Cs activity (Bq) and potassium content (gK).

$$^{137}\text{Cs}\ (\text{Bq/gK}) = -1614.243 + 9.307 \times (^{\circ}\text{N}) + 3.737 \times (^{\circ}\text{E}) + 539.87 \times (\text{gK/kg}_{\text{BW}}) \quad (5)$$

$$^{137}\text{Cs}(\text{Bq/kg}_{\text{BW}}) = -3034.159 + 17.152 \times (^{\circ}\text{N}) + 6.855 \times (^{\circ}\text{E}) + 1027.201 \times (\text{gK/kg}_{\text{BW}}) \quad (6)$$

The predicted values from these equations are plotted against the measured values from **Table 2** and displayed in **Figure 7b**.

The large spread in the values for males is probably due to large variations in body-weight and potassium content. While these quantities are less variable for the female group.

6. Whole body measurements of reindeer raising Sami

The high accumulation of fallout from nuclear bomb tests in the food chain lichen - reindeer meat—man (**Figure 2**), and large consumption of reindeer meat result in substantially increased ^{137}Cs levels in reindeer-raising Sami.

This relationship was explored during 1961 and 1962 by measurements of the ^{137}Cs body burden of the Same population in Jokkmokk. The Whole-body measurement program extended in 1963 to some other places in the reindeer-raising district. The results of these measurements which are summarised below have previously published by Kurt Lidén and collaborators [8–10].

Figure 8 shows the activity concentration (Bq/kg) of ^{137}Cs during the period 1961–1965 in old and young male Sami people in mountain Sami villages Jåkkåkaska, Tuorpon, Loukta-Mavas and Sirges (previously named Sirkas), as well in non-Sami rural people in the Jokkmokk district. The ^{137}Cs activity concentration increased rapidly between 1963 and 1964 due to the extensive atmospheric nuclear weapon testing in 1961. In 1965, average levels in the range of about 200–700 Bq/kg_{BW} were reached, with some individual males above 1000 Bq/kg_{BW}.

In **Figure 8** presents the variation of the ^{137}Cs activity concentration during the period September 1961–March 1965 for male elderly (age 20–70a) and youngster (age 11–19a) for both the Sami people and the control group at Jokkmokk. Between 1964 and 1965, the older group in rural non-Sami people in Jokkmokk district has a lower level of ^{137}Cs than the group in mountain Sami village Sirges (Si), which in turn was 40% lower than the groups Jåkkåkaska, Tuorpon, and Luokta-Mavas (JA). These differences are even higher for the younger group.

The activity concentration (Bq/kg) of ^{137}Cs during the period 1961–1965 in old and young male Sami people in various Sami villages (a and b) and rural non-Sami people in Jokkmokk district (c).

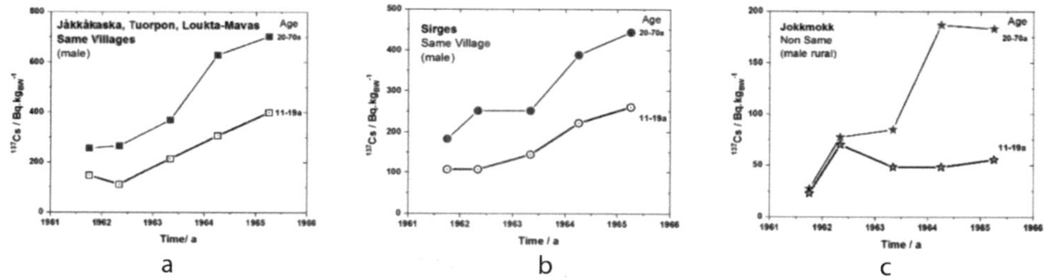

Figure 8.
(a) Mountain Sami villages people Jåkkåkaska, Tuorpon, and Loukta-Mavas. (b) Mountain Sami village Sirges. (c) Rural non-Sami in Jokkmokk district.

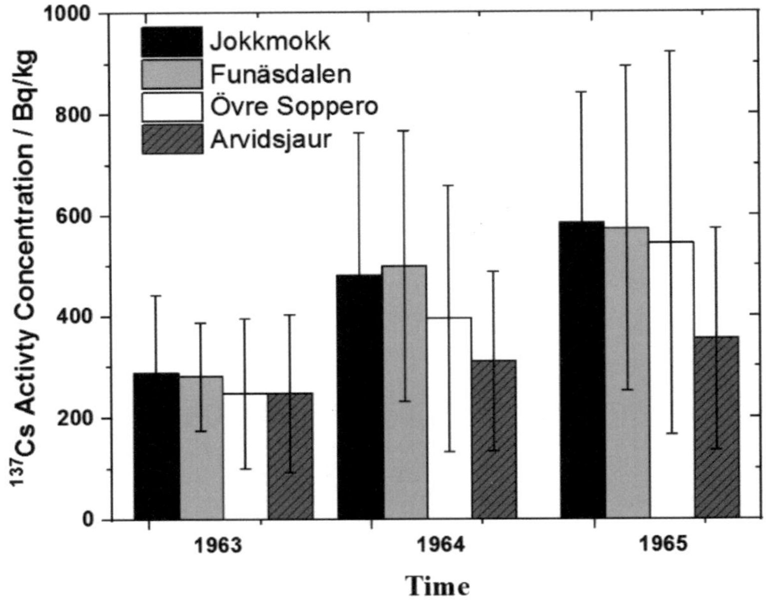

Figure 9.
The average values and the estimated standard deviation (SD) given for ^{137}Cs activity concentration during 1963–1965 in Sami people in four different reindeer raising districts. The Arvidsjaur district comprises three groups from forest villages. The other are from the mountain Sami villages in Jokkmokk municipality (Jåkkåkaska, Tuorpon, and Sirges) Funäsdalen and Övre Soppero.

Not shown in the figure, the differences in ^{137}Cs levels for the 20–70 age group of women in Sirges was lower than in JA with 10% in 1964 and 25% in 1965. The girls (age 11–19a) have about equal levels in 1964 in Sirges and JA. However, in 1965 the levels in Sirges was about 20% lower than in JA [10].

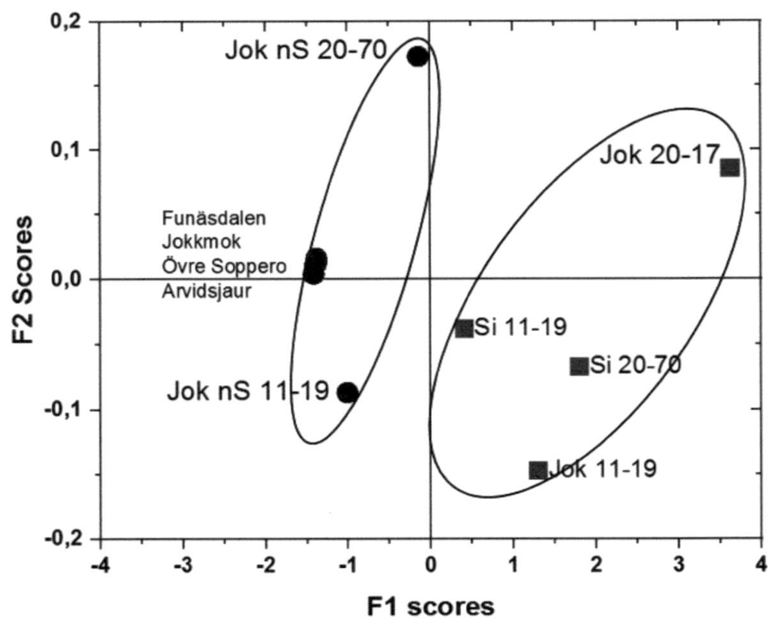

Figure 10.
The result of PCA analysis of all whole body measurements during 1963–1965.

In **Figure 9** are given the ^{137}Cs activity concentration during 1963–1965 in Sami people from four different districts. The Arvidsjaur district comprises three groups from forest villages, and the others are from mountain Sami villages in Jokkmokk municipality (Jåkkåkaska, Tuorpon, and Sirges), Funäsdalen and Övre Soppero. The average ^{137}Cs activity concentration during 1963–1965 increased from 300 to 600 Bq/kg body weight in Sami people living in mountain villages.

Although some individuals in 1965 showed ^{137}Cs levels above 1000 Bq per kg body weight, they expressed no fear about these high levels. Instead, individuals with high levels were respected and honoured, because high levels of radioactivity were a sign of hard and dedicated work with the reindeer.

Figure 10 shows the principal component analysis (PCA) results of the ^{137}Cs activity concentration during 1963–1965 in Saami people in four different districts displayed in **Figure 9** and the control groups given in **Table 2**. **Figure 10** clearly shows the difference between the Sami people and the people in the control groups. The main difference between the two populations is due to the higher consumption of meat and other products of reindeer by the Sami people.

7. Radiation dose contribution and health aspects

The annual radiation dose contribution to an adult person of a constant ^{137}Cs body-concentration of 1 Bq/kg_{BW} is estimated to be about 2.2 μSv per year. During 1995 the male reindeer-raising mountain Sami with an average ^{137}Cs body-concentration of about 500 Bq/kg received about 1.1 mSv per year.

The radiation-dose contribution to people who eat many foods with high concentrations of caesium-137 is estimated to be about 1–2 mSv/year. However, the radiation dose from a large intake of reindeer meat is also due to naturally occurring polonium-210 [11]. The total radiation dose from all environmental sources is estimated to be about 3–4 mSv/ year.

8. The Chernobyl accident

A catastrophic nuclear accident occurred on the 25–26 April 1986 in the No. 4 light water graphite moderated reactor at the Chernobyl Nuclear Power Plant near the now-abandoned town of Pripyat, in northern Ukraine, approximately 104 km north of Kiev. During several days the accident caused a large release of fission- and neutron-activation products to the atmosphere. Already the following day the atmospheric plume of released radioactivity reached Sweden. During April 28th a heavy rainfall deposited the radioactive dust in the plume over the central part of Sweden, leading to high surface contamination of ^{137}Cs (>100 kBq.m^{-2}) in Sami populated reindeer raising districts.

During 1991 and 1992 whole-body content of ^{137}Cs was measured in the Sami population of northern Sweden. The Radiation Physics department at Umeå University arranged measurements in three areas with various levels of ^{137}Cs deposition [12–14].

Two groups of individuals are randomly chosen, of which one group representing the urban population of the area, and the other the members of the Sami communities. The average whole-body content of ^{137}Cs in the general population varied between 1.1 and 2.0 kBq, and for the Sami population between 3.4 and 25 kBq. **Figure 11** shows the average ^{137}Cs activity concentration in the Sami and urban population in districts of various contamination level. In the most contaminated areas, there were some individuals with levels above 1000 Bq/kgBW.

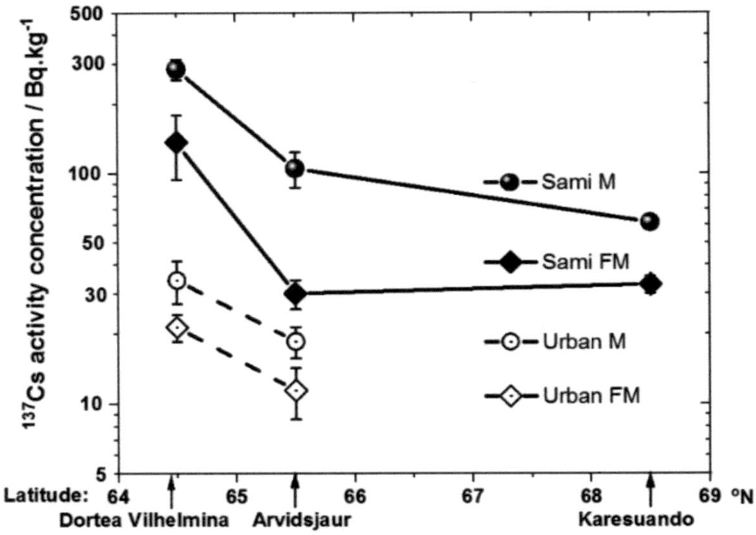

Figure 11.
The average ^{137}Cs activity concentration in the Sami and urban population in various districts [12].

9. Discussion and conclusion

Figure 12 displays the results of whole-body measurements of ^{137}Cs in the Swedish Sami population from nuclear weapons fallout during the 1960th and the Chernobyl fallout after the 1986 accident. For comparison, the 2006–2010 results of reported whole-body measurements of people living in the vicinity of Chernobyl are

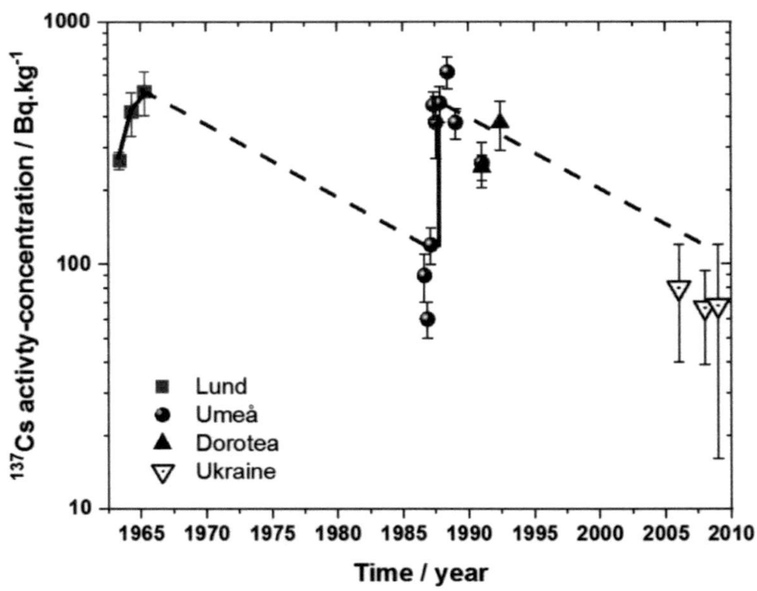

Figure 12.
Summary of the ^{137}Cs activity concentration in the Swedish Sami population measured by the research teams from Lund and Umeå universities. The ^{137}Cs activity concentration in people living in the vicinity of Chernobyl in Ukraine, (symbol Δ in the figure) derives from published data [15].

also given in **Figure 12** [15]. The extrapolated value of whole-body measurements of ^{137}Cs in the Swedish Sami population from 1994 to 2007 is about twice the Ukraine value. The Sami population living in Northern Sub Arctic regions is primarily affected by atmospheric deposition of the gamma emitting radionuclide ^{137}Cs.

Enhanced ^{137}Cs concentrations in reindeer raising people has also been reported from in Finland and Russia [16]. Eskimos and other inhabitants in Alaska consuming caribou (wild reindeer) show enhanced ^{137}Cs body concentrations [17].

In **Figure 13** is displayed values of the ratio of ^{137}Cs body activity (Bq) and potassium content (gK) in people living in Northern Sub Arctic regions and consuming reindeer or caribou (Alaska). The values in **Figure 13** derive from this work and **Table 3** in the 1964 Report of the United Nations Scientific Committee on the effects of atomic radiation [18].

The discovery of the massive increase of ^{137}Cs bodily activity (Bq) in the Sami population in Sweden and Finland and closely related populations of the Soviet Union and its steady increase in the 1960's actually forced the United States, United Kingdom and Soviet Union test bans in 1964 promising to refrain from test nuclear weapons in the atmosphere [7].

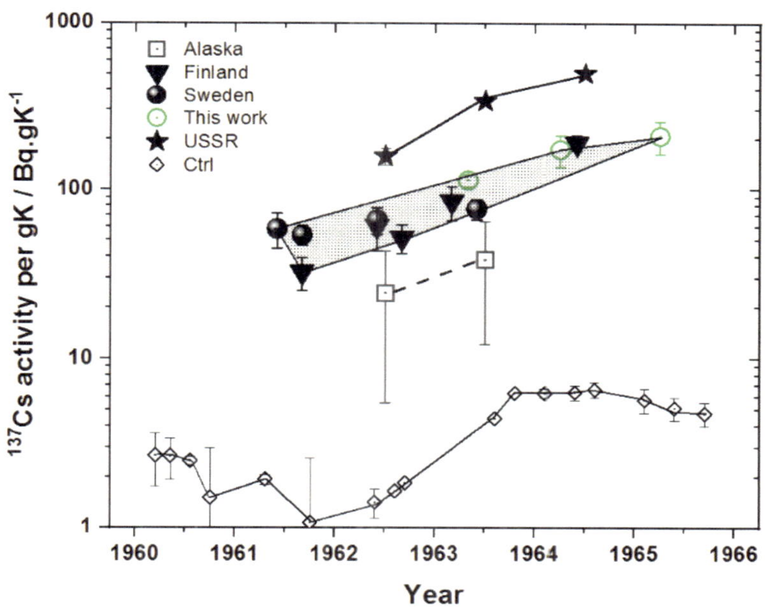

Figure 13.
The ratio of ^{137}Cs total body activity (Bq) and potassium content (gK) in people living in northern Sub Arctic regions consuming reindeer or caribou (Alaska). The lower curve is the average of the control groups in Lund [8].

Time	Jokkmokk	Funäsdalen	Ö Soppero	Arvidsjaur	AVE ± SD
1963.33	0.6 ± 0.3	0.6 ± 0.2	0.5 ± 0.3	0.5 ± 0.3	0.59 ± 0.05
1964.25	1.1 ± 0.6	1.1 ± 0.6	0.9 ± 0.6	0.7 ± 0.4	0.93 ± 0.19
1965.25	1.3 ± 0.6	1.3 ± 0.7	1.2 ± 0.8	0.8 ± 0.5	1.13 ± 0.24

Table 3.
The estimated annual radiation dose contribution to adults (mSv per year) of ^{137}Cs body-content at various places in Sweden and averages (AVE) with standard deviation (±SD) in the last column.

Acknowledgements

This chapter dedicates to professor emeritus Calle Carlsson on his 95th year of age, to honour his pioneering efforts in the project to measure the radioactivity of the Sami population living in Northern Sub Arctic regions, that forced the Test Prohibition Agreement pledging to refrain from testing nuclear weapons in the atmosphere.

Conflict of interest

The author declares no conflicts of interests.

Author details

Bertil R.R. Persson
Medical Radiation Physics, Lund, Lund University, Lund, Sweden

*Address all correspondence to: bertil_r.persson@med.lu.se

References

[1] Brune D, Forkman B, Persson B. Nuclear Analytical Chemistry. Lund: Studentlitteratur; 1984. 557 p

[2] Cederquist ES, Lidén KVH. Clinical use of NaI(Tl) counter. In: Proc Symp Whole-body Counting. Vienna; 1961

[3] Persson BRR. 55Fe, 90Sr, 134 Cs, 137Cs and 210Pb in the biosphere. In: Radiological Health Aspects of the Environmental Contamination from Radioactive Materials in Northern Sweden. Lund: Lund University; 1970

[4] Naversten Y, McCALL RC, Lidén K. Semi-portable whole-body counter for cesium 137 and other gamma-emitting isotopes. Acta Radiologica: Therapy, Physics, Biology. 1963;**1**(3):190-204

[5] DOE. United States Nuclear Tests: July 1945 through September 1992. (DOE/NV-209 REV15). Las Vegas, NV: Department of Energy, Nevada Operations Office. 2000-12-01; 2000. Contract No.: DOE/NV-209 REV15

[6] Mikhailov VN, editor. USSR Nuclear Weapons Tests and Peaceful Nuclear Explosions: 1949 through 1990: Sarov, Russia: The Ministry of the Russian Federation for Atomic Energy, and Ministry of Defense of the Russian Federation; 1996

[7] Carlsson C. Kurt Lidén: A pioneer in medical radiation physics (In Swedish: Kurt Lidén: En föregångare inom den medicinska strålningsfysiken). Strålskyddsnytt. 2001;**19**(2):20-21

[8] Lidén K, Gustafsson M. 137Cs levels of different population groups in Sweden. Acta Radiologica: Diagnosis. 1966;**4**(sup254):38-46

[9] Lidén K. ^{137}Cs burdens in Swedish Laplanders and reindeer. Acta Radiologica. 1961;**56**:237

[10] Lidén K, Naversten Y. The enhanced radiocaesium levels of people in northern Sweden. In: Assessment Od Radioactivity in Man. Vol. II. Wien: IAEA; 1964. p. 167

[11] Persson BRR, Gjelsvik R, Holm E. Radioecological modelling of Polonium-210 and Caesium-137 in lichen-reindeer-man and top predators. Journal of Environmental Radioactivity. 2018;**186**: 54-62

[12] Johansson L, Agren G. Cs-137 in the population of northern Sweden. Radiation Protection Dosimetry. 1994;**55**(2):131-142

[13] Johansson L, Bjoreland A, Wickman G, Eriksson A. Distribution of radioactive caesium in the population of northern Sweden: A follow-up study. Radiation Protection Dosimetry. 1999; **86**(1):59-62

[14] Johansson L, Wickman G, Agren G, Eriksson A, Jonsson H, Tavelin B. Distribution of radioactive caesium in the population of northern Sweden 1988–1993. Radiation Protection Dosimetry. 1995;**62**(4):203-212

[15] Lindgren A, Stepanova E, Vdovenko V, McMahon D, Litvinetz O, Leonovich E, et al. Individual whole-body concentration of (137)cesium is associated with decreased blood counts in children in the Chernobyl-contaminated areas, Ukraine, 2008-2010. Journal of Exposure Science & Environmental Epidemiology. 2015; **25**(3):334-342

[16] Miettinen JK, Jokelainen A, Roine P, Liden K, Naversten Y. ^{137}Cs and potassium in people and diet a study of Finnish Lapps. Annal Academia Scientia Fennicae. Ser. A11, Chemia. 1963:120

[17] Palmer HE, Roesch WC, Griffin BI, Hanson WC. Cesium-137 in Alaskan Eskimos. Science. 1963;**142**(3588):64

[18] UNSCEAR. Report of the United Nations Scientific Committee on the Effects of Atomic Radiation. New York: United Nations; 1964

Section 2

Radiation Detection and Measurement

Chapter 5

High Purity Germanium: From Gamma-Ray Detection to Dark Matter Subterranean Detectors

Nicolas Fourches, Magdalena Zielińska and Gabriel Charles

Abstract

High purity germanium remains the material of choice for the detection of photons in the range of MeV or higher, down to the hard X-ray range. Since the operation of HPGe-based detectors is possible only at or below the liquid nitrogen temperature, their advantage is mainly the resolution, which matches the Fano factor if appropriate cooled electronic readout is used. We focus here on present-day applications of HPGe detectors, which are now broader than ever despite the recent development of room-temperature photon detectors based on binary compounds. We present in particular dark matter detectors and γ-ray trackers as examples of the recent applications of HPGe as a detecting medium. More generally, we discuss the future of γ-ray detectors and the role that the semiconductor detectors will keep with respect to alternative detection materials. This chapter is an introduction to this general topic, and the reader is encouraged to refer to research and review articles on this subject published in the past or recently.

Keywords: high purity germanium, traps and defects, gamma-ray detection, dark matter, nuclear spectroscopy, nuclear material monitoring

1. Introduction

This chapter provides a technical overview of HPGe detectors and their applications, both in science and day-to-day life. This overview covers the applications of HPGe as a material for γ-ray detection and its other more recent use in particle physics. It represents an introduction rather than a complete and exhaustive description of possible detector applications of HPGe.

Since the 1970s, photon detectors (γ and X) have been developed from high purity germanium [1]. The reason why HPGe has remained in use for such a long time as high-resolution γ- and X-ray detector material is mainly because it contains a very low concentration of electrically active defects [2, 3] which may be lower than 10^9 cm^{-3}. Such value is difficult to reach in compound semiconductors. Even for detector-grade silicon, the doping level is slightly higher. High electron density (Z = 32), together with low average energy, is needed for e-h pair generation. Above 1 keV of initial particle energy, germanium is a good choice for the detection of photon or ionizing particles. See **Table 1** and Refs. [4–10] for studies, both

Material	Si	Ge	CdTe	GaAs	SiC β	(diamond) C
Density in gcm^{-3}	2.33	5.33	5.85	5.32	3.21	3.5
Bandgap	1.1 eV	0.67 eV	1.44 eV (dir)	1.4 eV (dir)	2.3 eV	5.47 eV
Breakdown field (MV/cm)	0.3	0.1	0.4	0.4	2	20
ε or Eth	3.6 eV	2.98 eV	~4.5 eV	~4.5 eV	8.8 eV	12 eV
LET(MeV/cm)	3.6	7.5	7.3	7.5	5.5	6.3
LET($MeVg^{-1}cm^2$)	1.6	1.4	1.25	1.4	1.7	1.8
Number of electron-hole pairs generated	<~105 e-h / µm	<~ 250 e-h / µm	<~ 162 e-h / µm	<~167 e-h / µm	<~63 e-h / µm	<~52 e-h / µm

Table 1.
Semiconductors that may be used for the direct detection of γ rays and their average energy for electron hole creation Ehn and other quantities, the linear energy transfer is indicated for high-energy charged-particle detection (at Minimum Ionization).

theoretical and experimental, on ionization in semiconductors. In the early days of germanium γ-ray detection, less purified material was used. The compensation method for net doping density reduction was based on the lithium drifting technique [11]. With an applied voltage bias, the drift of the interstitial Li+ ions through the detector is made possible leading to the neutralization of acceptors by the creation of Li+ acceptor pairs. However, progress in Ge purification led to the obsolescence of compensation techniques. Lithium is still used as a doping ion for the creation of n+ contacts, even though ion-implantation is now a mainstream technique for this purpose. Room-temperature γ-ray detection is still a challenge as the best candidate material exhibits higher defect concentrations and/or is difficult to grow into large crystals. Cadmium telluride is a material of choice for photo-detection, and gallium arsenide is another option. However, considerable progress would be necessary to match the easiness of use of HPGe in its present-day applications, despite the need for cryogenic apparatus.

The main contributions to the development of high-purity germanium have been made already in the 1970s by the LBNL [1] and other US laboratories as well as industrial companies. There are presently a few corporations in Europe which supply such materials, without mentioning those from Eastern Europe. High-purity single crystals are usually grown in hydrogen atmosphere in a silica crucible [9, 12]. This prevents the introduction of oxygen and other electrically active impurities. However, this process introduces substitutional silicon, which is electrically inactive, similar to carbon. These crystals exhibit high levels of interstitial hydrogen, which may form complexes with some metallic impurities such as copper [13]. Good quality crystals (despite being small) may be fabricated in University labs [11, 14] which show that nowadays the availability of such crystals is better than ever. Gamma-ray detector manufacturers usually use commercially available crystals, which are subsequently processed, or in some cases fabricate the material itself. It should be noted that except for isotopically enriched crystals (for physics experiments), most crystals are made of natural germanium with a proportion of isotopes indicated in **Table 2**.

Isotope	^{72}Ge	^{73}Ge	^{74}Ge	^{76}Ge	^{70}Ge
Nuclear Spin	0	1/2	0	0	0
Abundance	27.45 %	7.76 %	37.562%	7.76%, Double beta decay. Very long half life	20.52 %

Table 2.
Stable isotopes found in natural germanium with the nuclear moments and abundance.

2. Gamma-ray detection and spectroscopy

In this chapter, we introduce the principle of γ-ray detection and spectroscopy. Most high-energy photons (100 keV–10 MeV) interact with the electrons in the HPGe material [15]. As the density of electrons is proportional to Z, Ge is a material close to the optimum among well-characterized semiconductors, together with CdTe and GaAs. As the absorption coefficient increases along with Z, this results in a good detection efficiency for a given detector volume compared for instance with silicon. Here are the three main processes by which gamma rays lose energy in the detecting media: the photoelectric effect, Compton scattering, and e+e− pair creation. The photoelectric process [16] is dominant at low energies (<100 keV approximately) and is related to emission of electrons from the atomic shells. This process depends on the energy of the photons and the atomic number of the detecting media. The energy of a photon is absorbed by an inner-shell electron and leads to its emission from the atom. Subsequently, the photoelectrons lose their kinetic energy in the semiconductor by electron-hole pair generation. This first term is the photoelectron energy ($E_{photoelectron}$), and the second is the difference between the gamma energy (E_γ) and the electron binding energy ($E_{bindingenergy}$).

$$E_{photoelectron} = E_\gamma - E_{bindingenergy}, \tag{1}$$

The electron binding energy for K-shell valence electrons is of the order of 11 keV in Ge, compared to around 1.8 keV in silicon. Other, shallower shells may also be excited, contributing to the signal, which means that as the photon energy increases, inner shells can be excited gradually and the absorption will exhibit abrupt increase. This does not have a direct influence on the average number of electron-hole pairs generated per unit of energy deposited; we will discuss this in the following paragraph [8]. The absorption coefficient is proportional to:

$$\frac{Z^4}{E^3} \tag{2}$$

This means that the contribution of the photoelectric effect vanishes at high energies (E), when the other two processes of interaction of gamma rays with matter become dominant. In the intermediate energy range, Compton effect dominates, hence the absorption of the photon becomes a multistep process with Compton electrons (<1 MeV) being absorbed after traveling a short distance, while Compton photons are created and absorbed in subsequent steps. One should note that some photons can escape the detecting medium and consequently do not contribute to a full-energy peak. These induce a signal background (Compton background) that is particularly large in low volume detectors. Techniques exist to mitigate this drawback using a secondary detector surrounding the primary one. This secondary detector is referred to as a Compton shield [17]. This anticoincidence scheme with these two detectors eliminates the un-absorbed photon events. This shows that for high detection efficiency, a relatively large volume of the detector is necessary for γ-ray spectroscopy. This is much different for γ-ray tracking is, when detector granularity is required. Time coincidence between events in neighboring detectors is the way to proceed to identify the particles. The probability of interaction with the electrons of the solid through Compton effect increases when the density of electrons n increases, being roughly proportional to it. If ρ is the mass density of the solid and A the mass number, the average electron density is proportional to ρ/A, which is the atomic density, multiplied by Z, the number of electrons per atom [16].

$$n = Z \times \frac{\rho}{A} \tag{3}$$

This shows that as the Z/A ratio is constant, it is reasonable to use very dense media. Hence, it proves that germanium is a better choice than silicon for this purpose, as would be CdTe or GaAs, if the defect concentration could be reduced to an acceptable level. We will discuss the problem of defects in the following paragraph. Present-day experiments benefit from simulation codes such as GEANT4 or others [18] for their design. HPGe γ-ray detectors are no exception. The other process that becomes important at higher photon energy is the electron-positron pair creation, which may appear at energies above 1022 keV. This kind of interaction is related to γ-nuclear coupling, where the γ-photon interacts with the nucleus resulting in a creation of a e−/e+ pair with a total kinetic energy equal to $E_k = E_\gamma - 1.022$ MeV (E_k is the kinetic energy of the e = e− pair, and E_γ is the energy of the incident gamma photon). Subsequently, the positron may annihilate with an electron in the material, which leads to the production of two γ-rays of 511 keV that may (but not have to) be absorbed in the detecting material through Compton or photoelectric effect processes. As one or both 511-keV photons may escape from the detector with no energy deposition, satellite peaks appear in the measured energy spectrum, called escape peaks and separated from the total absorption peak by a 511 keV energy difference (or 1022 keV for double escape). Again, some discrimination is possible to avoid this effect, although one should be interested in visualizing the different peaks to investigate how the incident particle has interacted with the detector medium. The e + e- pair creation is most important at energies above 10 MeV [16], which shows that HPGe is not a good choice for very energetic photons such as cosmic radiation or those generated by high-energy accelerators (~1 GeV) since huge monocrystals would be required to provide a significant detection efficiency. This is hardly possible, with the largest commercially available HPGe crystals weighing 1–2 kg (larger crystals have been grown for specific applications) and having a diameter of a few centimeters. If we consider photofission [19], it has been reported that it occurs in natural lead for 10 MeV range photons. However, as the dependence on the fissility parameter is sublinear in logarithmic scale, it should be considered as being very weak for a germanium nucleus. Let us compare with lead, the fissility parameters read:

$$Z^2/A = 32.5 \text{ for Pb} \quad Z^2/A = 14 \text{ for Ge} \tag{4}$$

Hence, according to these figures, photofission should be much lower for natural Ge than for natural Pb. We therefore can consider this contribution as negligible in the energy range a few tens of MeV for the photons that are usually analyzed using HPGe spectrometers. This is the case in nuclear sciences for radioisotope identification and monitoring. Hence, for nuclear physics, HPGe is an adequate choice and widely used for detection of low energy and medium energy photons. In high-energy physics experiments, electromagnetic calorimeters are mostly based on dense liquids or solids and exhibit some granularity [20], while the energy resolution at high energy is not as good as for HPGe at low energies. Up to now, no high energy physics experiment has ever introduced HPGe as a semiconductor detector in calorimeters. Instead, silicon cells with absorbers are used in calorimeters and are proposed for several future detectors with the advantage of operation at room temperature, which is possible with semiconductors exhibiting a larger band gap, such as silicon.

3. Technology, material, geometry, and performance of HPGe gamma-ray detectors

Focusing on photon energy resolution criterion, HPGe still provides one of the best results. With a good quality material, the energy resolution is very close to the Fano limit [21, 22]. To obtain this, the readout electronics must be low noise. Historically, the front-end transistor was a JFET cooled to the temperature of the detector (77 K, liquid nitrogen temperature) [9, 23]. This reduced the energy resolution to less than 1 keV for 1 MeV photons. It would now be possible to use very low-noise room temperature CMOS μ electronic circuits [24] to match the low-noise specifications required for γ-ray spectrometry. The front-end readout electronics for these detectors is usually a charge-sensitive device, which is necessary for spectrometric measurements for which the generated charge is proportional to the energy deposited in the detectors. These CSA (charge-sensitive amplifiers) have an integrating pole followed by filtering stage (usually based on derivation-integration schemes for optimal filtering). This is done by a so-called shaper in order to optimize the signal/noise ratio and to reduce event pile up through fast operation. This channel is however slow (a few μs), so it may be supplemented by a fast current-sensitive channel, for coincidence, veto (anticoincidence) or timing purposes.

Because a typical size of a HPGe detector is not optimum for timing measurements, the resolution obtained with a CFD (Constant Fraction Discriminator) is close to 400 ps [25], which is a high value compared with other fast detectors (PM or APD). In fact, this value was measured for a planar germanium detector of a volume of a few cubic centimeters, and it would be even higher for standard coaxial HPGe detectors.

3.1 Starting material

In order to reach optimum resolution of semiconductor detectors, the crystal defect density should be decreased, in particular for those that are electrically active. This has become possible when defect and impurity control in the process of crystal growth has reached a sufficient level of reliability. In addition to point defects and impurities, dislocations play a major role as they behave like a sink for impurities [1, 9, 12]. They can currently be revealed by chemical crystal etching. The pits observed using optical microscopes are related to dislocations, which allows determining the etch-pit density. These dislocations induce a broad DLTS signal in n-type high purity material [3]. These are moderately deep donor states with carrier capture cross section of the order of 10^{-13} cm^{-2} or more. The peak is at 50–60 K for a large emission rate window (56 s^{-1}) [3]. This means that at 77 K, the emission rate is large enough to have no marked effect on carrier trapping as the electrons are released with a time constant that is low compared with the drift time. With σ_n higher than 10^{-13} cm^{-2}, which is the case with a filling pulse shorter than 1 ms in duration (with a 10^{10} cm^{-3} carrier concentration, and 10^6 cm s^{-1} of thermal velocity at 50 K, in our DLTS measurements), the emission rate at 77 K exceeds the capture rate for an activation energy of 100 meV. The other fact is that the DLTS signal fades away above 65 K, so the contribution of the dislocation band should be low as long as the concentration remains at a reasonable level. This gives a rule of thumb [2] for the dislocation density that should not exceed 10^4 cm^{-2} and should be above 10^2 cm^{-2} for detector-grade material, at least for the material used in the 1980s–1990s. If the dislocation density is too low, the deep impurities such as those related with copper (substitutional or bound with hydrogen) will be higher, as they cannot precipitate onto the dislocation lines and have a higher density than isolated impurities. They give rise to deep hole traps that affect greatly the hole transport even at 77 K. These traps have an activation energy higher than 0.160 eV and capture cross sections above 10^{-13} cm^{-2}, (similar to coulombic/attractive centers) with concentrations reaching 10^9 cm^{-3}.

3.2 Geometry and process

The structures that have been used from the early days of Ge detectors are p+n–n+ and p+p–n+, usually with a Li-diffused n+ contact and a boron-implanted p+ contact, and a thin metallization of sputtered aluminum or electrolytically plated gold in old detectors. Following an appropriate surface treatment (etching with CP4 mixture after cleaning with solvents such as acetone) and rinsing with a slightly oxidizing mixture, the detectors are placed in secondary vacuum, and after surface desorbing, the leakage current can be stabilized at low values. The leakage dark current at 77 K with a reverse bias of 1 kV or more for large coaxial detectors can be reduced to the order of a few pA or less. Usually a thermal treatment above room temperature will allow surface desorbing, and consequently impurities on the surface will be eliminated. The question of passivation is still open, as germanium does not have the self-passivation properties of silicon, for which oxygen creates a stable oxide layer of a few nm at room temperature. Germanium oxides are not stable [26–29] and react with water. More recent passivation methods are based on amorphous layer deposition such as a-Ge-H (amorphous hydrogenated germanium). These constitute a coating, rather than standard passivating layers. In spite of this, their effect is to stabilize leakage currents at 77 K at reasonable values. However, these values are higher than those for bare Ge material in vacuum.

As the presence of defects mainly alters the transport of holes, efforts were made to reduce the mean-drift length of holes by an adequate electrode configuration [30]. For coaxial detectors, the hole-collecting electrode can be placed at the outer surface of the detectors. This is usually a p+ outside implant, which is shallower than the n+ electrode, which is placed on the axis of the detector. Holes have a shorter drift distance than the electrons, the p+ electrode being negatively biased. This configuration is less sensitive to cumulative nonionizing radiation effects. N-type material is used contrary to p-type material with opposite electrode configuration in the conventional detectors. The same is true for planar detectors, for which the collecting length is reduced to values well below 1 cm. The cumulative radiation effects are mainly deep defects introduced by irradiation by nonionizing particles, such as hadrons (particularly neutrons and protons), but also energetic electrons. These particles, particularly at energies in the MeV range, induce displacement cascades in the detecting crystalline material and therefore produce defects of various nature, point-like or clusters. The temperature of operation and irradiation has a great influence on the outcome [25]. The result is a degradation of the energy resolution, which can be observed as a broadening of the photo-peaks, with a tail at low energy giving them an asymmetric aspect. As the detector can be compared to an ionization chamber with two electrodes polarized at different potentials, the current integrated by the readout electronics is a displacement current. The total charge collection is achieved when all holes and electrons are collected by the n+ (positively biased) and p+ (negatively biased) electrodes (cathode and anode). If carriers become trapped during the transport process, a loss of charge occurs leading to a peak tail at low energy. The collected charge is lower than the photo generated charge. Statistically, there is a certain charge deficit in the signal corresponding to one event, so the high-energy side of the peaks is not much affected [30], contrary to the low energy side. This leads to a loss of energy resolution. The charge loss is often followed by charge reemission with a large time constant, which has no real influence of the spectrum. An analytical model of radiation-induced defects has been proposed in Ref. [25] and developed in later papers.

3.3 Defects

The radiation-induced effects have been widely studied in HPGe detectors [30, 25, 31], and, in the beginning, without a quantitative relation to crystalline defects introduced by irradiation in HPGe. Since these effects result in resolution degradation, their characterization is of utmost importance. We can cite a few extensive works on this subject, mostly using electrical measurements [12, 13]. A powerful characterization technique called photoelectric spectroscopy [32] or alternatively photo-thermal ionization spectroscopy (PTIS) has been successfully applied to HPGe, but this technique is limited to shallow hydrogenoid levels with low concentration and not operation-detrimental deep levels. It applies a two-step process, phonon + photon (far infrared) and requires low temperature operation (LHe) below the ionization temperature of impurities and dopants. No overlapping of the bound electron wavefunction is required, so it can only be used when the impurities have a low concentration. For deep-defects, deep level transient spectroscopy (DLTS) became the technique of choice. In particular, it helped to establish that deep traps that are created by a temperature increase above 77 K are more detrimental to detector operation than the primary defects that are created by irradiation at low temperature [3]. Neutron-induced defects are thought to be vacancy and interstitial related at 77 K. After annealing above this temperature, divacancies and impurity-vacancy defects are the most numerous centers observed that are electrically active. They give rise to numerous deep hole-traps with capture cross sections of the order of 10^{-13} cm^{-2}. They only anneal out above 420–470 K, where less electrically active defects are created. The stable defects at 100 K and less are disordered regions containing a large concentration of vacancies and interstitials. These act as hole-traps but are less effective in degrading resolution than secondary defects observed at room temperature. Numerous studies have been devoted to the degradation of the resolution, and most of them identify the causes as being related to the defects with capture cross sections of the order of 10^{-11} cm^{-2} [2]. We know from experiments and simulations that these are due to zones with a high local defect density, which enhances capture probability [33] through an electrostatic effect.

Street [34] has found that the presence of disorders in amorphous silicon enhances the cross section for the capture of carriers by the defects. Later, an analytical model has been developed that clearly explains this effect using simple assumptions [33]. This mitigates the direct role of disordered regions as being the sole origin of carrier capture at 77 K. The sizes of these disordered regions areof the order of the range of primary recoil atoms (**Figure 1**). Isolated defects should contribute greatly to the trapping process. The recoil of an atom is induced by the collision with the impinging particle. SRIM simulations show that its range is of the order of 10 nm at 10–30 keV, with around a few hundred vacancies being created on its trajectory. The recoil energy has been computed for neutrons in the MeV energy range, see **Figure 1**. In most cases, a thermal treatment above room temperature is used to remove radiation damage. In Ref. [31], recombination-enhanced annealing using minority-carrier injection was applied but with no significant results, at least at room temperature. At low temperature, when the defects are not stable, no improvement could be observed with this method. However, a strong dependence of the annealing stages on the material type (p or n) was observed in Refs. [3, 25] and other detector studies.

3.4 Alternatives to HPGe

For a long time, materials alternative to HPGe have been proposed, and corresponding detectors were fabricated. GaAs is one example (mostly for X-ray detection), and, more importantly, CdTe. The goal was to eliminate the need for cryogenic operation by the use of large band gap semi-insulating material (diamond is also considered). Many difficulties still exist, mostly related to the defect density that is much higher in binary materials [9]. The other drawback is the possibility to grow large crystals, which proves to be more difficult for alternative materials. Large diamond single crystals with a low nitrogen content are difficult to grow as they need high-pressure processing. However, some CdTe photon detectors using segmented crystals for photon identification have been successfully implemented on space missions (INTEGRAL) [24]. Segmentation allows reducing drift length and therefore trapping, so that the resolution can be maintained at an acceptable level.

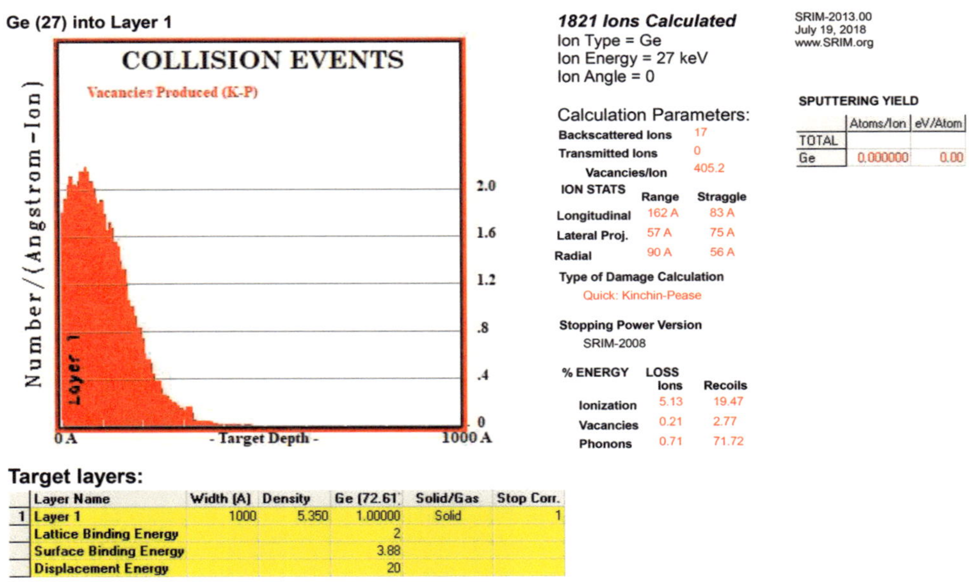

Figure 1.
SRIM simulation showing the vacancy distribution for a Ge recoil of 30 keV.

3.5 Particle identification

Particle identification through pulse shape discrimination is one of the developments that are used mainly in nuclear physics [35]. These techniques can be adequately used to determine the region of the detector where the interaction took place. There have been early reports for methods of particle identification in germanium and silicon detectors [36, 37]. The detection of photons is basically through ionization, and there is no important interaction with the nucleus, which could transfer momentum to the nucleus. If we consider other particles with a significant mass, their interaction with the nucleus may induce the recoil, which in turn is slowed down in the detector material. The slowing-down process results in ionization (electron-hole pair generation), defect creation (vacancies and interstitials), and phonon creation. One should note that according to SRIM simulations, in the typical energy range for the recoil (20–30 keV), the most important contribution (73%) is from phonon emission. The energy used for vacancy creation is of the order of 3%, which indicates that defect monitoring [38] could provide an alternative way to estimate the total integrated flux of interacting particles, since when stable defects are created, their concentration is proportional to the total number of interacting particles in a given time interval, which can be very long. The rest of the energy is used for ionization (25%). This result is close to the experimental result obtained for cryogenic detectors, as it will be discussed in the paragraph on dark matter detection.

3.6 Simulation

Simulations of the interaction of photons with matter have been given a certain attention. Monte Carlo codes have been developed. Recently, the NWEGRIM code [18] from Pacific Northwest laboratories has been used to simulate the interaction of photons in silicon, but these simulations could also be applied to germanium. GEANT4 has been used for simulation of charged particles interacting with germanium [39].

4. Gamma-ray tracking arrays

High-purity germanium γ-ray tracking arrays, such as AGATA (Advanced GAmma Tracking Array) [40] and GRETINA/GRETA (Gamma Ray Energy Tracking Array) [41] represent the state-of-the-art in high-resolution γ-ray spectroscopy for nuclear physics experiments. These spectrometers are composed of highly segmented large-volume HPGe crystals. Pulse-shape analysis, applied to the recorded signals from the segments, yields three-dimensional interaction positions with a typical precision of about 2 mm [40, 41]. Subsequently, a γ-ray tracking algorithm is applied to the determined interaction points in order to group and order them in sequences corresponding to individual γ-rays. In this procedure, the geometrical criteria and the Compton scattering formula are used, and the full energy of a γ-ray is determined as the sum of the energies of the interactions ascribed to the same trajectory.

The γ-ray tracking arrays provide improved energy resolution for in-beam nuclear physics studies, thanks to much reduced Doppler broadening as compared to standard γ-ray spectrometers. The use of γ-ray tracking also eliminates the need for Compton-suppression shields, commonly used with HPGe crystals in order to improve the peak-to-total ratio. Consequently, the entire 4π solid angle can be surrounded with Ge crystals, which leads to significantly increased detection efficiency. The resolving power of a 4π γ-ray tracking array is estimated to be up to two orders of magnitude better than that of the existing conventional γ-ray spectrometers, depending on the physics case [41]. This is particularly important for studies of very exotic nuclei far from stability, employing weak radioactive-ion beams at intermediate energies (up to several hundred MeV/u), which leads to recoil velocities that may exceed 30% of the speed of light [42].

5. Dark matter direct detection and related studies

The last two decades have seen an important effort devoted to search for dark matter, using underground direct detection apparatus [36, 37, 43–47]. The first stage was to design a detector that is able to discriminate between particles in order to identify the so-called weakly interacting massive particles that are thought to be a constituent of dark matter. The developed detectors include two channels: a "thermal" channel which is based on the thermalization of phonons and a ionization channel which is proportional to the number of carriers collected. In reality, the so-called Luke-Neganov effect [48, 49] affects the properties of the detector. The Luke-Neganov effect consists in the amplification of the phonon signal by the electron (carrier) drift [50].

The charge signal may be expressed as:

$$s_c = \frac{E}{\varepsilon} q, \tag{5}$$

where E is the average energy for electron-hole pair creation, and

$$s_T = \frac{E}{\varepsilon} qV + E \tag{6}$$

The S_T term is proportional to the charge multiplied by the voltage drop, so we can write that if t is the time necessary for the charge to be collected,

$$I = Sc/t \tag{7}$$

First term: ItV, which is a Joule-like term. Energy = power × time = current × voltage × time = charge × voltage.

These relations indicate the need to operate at weak field to achieve a reduction in the contribution of the phonon signal. At high fields, the phonon signal tends to grow linearly with the applied voltage. Of course, the heat signal can only be detected if the calorific capacity is low enough, and so the operating temperature should be very low. Germanium detectors are also good cryogenic (mK range) bolometers. Another reason for the choice germanium is that its nucleus is nucleon rich (for instance compared with silicon). This should enhance interaction cross sections of WIMPS with the detecting medium. If we consider the interaction of fast neutrons with germanium (in the MeV range) as an example, the elastic-scattering cross section is much higher than the inelastic cross section [31] and is of the order of 3 barns [5]. Hence, the mean free path of these fast neutrons is close to a few centimeters. The recoil nucleus with an energy of a few tens of keV dissipates 25% of its energy into ionization, which amounts to a few keV (5 keV). The ionization channel monitors this fraction of energy. A similar phenomenon was observed, but not published, in planar HPGe 77 K detectors, exposed to neutrons, in the MeV range [51]. The heat channel monitors almost entire energy deposited by the recoils, except for the fraction needed for vacancy-interstitial (defect) creation. If a neutron is fully absorbed in the bolometer following multiple scattering, the total energy of all recoils matches the initial energy of the impinging neutron, and therefore the heat channel signal reflects the total energy deposited.

The calibration runs yielded the ratio of the ionization signal to the phonon energy equal to 30%. This is very close to the value of 25%/73% = 0.35 determined in a SRIM simulation. When the particle interacts with matter purely via ionization, as it is the case for photons, the ratio of ionizing energy to the total energy is close to one, if no Luke-Neganov amplifying occurs. Therefore, this configuration can be used for particle discrimination [46]. For the WIMPS experiments, if we consider that these particles interact with the nuclei of the detecting media, this provides a way of discriminating the photons from other events. At this cryogenic energy (a few mK), the shallow and deep levels may disturb the charge transport through the detector (**Figure 2**). This is the reason why the detector is saturated with photo-generated carriers prior to data acquisition. In spite of this, more detailed defect studies on the starting material should be made [52]. In particular, as very shallow levels may have an impact, the nonthermal carrier emission or capture should be studied. Additionally, a method using alternately biased electrodes on each surface of the detectors, including the sides, has enabled the measurement of volume-only events, eliminating near-surface events related to low-energy particles strongly interacting with the medium [46]. Other techniques have been developed [53] to solve this problem.

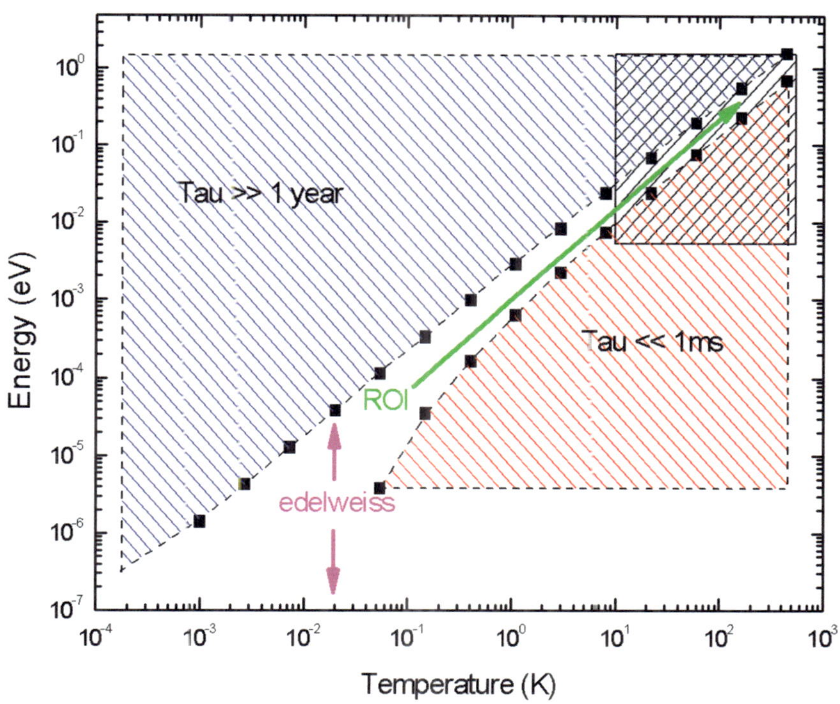

Figure 2.
Diagram showing the energy of the trapping level in Ge band gap versus temperature, using a simple Boltzmann factor. This is valid at high temperature for thermal emission with a time constant Tau. The region investigated by DLTS is delimited by the upper square on the right. The purple arrow shows the operation of DM cryogenic experiments such as EDELWEISS.

6. Applications in industry and daily life

As many particle detectors, HPGe detectors are now also used for other purposes than fundamental research. HPGe detectors can now be found in different industries such as environment control, medical imaging, and of course radioactive material control.

One of the most common uses of HPGe is the monitoring of atmospheric radiations. Stations around the world check the air quality and collect air content samples or solid samples, such as rocks, water, or plants. The samples are then analyzed in a laboratory. Thanks to their excellent resolution and low background once placed in a suitable environment, and the radio-elements contained in the samples can be determined. It is, for example, possible to detect radio-elements with concentrations down to 0.1–1 Bq/kg in solids and 0.1 Bq/L in water. The system can be improved by adding scintillator array around the germanium detectors to eliminate signals from cosmic rays. Such system is, for example, used to obtain a sensitivity of millibecquerels per cubic meter of an air sample.

Another practical example of use of HPGe to study radioactivity of liquids can be found in the study of old wines. Take a 1928 vintage bottle of Bordeaux wine; its cost can approach $10,000. However, how sure is the buyer that this bottle is really from 1928? To check this, the bottle was placed in an array of germanium detectors. The analysis immediately showed that the bottle contained traces of Cs137. Cs137 is a radioactive element presents in the atmosphere due to nuclear bombs..., which means that it cannot be found in a bottle of 1928. The quantification of Cs137 is now used to date wine produced after 1950 without opening the bottle.

So far, HPGe detectors for medical purposes were mainly used for nuclear waste control produced by hospital. Here, the procedure is the same than for any nuclear waste collected from a nuclear power plant for example. The contaminated volume is placed near the detector. The analysis of the spectrum will reveal radioactive elements present and their quantity. The cost and mechanical constraints related to the use of liquid nitrogen limited the deployment of HPGe in hospitals for medical imaging. However, novel imaging methods, combining HPGe and silicon detectors are envisioned for future scanners such as ProSPECTus. Such system could reduce the dose to the patient, improve image quality, and be faster to acquire.

The list of industrial applications of HPGe will become longer as we progress toward more reliable, easier to use, and cheaper detectors. Originally only pushed by fundamental research, the development of new HPGe is now also driven by the needs of companies using them. New applications, not yet considered, will then emerge.

7. Conclusions

The future of HPGe detectors depends on the availability of other materials that would match the resolution performance of HPGe at 77 K, while operating at room temperature. Segmentation of the detectors provides a way to reduce carrier collection lengths and hence to mitigate the effects of electrically active deep defects. With integrated microelectronics, the noise can be as low as a few tens of electrons. With on detectors CMOS chips and low capacitance detectors, the electronic noise can be reduced to less than 100 e-h pairs. With ε equal to 4.5 eV (GaAs, CdTe), this corresponds to a resolution of the order of 450 eV. This could also be applied to HPGe detectors that are being used for double beta decay or dark matter experiments. The multielectrode scheme that is implemented in EDELWEISS III [46] is a first step toward a time-projection chamber HPGe detector, allowing an improved discrimination ability. Each electrode could be a separate channel, which would be easy for the detector operated at low field and low voltage. It seems clear that the use in more routine applications such as, for instance, high-precision radioactive material characterization and radioactive material tracing to avoid nuclear dissemination will remain a field where HPGe is the most competitive despite the need for LN2 cryogenic installations. Development of cryogenic cooling fridges will eliminate this specific constraint. In scientific applications, HPGe, being one of the chemically purest materials ever fabricated, will remain needed.

Acknowledgements

The contribution of researchers from IRFU and other institutions, with whom the first author had fruitful discussions while working on dark matter instrumentation, is gratefully acknowledged.

Conflict of interest

The authors declare having no conflict of interest.

Author details

Nicolas Fourches[2]*, Magdalena Zielińska[2] and Gabriel Charles[1]

1 Institut de Physique Nucleaire d'Orsay, Orsay Cedex, France

2 Universite Paris-Saclay, CEA/IRFU, Gif/Yvette Cedex, France

*Address all correspondence to: nicolas.fourches@cea.fr

References

[1] Baertsch RD, Hall RN. Gamma ray detectors made from high purity germanium. IEEE Transactions on Nuclear Science. 1970;**17**:235-240

[2] Haller EE, Hansen WL, Goulding FS. Physics of ultra-pure germanium. Advances in Physics. 1981;**30**:93-138

[3] Fourches N, Walter G, Bourgoin JC. Neutron-induced defects in high-purity germanium. Journal of Applied Physics. 1991;**69**:2033-2043

[4] Alig RC, Bloom S. Electron-hole-pair creation energies in semiconductors. Physical Review Letters. 1975;**35**(22):1522-1525

[5] Alig RC, Bloom S, Struck CW. Electron-hole-pair creation energies in semiconductors. Bulletin of the American Physical Society. 1980

[6] Alig RC, Bloom S, Struck CW. Scattering by ionization and phonon emission in semiconductors. Physical Review B. 1980;**22**:5565

[7] Alig RC. Scattering by ionization and phonon emission in semiconductors. II. Monte Carlo calculations. Physical Review B. 1983;**27**:968

[8] Klein CA. Bandgap dependence and related features of radiation ionization energies in semiconductors. Journal of Applied Physics. 1968;**39**:2029-2038

[9] Bertolini G, Cappellani F, Restelli G. Current state of the art in semiconductor detectors. Nuclear Instruments and Methods. 1973;**112**:219-228

[10] Campbell L, Gao F, Devanathan R, Weber WJ. Model of plasmon decay for electron cascade simulation. Nuclear Instruments and Methods in Physics Research Section A: Accelerators, Spectrometers, Detectors and Associated Equipment. 2007;**579**:454-457

[11] Miller GL, Pate BD, Wagner S. Production of thick semiconductor radiation detectors by lithium drifting. IEEE Transactions on Nuclear Science. 1963;**10**:220-229

[12] Hall RN. Characterization of pure germanium for detector fabrication. IEEE Transactions on Nuclear Science. 1972;**19**:266-270

[13] Pearton SJ. Hydrogen passivation of copper-related defects in germanium. Applied Physics Letters. 1982;**40**:253-255

[14] Wang G, Mei H, Mei D, Guan Y, Yang G. High purity germanium crystal growth at the University of South Dakota. Journal of Physics: Conference Series. 2015

[15] Olive KA et al. Review of particle physics. Chinese Physics C. 2014;**38**:090001

[16] Nelson G, Reilly D. Gamma-ray interactions with matter. In: Passive Nondestructive Analysis of Nuclear Materials. 1991. pp. 27-42

[17] Sakai E. Recent Topics in Semiconductor Detector Development. 1978

[18] Gao F, Campbell LW, Devanathan R, Xie Y, Corrales LR, Peurrung AJ, et al. Monte Carlo method for simulating γ-ray interaction with materials: A case study on Si. Nuclear Instruments and Methods in Physics Research Section A: Accelerators, Spectrometers, Detectors and Associated Equipment. 2007;**579**:292-296

[19] Cetina C, Berman BL, Briscoe WJ, Cole PL, Feldman G, Heimberg P, et al. Photofission of heavy nuclei at energies

up to 4 GeV. Physical Review Letters. 2000;**84**:5740

[20] Behnke T. The International Linear Collider Technical Design Report. Detectors. Vol. 4. 2013

[21] Fano U. Ionization yield of radiations. II. The fluctuations of the number of ions. Physical Review. 1947;**72**:26

[22] Fano U. On the theory of ionization yield of radiations in different substances. Physical Review. 1946;**70**:44

[23] Upp DL, Keyser RM, Twomey TR. New cooling methods for HPGE detectors and associated electronics. Journal of Radioanalytical and Nuclear Chemistry. 2005;**264**:121-126

[24] Lebrun F, Leray JP, Lavocat P, Crétolle J, Arques M, Blondel CE, et al. ISGRI: The INTEGRAL soft gamma-ray imager. Astronomy & Astrophysics. 2003;**411**:L141-L148

[25] Fourches N, Huck A, Walter G. The role of secondary defects in the loss of energy resolution of fast-neutron-irradiated HPGe gamma-ray detectors. IEEE Transactions on Nuclear Science. 1991;**38**:1728-1735

[26] Bernstein RB, Cubicciotti D. The kinetics of the reaction of germanium and oxygen. Journal of the American Chemical Society. 1951;**73**:4112-4114

[27] Onsia B, Conard T, De Gendt S, Heyns M, Hoflijk I, Mertens P, et al. A study of the influence of typical wet chemical treatments on the germanium wafer surface. Solid State Phenomena. 2005;**103**:27-30

[28] Kobayashi M, Thareja G, Ishibashi M, Sun Y, Griffin P, McVittie J, et al. Radical oxidation of germanium for interface gate dielectric GeO_2 formation in metal-insulator-semiconductor gate stack. Journal of Applied Physics. 2009;**106**:104117

[29] Delabie A, Bellenger F, Houssa M, Conard T, Van Elshocht S, Caymax M, et al. Effective electrical passivation of Ge(100) for high-k gate dielectric layers using germanium oxide. Applied Physics Letters. 2007;**91**:082904

[30] Pehl RH, Madden NW, Elliott JH, Raudorf TW, Trammell RC, Darken LS. Radiation damage resistance of reverse electrode GE coaxial detectors. IEEE Transactions on Nuclear Science. 1979;**26**:321-323

[31] Fourches N. Etude des defauts dus a l'irradiation aux neutrons rapides dans le germanium de haute purete. 1989

[32] Kogan SM, Lifshits TM. Photoelectric spectroscopy—A new method of analysis of impurities in semiconductors. Physica Status Solidi (a). 1977;**39**:11-39

[33] Fourches N. High defect density regions in neutron irradiated high-purity germanium: Characteristics and formation mechanisms. Journal of Applied Physics. 1995;**77**:3684-3689

[34] Street RA. Disorder effects on deep trapping in amorphous semiconductors. Philosophical Magazine B. 1984;**49**: L15-L20

[35] Pullia A, Pascovici G, Cahan B, Weisshaar D, Boiano C, Bassini R, et al. The AGATA charge-sensitive preamplifiers with built-in active-reset device and pulser. In: 2004 IEEE Nuclear Science Symposium Conference Record; 2004

[36] Shutt T, Ellman B, Barnes PD Jr, Cummings A, Da Silva A, Emes J, et al. Measurement of ionization and phonon production by nuclear recoils in a 60 g crystal of germanium at 25 mK. Physical Review Letters. 1992;**69**:3425

[37] Gerbier G, Lesquoy E, Rich J, Spiro M, Tao C, Yvon D, et al. Measurement of the ionization of slow silicon nuclei in silicon for the calibration of a silicon dark-matter detector. Physical Review D. 1990;**42**:3211

[38] Budnik R, Cheshnovsky O, Slone O, Volansky T. Direct detection of light dark matter and solar neutrinos via color center production in crystals. Physics Letters B. 2018;**782**:242-250

[39] Fourches NT, Chipaux R. Electrical-modelling, design and simulation of cumulative radiation effects in semiconductor pixels detectors: Prospects and limits. Journal of Instrumentation. 2015;**10**:C01036

[40] Akkoyun S, Algora A, Alikhani B, Ameil F, De Angelis G, Arnold L, et al. Agata—Advanced gamma tracking array. Nuclear Instruments and Methods in Physics Research Section A: Accelerators, Spectrometers, Detectors and Associated Equipment. 2012;**668**:26-58

[41] Paschalis S, Lee IY, Macchiavelli AO, Campbell CM, Cromaz M, Gros S, et al. The performance of the gamma-ray energy tracking in-beam nuclear array GRETINA. Nuclear Instruments and Methods in Physics Research Section A: Accelerators, Spectrometers, Detectors and Associated Equipment. 2013;**709**:44-55

[42] Weisshaar D, Bazin D, Bender PC, Campbell CM, Recchia F, Bader V, et al. The performance of the γ-ray tracking array GRETINA for γ-ray spectroscopy with fast beams of rare isotopes. Nuclear Instruments and Methods in Physics Research Section A: Accelerators, Spectrometers, Detectors and Associated Equipment. 2017;**847**:187-198

[43] Creswick RJ, Avignone III FT, Farach HA, Collar JI, Gattone AO, Nussinov S, et al. Theory for the direct detection of solar axions by coherent primakoff conversion in germanium detectors. Physics Letters B. 1998;**427**:235-240

[44] Essig R, Mardon J, Slone O, Volansky T. Detection of sub-GeV dark matter and solar neutrinos via chemical-bond breaking. Physical Review D. 2017;**95**:056011

[45] Abt I, Caldwell A, Gutknecht D, Kröninger K, Lampert M, Liu X, et al. Characterization of the first true coaxial 18-fold segmented n-type prototype HPGe detector for the gerda project. Nuclear Instruments and Methods in Physics Research Section A: Accelerators, Spectrometers, Detectors and Associated Equipment. 2007;**577**:574-584

[46] Armengaud E, Arnaud Q, Augier C, Benoît A, Bergé L, Bergmann T, et al. Performance of the EDELWEISS-III experiment for direct dark matter searches. Journal of Instrumentation. 2017;**12**:P08010

[47] Armengaud E, Augier C, Benoit A, Bergé L, Bergmann T, Blümer J, et al. Search for low-mass WIMPs with EDELWEISS-II heat-and-ionization detectors. Physical Review D. 2012;**86**:051701

[48] Luke PN. Voltage-assisted calorimetric ionization detector. Journal of Applied Physics. 1988;**64**:6858-6860

[49] Neganov BS, Trofimov VN. USSR Patent No 1037771; 1985

[50] Di Stefano P. Recherche de matière sombre non-baryonique au moyen d'un bolomètre à ionisation dans le cadre de l'expérience EDELWEISS; 1998

[51] Huck A. CRN, Strasbourg, Private Communication (Nicolas Fourches); 1989

[52] Domange J, Olivieri E, Fourches N, Broniatowski A. Thermally-stimulated current investigation of dopant-related D- and A+ trap centers in germanium for cryogenic detector applications. Journal of Low Temperature Physics. 2012;**167**:1131-1136

[53] Yang LT, Li HB, Wong HT, Agartioglu M, Chen JH, Jia LP, et al. Bulk and surface event identification in p-type germanium detectors. Nuclear Instruments and Methods in Physics Research Section A: Accelerators, Spectrometers, Detectors and Associated Equipment. 2018;**886**:13-23

Chapter 6

Gamma Ray Measurements Using Unmanned Aerial Systems

Monia Kazemeini, John Vargas, Alexander Barzilov and Woosoon Yim

Abstract

Gamma ray measurements involved in monitoring technologies of field conditions are of vital importance for environmental safety and radiation protection. This chapter addresses the method of cooperative gamma sensing using multiple unmanned aerial systems. Section 1 provides an introduction. The design of semiconductor and scintillation gamma ray sensors integrated into aerial robotic platforms is discussed in Section 2, along with the fusion of time-stamped radiation data with position information using the real-time kinematic positioning technique. Section 3 addresses the multirobot contour mapping of radiation fields. Computer simulation of radiation contour mapping is discussed in Section 4. Experimental verification of the contour mapping and source-seeking algorithm is described in Section 5. Section 6 summarizes results of the project.

Keywords: gamma rays, remote sensing, unmanned aerial system, robot operating system, contour mapping

1. Introduction

Technologies of gamma ray sensing in radiation zones are vital for environmental safety and radiation safety. Hazardous radiological wastes can be a result of large nuclear projects; radiological materials can be displaced, lost, or smuggled [1–4]. Additionally, they can be released into the environment due to man-made accidents or natural disasters. One example is the Fukushima Daiichi nuclear power plant disaster [5] that led to radiological contamination of the plant's infrastructure and territory adjacent to the plant. The use of Unmanned Aerial Systems (UAS) allows for the remote radiation sensing, mapping, and wide area search operations while keeping users away from the risk of exposure.

Small scale, multiuse UAS platforms equipped with navigation and radiation sensing capabilities permit for the surveillance in hard to reach, hazardous areas while allowing for the measurements to be dynamically tracked and mapped [6]. The measured data could be used for situational awareness and further analysis of radiation fields in temporal and space domains. Furthermore, based on the cooperative sensing algorithms, the UAS swarm can be programmed to search for unattended radiation sources. In order to accomplish radiation monitoring tasks using UAS, gamma ray sensors should be integrated into a robotic aerial platform. The sensor's data analysis should be automated and carried out onboard taking into account limited computational resources of flight computers.

IntechOpen

2. Radiation sensors

Since UAS has weight and size limitations for onboard components so that the flight efficiency is not reduced due to the payload, the choice of radiation sensors is important. Radiation sensors should be swappable (attached and detached from the robotic platform) in the field conditions. The sensors should have low power requirements so that the integration into the platform does not affect the battery power. The data processing limitations should be also taken into account. To address this necessity, two types of flight ready ambient temperature radiation sensors were developed: a Cadmium Zinc Telluride (CZT) semiconductor sensor for high-resolution gamma spectrometry and the elpasolite scintillation sensor $Cs_2LiYCl_6:Ce^{3+}$ (CLYC) for neutron and gamma measurements. Both sensors were designed as plug-and-play interchangeable modules.

2.1 CZT sensor

A CZT is a wide bandgap semiconductor [7]. Gamma rays that interact with this detection medium deposit their energy into the CZT operating in a direct-conversion mode at an ambient temperature. Therefore, no cooling is needed which significantly simplifies the sensor's integration into the robots and their practical use. These semiconductors are able to process more than 10 million photons per square millimeter per second.

The CZT module GR1-A (Kromek) [8] was integrated into a multicopter aerial platform. The GR1-A includes a 1 cm^3 cubic CZT crystal. Electric signals generated by the CZT proportionally to energy of the incident gamma ray are amplified by a preamplifier and then processed by a shaping amplifier. A 4096 channel analyzer produces the discrete data array of the gamma spectrum that can be processed further. The scheme of the CZT sensor operation is shown in **Figure 1**.
The USB interface is used for the data communication as well as power. The power consumption of this module is about 250 milliwatts. The module's volume is $2.5 \times 2.5 \times 6.1$ cm^3 and the weight is about 50 grams. The sensor's energy range is from 30 keV to 3 MeV with the Full Width at Half Maximum of the peak (FWHM) energy resolution of less than 2% at photon energy of 662 keV. The module's

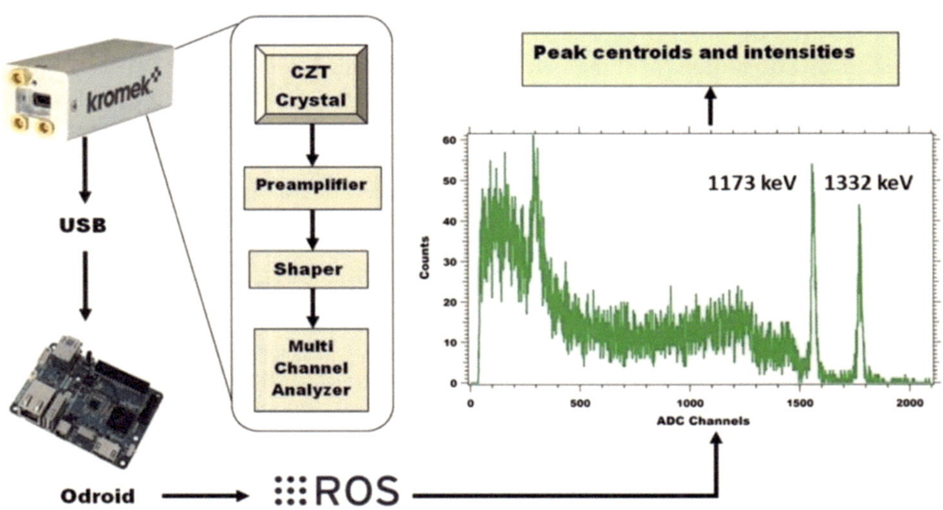

Figure 1.
CZT sensor operation.

temperature range is from 0 to 40°C. The CZT sensor mounted onto the UAS platform (the DJI S1000+ octocopter) is shown in **Figure 2**.

Robot Operating System (ROS) [9] was used for fusion of the data sent from multiple sensors connected to the robotic platform (radiation sensors and a GPS sensor). ROS is an open source tool, consisting of libraries used for robot applications. It allows including a number of independent nodes which communicate to each other, using a publishing and subscribing messages to the topics. Once a data array was sent from the CZT sensor via USB to an onboard Linux minicomputer (the Odroid model) and processed using ROS, a gamma ray spectrum is generated. The example of a measured spectrum using a ^{60}Co gamma-ray source is shown in **Figure 1**.

The gamma ray spectrum is then analyzed using a code to extract the peaks of interest along with their intensities. This allows for identification of isotopes (sources) that emit these gamma rays, and also for estimation of source strength. Computational algorithms are widely used for photon spectral analytics for various applications [10–14]. For the rapid spectral analysis on board of robotic platforms with a limited computing power, a simple, but robust algorithm is needed. The algorithm based on Mariscotti's technique [15] was developed for the peak identification in the presence of a background in the measured spectrum. In this technique, peaks are approximated using a Gaussian function.

Within a small interval of several peak's widths, which is applicable for CZT detectors and scintillation detectors, the spectrum's background is represented as a linear polynomial. In this interval, number of counts in the spectrum channel i is reconstructed as $Sp(i) = Gauss(i) + B_1 + i \cdot B_2$, where $Gauss(i)$ is a Gaussian function, and constants B_1 and B_2 are associated with the background. This method locates a peak where the 2nd derivative of the function $Sp(i)$ is not zero.

The function $Gauss(i) = I \cdot exp\left(-[i-p]^2/[2\sigma^2]\right)$ includes the height I of the peak that is centered at a channel p. The value of σ is found as $FWHM_{peak}/2.355$. For a discrete data $y(i)$, the 2nd derivative of $Sp(i)$ was approximated as a centered finite-divided difference: $Sp''(i) = \left[y(i+1) - 2 \cdot y(i) + y(i-1)\right]/h^2$, where h is the channel's width.

Since $y(i)$ carries statistical errors, the $Sp(i)$ data will also exhibit statistical fluctuations about expected values at each channel. Hence, a peak cannot be resolved in case of an expected value of $Sp(p)$ which is equal to the standard deviation. For $\sigma = 4$ it corresponds to the minimum resolvable I value of 600. To resolve much smaller peaks, function $Sp(i)$ should be 'smoothed', consequently reducing the standard deviation. The smoothing was done by calculating an average of $Sp(i)$ values over w

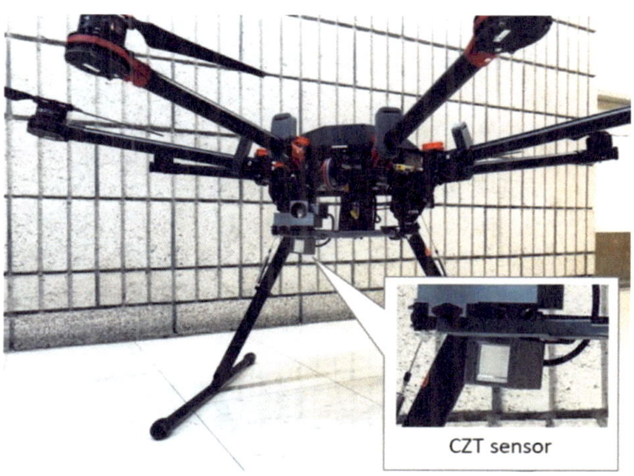

Figure 2.
CZT sensor mounted on the UAS platform.

adjacent channels near the channel i. This smoothing can be repeated z times yielding $S_i(z,w) = \sum_j C_{ij}(z,w) y_j$, where C_{ij} are weighting factors. Its standard deviation is $D_i(z,w) = \left[\sum_j C_{ij}^2(z,w) y_j\right]^{1/2}$. Values z and w were determined for the CZT sensor based on the desired I value.

This algorithm makes it possible to distinguish photopeaks from Compton shoulders in the spectral data. The peak centroid was found at the center of the Gaussian function. The intensity of the peak was calculated using the area under the Gaussian. The C code implementing this algorithm was written as a function within ROS. The spectrum data array can be processed using this function returning the energies of found peaks and their intensities, as a result minimizing the data transfer from the UAS platform to a ground control station.

2.2 CLYC sensor

A 2.54-cm diameter, 2.54-cm long cylindrical CLYC [16, 17] crystal was utilized in the design of a gamma-neutron sensor. The CLYC's density is 3.31 g/cm^3. Its scintillation light wavelength range is from 275 to 450 nm (the peak is at 370 nm). The crystal's refractive index is 1.81 at 405 nm. This crystal was coupled with a super bialkali photomultiplier tube matching the CLYC's emission wavelength range, a miniature digitizer, and a high voltage generator; all packaged in a custom housing (**Figure 3a**). The ^6Li isotope enrichment of the CLYC was 95%. Neutron detection was achieved via $^6Li(n,\alpha)t$ reaction. The scintillation light yield of CLYC is 20,000 photons per 1 MeV for gamma rays and 70,000 photons per thermal neutron. CLYC's scintillation properties allow for gamma spectrometry. This sensor operates without cooling. The measured gamma-ray FWHM energy resolution was 5% at 662 keV and 3.3% at 1332 keV.

Moreover, CLYC exhibits neutron-photon pulse shape discrimination (PSD) properties. The photon-induced excitation in the CLYC medium leads to a fast core to valence (CVL) decay and prompt cerium decay with 1 and 50 ns decay constants, respectively. A neutron event in CLYC causes a slow cerium self-trapped excitation (Ce-STE, 1000 ns decay constant) [18].

The digitized neutron and photon signals of the CLYC sensor are shown in **Figure 3b**. Each signal was analyzed using the eMorpho digitizer, generating three values saved in a list mode: the time stamp; the integral under the front of the signal's curve assessed using the partial integration time; and the integral under the entire curve that is proportional to the energy of measured radiation. In order to segregate neutron signals from gamma-ray signals, a PSD value (calculated as ratio of the area under the tail part of the signal to the front part of the signal) was used. Because neutron signals have longer tails, it leads to larger PSD values than in the case of gamma-ray signals. The experimental plot of neutron-gamma PSD for the CLYC sensor using a PuBe source is shown in **Figure 3c**. Neutron and gamma ray signals are well separated in this plot. The neutron signals appear at the value of 3200 keV, electron equivalents (keVee). The figure of merit of neutron-gamma PSD for CLYC was evaluated as 2.3 using an approach described in [19].

2.3 Sensor integration with UAS

The 'plug-and-play' concept was used to integrate the radiation sensors into the UAS using ROS. This approach supports 'hot swapping' of the sensors into the UAS platform, meaning that the user does not need to set up sensor's parameters each time the UAS is powered on. When the CZT or CLYC sensor is plugged, the operating system recognizes it, and then installs the associated driver to read and process the sensor's data. The data is published and the message appears. Similarly, when the

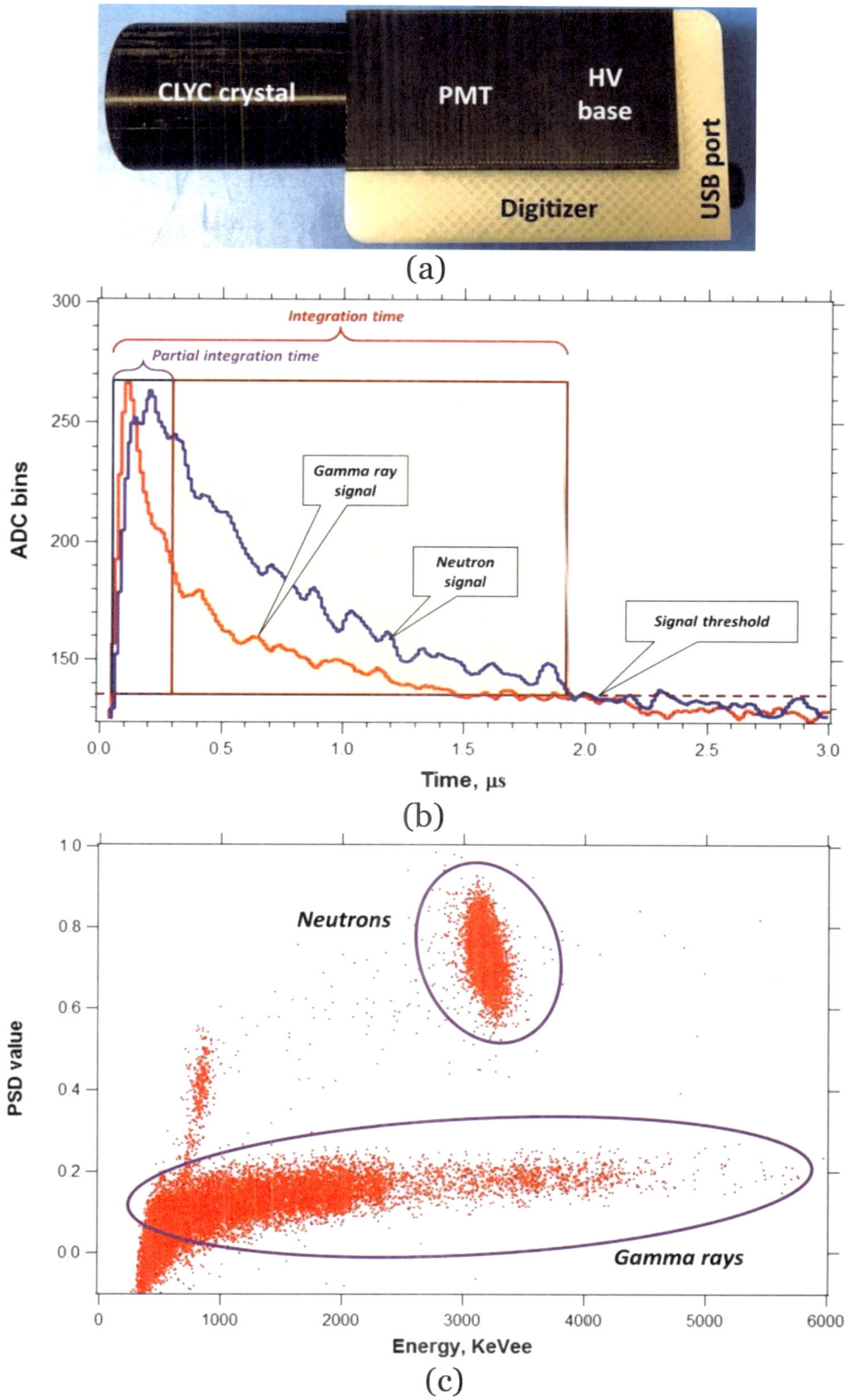

Figure 3.
(a) CLYC sensor, (b) CLYC signals, and (c) neutron-gamma PSD plot.

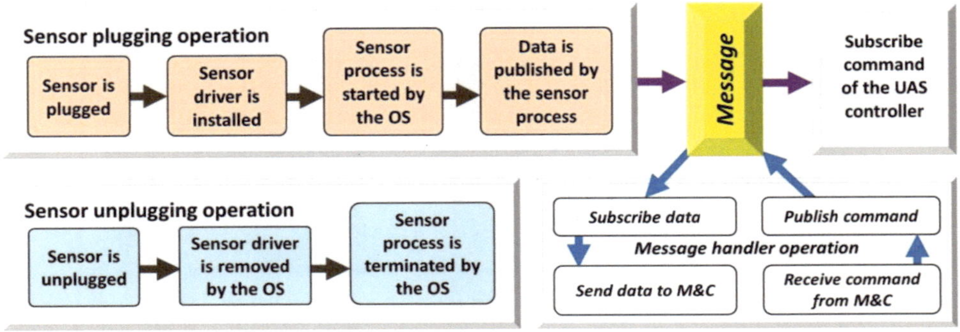

Figure 4.
Plug-and-play operation scheme for radiation sensors.

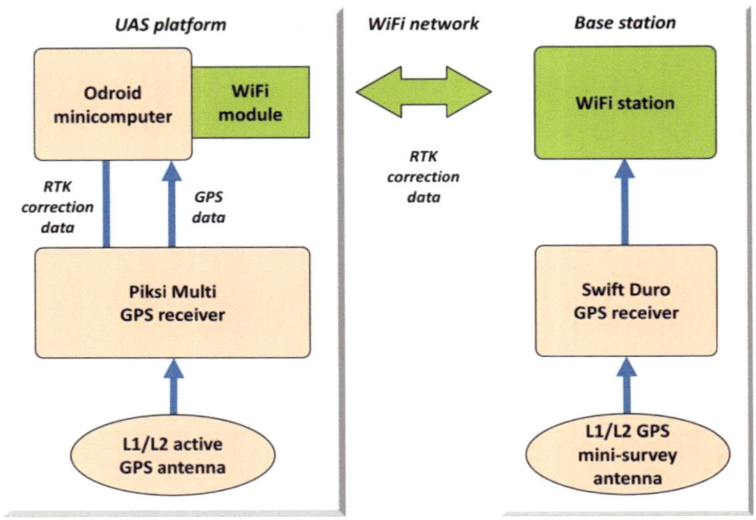

Figure 5.
Scheme of the RTK GPS technique.

sensor is unplugged, the operating system terminates the process and deletes the sensor's driver. The scheme of the plug-and-play operations with CZT and CLYC sensors is shown in **Figure 4**.

The ROS programming environment was also utilized for fusion of the radiation sensor's data with time and position data to enable the spatiotemporal analytics of measured radiation fields. To determine the UAS coordinates at a time of the gamma-ray measurement, the real-time kinematic (RTK) positioning technique was used. This navigation method enhances precision of the position data obtained using satellite positioning systems. Based on the measurements of the phase of the signal's carrier wave, it employs a single reference 'base' station for real-time corrections of the UAS position. The scheme of the RTK GPS technique is shown in **Figure 5**.

A base station contains a fixed Swift Duro RTK GPS receiver, having known coordinates. Multiple UAS platforms carry the RTK GPS and L1/L2 GPS antenna. Correction data are transmitted from the base station to the UAS's RTK Piksi Multi GPS receiver thus allowing calculating the ionospheric error. This method permits raw data measurements with a frequency of 20 Hz. Accuracy of the UAS position estimation using the RTK technique is 10 mm horizontally and 15 mm vertically.

It was experimentally verified using a single base station and four UAS platforms. Such precision of position estimation for multiple UAS platforms enables to use the RTK GPS in cooperative multirobot radiation monitoring tasks including contour mapping and source search.

3. Multirobot contour mapping of radiation fields

One of the approaches of cooperative multirobot sensing is the method of contour mapping of radiation areas. It is based on the use of several UAS platforms (the 'swarm') equipped with radiation sensors. This approach allows for the automatic determination of the contour in space that corresponds to a preset radiation dose. Thus, the multirobot system could locate and follow a boundary of the hazardous zone.

In this section, the contour mapping algorithm is presented along with a gradient direction estimation and heading angle calculation scheme for the swarm consisting of three UAS that are positioned in a circular formation in the two dimensional space. It is assumed that a gamma-ray sensor, CZT or CLYC, is mounted on each UAS platform. The gamma-ray data are time stamped and merged with the position data as discussed in the previous section.

3.1 Gradient direction estimation

The contour mapping is based on two components: the gradient direction estimation and the average radiation level calculated using the radiation measurement data from sensors mounted on the UAS platforms of the swarm. The average of a scalar field is estimated over a circular area of radius r centered at a point c as shown in **Figure 6**. T_{avg} is the average radiation level calculated using the data from sensors of UASs flying in a circular formation:

$$T_{avg}(x,y) = \frac{\sum_{i=1}^{N} T_i(x,y)}{N} \tag{1}$$

Here, N is the number of UAS platforms, T_i is the intensity of the measured gamma peak of interest at a point(x,y) by ith UAS. The formation center moves toward the increasing (a source-seeking method) or the constant (a contour mapping method) value of the average of sensor readings. To find the required direction

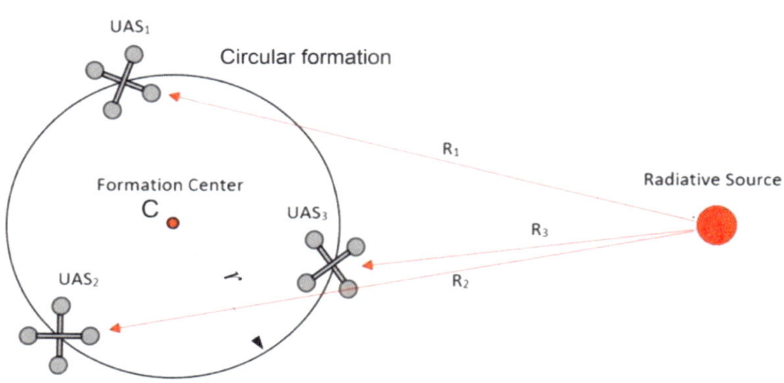

Figure 6.
Formation of three UAS platforms around a circle of radius r. Radiation measurements T_n by three UAS are based on $1/R^2$ model.

of motion of the swarm's center, the gradient of T_{avg} should be determined using multiple readings T_n from the UAS's sensors. It is assumed that N measurements are taken out of the readings distributed inside the circle according to a known distribution (e.g., uniform or based on a $1/R^2$ model). Using the composite trapezoidal rule, the gradient is estimated [20, 21] as:

$$\nabla_c T_{avg} \approx \frac{2}{\pi r^2} \sum_{i=1}^{N} T_n(p_i)\, p_i\, \Delta s \tag{2}$$

Here, $p_i = x_i - c$ and $\Delta s = 2\pi r / N$. It should be noted that the origin is moved to the center of the circle, and the integral is approximated by a finite number of measurements.

Three UAS platforms of the swarm ($n = 3$) are equally distributed around a circle circumference. Horizontal and vertical components of the gradient can be calculated using Eq. (2). Then the formation center c can be moved relative to its current position using the estimated direction of gradient for the source-seeking behavior for the swarm. In this technique, any number of UAS platforms can be used in the swarm. For accurate estimation of the gradient, these drones should be evenly distributed around a circle. In the gradient estimation algorithm, there is an inherent error in a $1/R^2$ scalar field. A relatively small change in distance can have a large effect on gamma measurement for each UAS depending on relative orientation of the swarm to the source as well as its distance from the source. In the contour mapping technique, three UAS platforms rotate about the swarm's center to improve the gradient estimation by changing the relative direction of the radiation source with respect to the UAS. **Figure 7a** illustrates how a gradient estimation error ε is defined. It should be noted that ε increases as a distance to the source relative to the swarm's center decreases. Therefore, the estimation error decreases when the source is located far away from the swarm (**Figure 7b**). The number of UAS in the swarm affects the gradient error as well. **Figure 7c** shows that the gradient estimation improves with more UAS in the swarm.

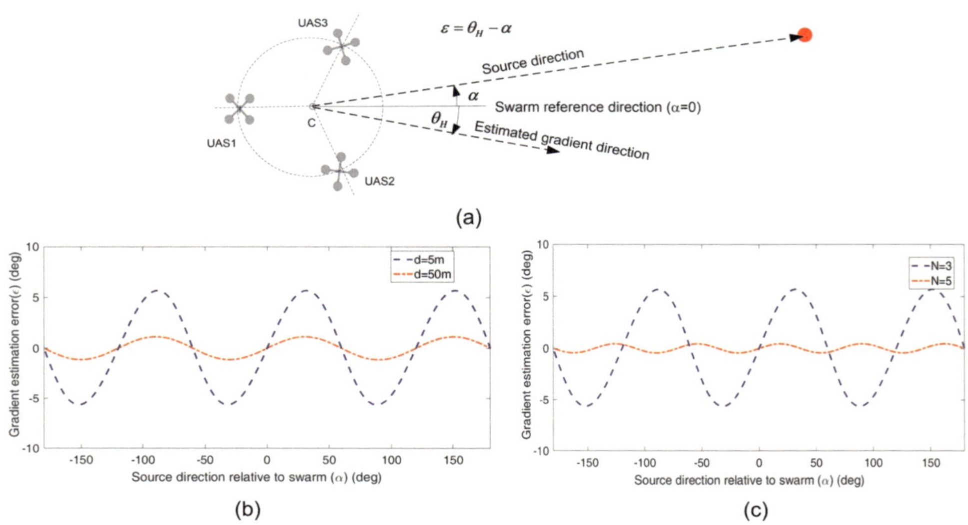

Figure 7.
(a) Gradient estimation, (b) effect of source distance on the gradient estimation error, and (c) effect of number of UAS on the gradient estimation error.

3.2 Heading angle

The bulk heading angle of the UAS swarm ψ depends on how far this swarm is from the desired radiation contour to be mapped. As the swarm approaches the desired contour, the heading angle must be directed to a tangential direction for this contour. When the swarm is far away from the contour, the heading angle will be directed toward the source, which is the source-seeking behavior of the swarm as shown in **Figure 8**. Here, θ_H is an estimated steepest gradient direction, and ϕ is a control angle determining a bulk heading angle Ψ [21]. The angles are measured with respect to a positive x-axis. The control angle ϕ is determined by how far the desired contour is located based on the average radiation measurement: $e_s = T_r - T_m$, where T_r is the desired radiation intensity of the contour to be mapped and T_m is an average radiation intensity measured using three UAS platforms. Here, an arbitrary constant of small magnitude R is used along with R_c to calculate a heading angle. Value of R_c is determined based on the PID control action from the measurement difference e_s with respect to the reference contour value T_m:

$$R_c = K_p e_s + K_d \frac{de_s}{dt} + K_i \int e_s dt,\ \phi = \tan^{-1}\frac{R_c}{R} \tag{3}$$

where K_p, K_i, and K_d are the proportional, integral, and derivative gains, respectively. As shown in Eq. (3), a heading angle ψ becomes 90 degrees for a large value of R_c when the swarm is far away from the reference contour.

For the swarm that is near the reference contour, R_c becomes small and ψ is close to zero. This leads to the equation for determining the heading angle:

$$\psi = \theta_H - \frac{\pi}{2} + \phi \tag{4}$$

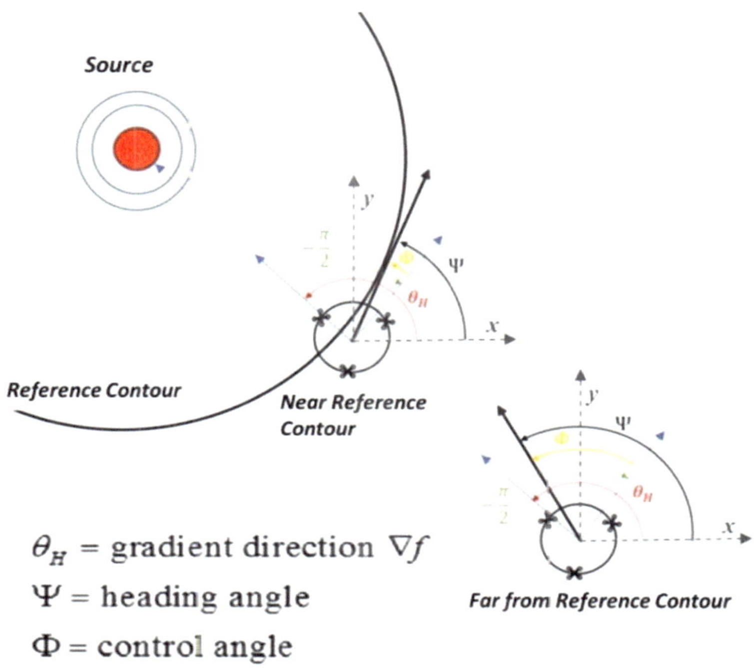

Figure 8.
Bulk heading angle of the swarm Ψ and the gradient angle θ_H.

Based on Eq. (4), the swarm will move in a tangential direction near the reference contour, and demonstrate a source-seeking behavior when it is far away from the source or the reference contour.

4. Computer simulation of radiation contour mapping

4.1 UAS swarm simulation in $1/R^2$ field

To validate the contour mapping algorithm, two types of radiation fields were used. The first is based on a $1/R^2$ model for a radioactive source located at the distance R from a sensor. This simplified model was employed to properly tune and improve contour mapping and source-seeking behaviors of the swarm. The algorithm was improved by including the UAS dynamics and adaptive spinning of the swarm (for the reduced flight trajectory of each UAS). The improved algorithm was validated using a realistic gamma field formed by multiple sources placed in the area with physical obstacles that was computed using the Monte Carlo Neutron and Particles code (MCNP) [22]. MCNP is widely used for simulation of coupled photon, neutron, electron and particle transport in complex geometries [23, 24].

A stochastic nature of radiation sensing was taken into account in the sensor's data. A random noise was introduced to the radius value of the $1/R^2$ model in the simulation: a noise of ±2.5 m was added, causing erratic gradient estimation and heading generation, as expected. To reduce this effect, a moving average filter was applied to both the gradient estimation and the heading angle generation. The simulation scheme is shown in **Figure 9**, including the UAS spinning formation used to estimate the steepest gradient. The UAS dynamics was incorporated for the realistic simulation of flight trajectories of aerial platforms during contour mapping and source seeking. Instead of using just kinematic motions of each UAS for this simulation, their dynamics was used to calculate their flight trajectories. The effect of inclusion of dynamics of UAS platforms on their trajectories that the swarm is able to follow is shown in **Figure 10**. A double integrator was incorporated in MATLAB to approximate the UAS dynamics. This provides more realistic simulation comparing to a case when only kinematics is considered.

A spinning formation of the swarm improves the estimation of radiation gradient direction [21, 25]. Spinning occurs around a virtual center of the UAS formation while the center of the formation moves along the desired vector to accomplish the contour mapping or source seeking. **Figure 11a** illustrates the swarm of three UAS platforms spinning around a virtual center to counteract error from the gradient estimation algorithm. A bold line shows a path traveled by the swarm's center in mapping a contour without spinning (**Figure 11b**) and with spinning (**Figure 11c**). The non-spinning formation exhibits poor mapping performance primarily due to large gradient estimation errors which depend on the relative direction to the source with respect to the swarm.

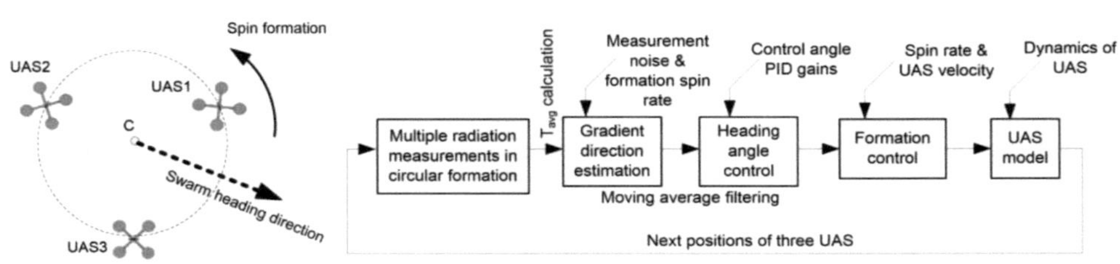

Figure 9.
Simulation scheme.

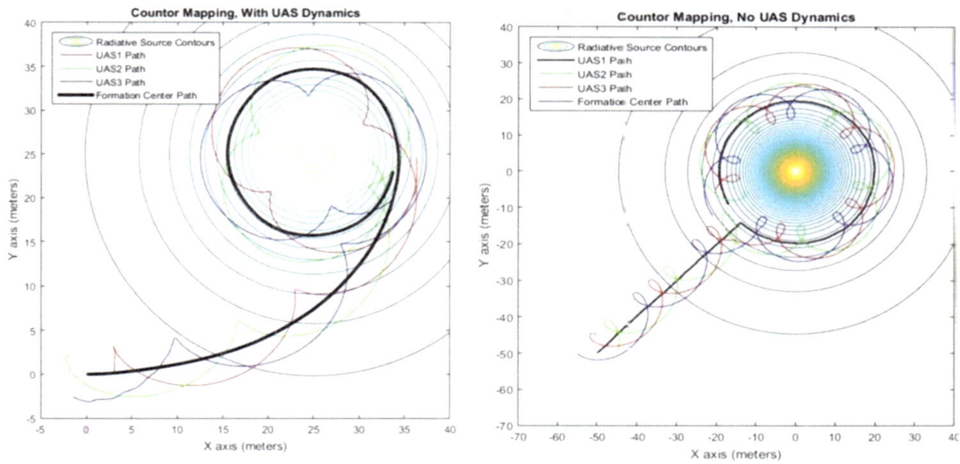

Figure 10.
Contour mapping with UAS dynamics and without it.

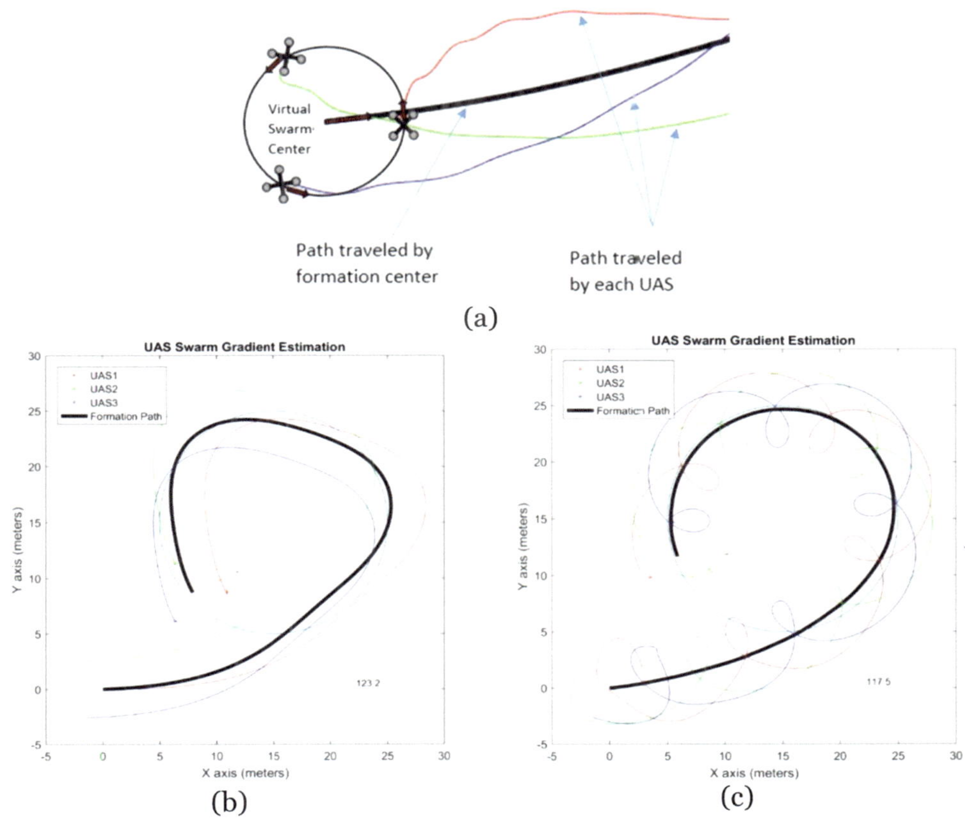

Figure 11.
(a) UAS swarm spins around a virtual center to counteract error from the gradient estimation algorithm (b) without spinning and (c) with spinning in mapping a contour.

Having the swarm circling around a virtual center, while this center travels in the direction dictated by the swarm heading algorithm, increases the total flight paths significantly. This led to development of an adaptive spin rate adjustment scheme to avoid unnecessary spinning formation when it is not required.

When the UAS swarm is far away from the source or the contour to be mapped, the direction to them is nearly fixed, thus it is not necessary to spin the formation. When the swarm is near the contour, it is necessary to spin the formation since most contours have a curvilinear shape. The criterion for spinning is chosen as a radius of curvature of the formation center's path.

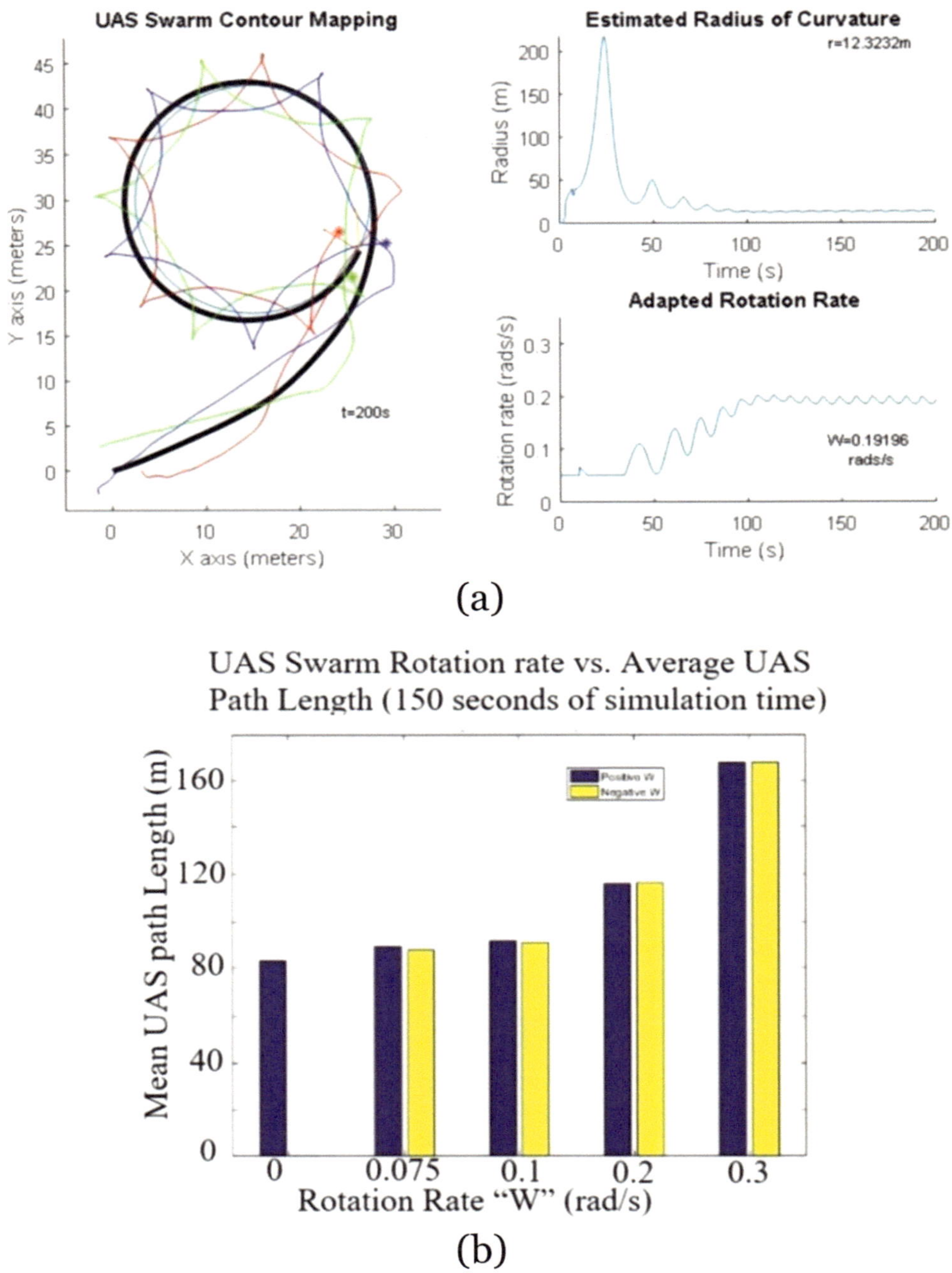

Figure 12.
(a) Adaptive tuning of spin rate based on radius of curvature of a formation center path. In this example, as an estimated radius of curvature approaches the target contour located at 10 m, a spin rate converges to the desired 0.25 rad/s and (b) plot of average path lengths in terms of spin rates.

As shown in **Figure 12a**, the swarm starts to spin when it approaches the contour to be followed. In this simulation, a spin rate control was bound with a lower value of 0.05 rad/s and an upper value of 0.3 rad/s. There is a big spike in a radius of curvature plot due to the nearly linear path the swarm has to follow. During the source-seeking stage, a minimum spinning was maintained to save on the total flight paths. Total path length in terms of spin rates is shown in **Figure 12b** which demonstrates a nearly negligible difference in path length if the swarm spins clockwise or counterclockwise. Rotation rates of 0, ±0.075, ±0.1, ±0.2, and ± 0.3 rad/s were used to compare the average of the flight path lengths.

Multiple radiation sources were also considered. **Figure 13** shows contour mapping simulation for three sources; the algorithm was capable to trace the desired

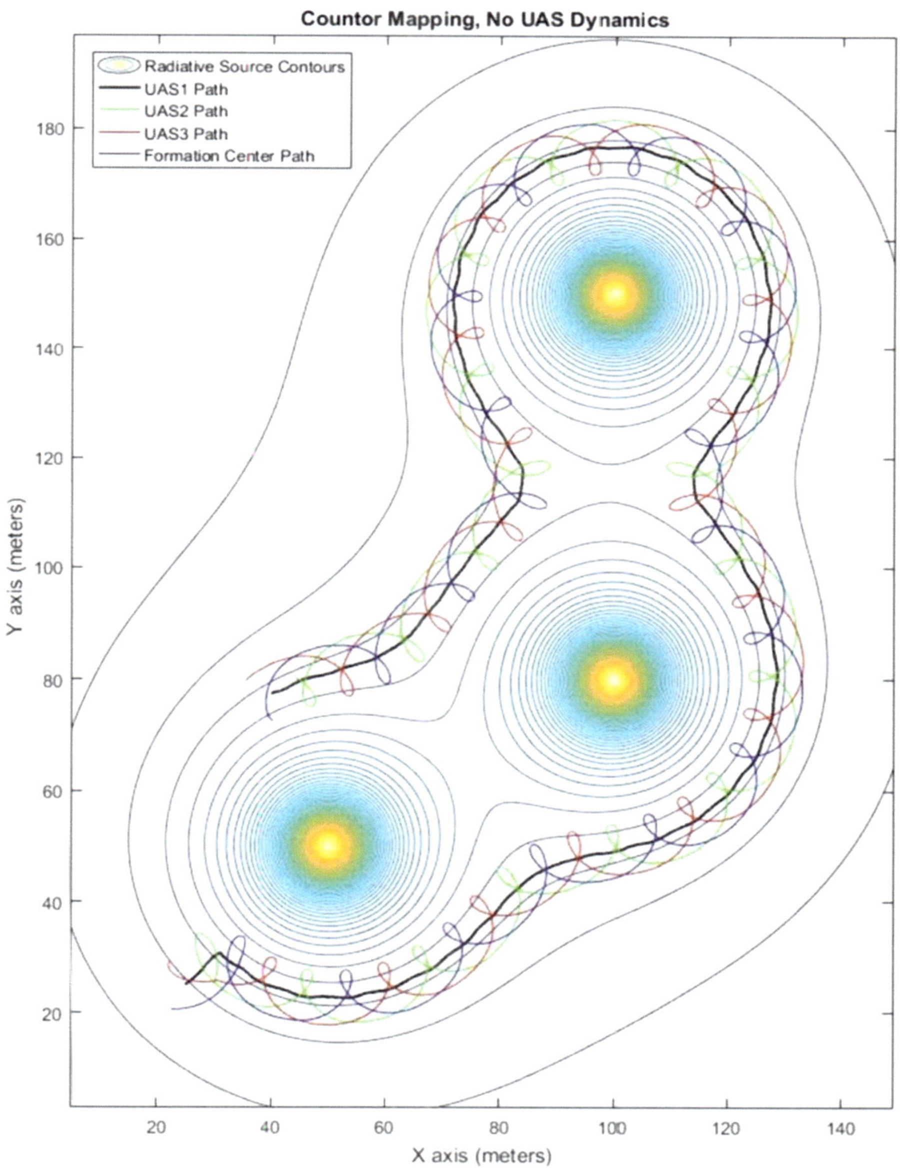

Figure 13.
Contour mapping for three radiation sources in the $1/R^2$ model.

contour reasonably well. Simulation was also done for a moving radiation source. The algorithm can accomplish tracing of the desired radiation contour well if the source moves reasonably slow. **Figure 14** shows a moving source traveling at 0.07 meters per second along the line from a point at (10, 40) m to a point (40, 10) m.

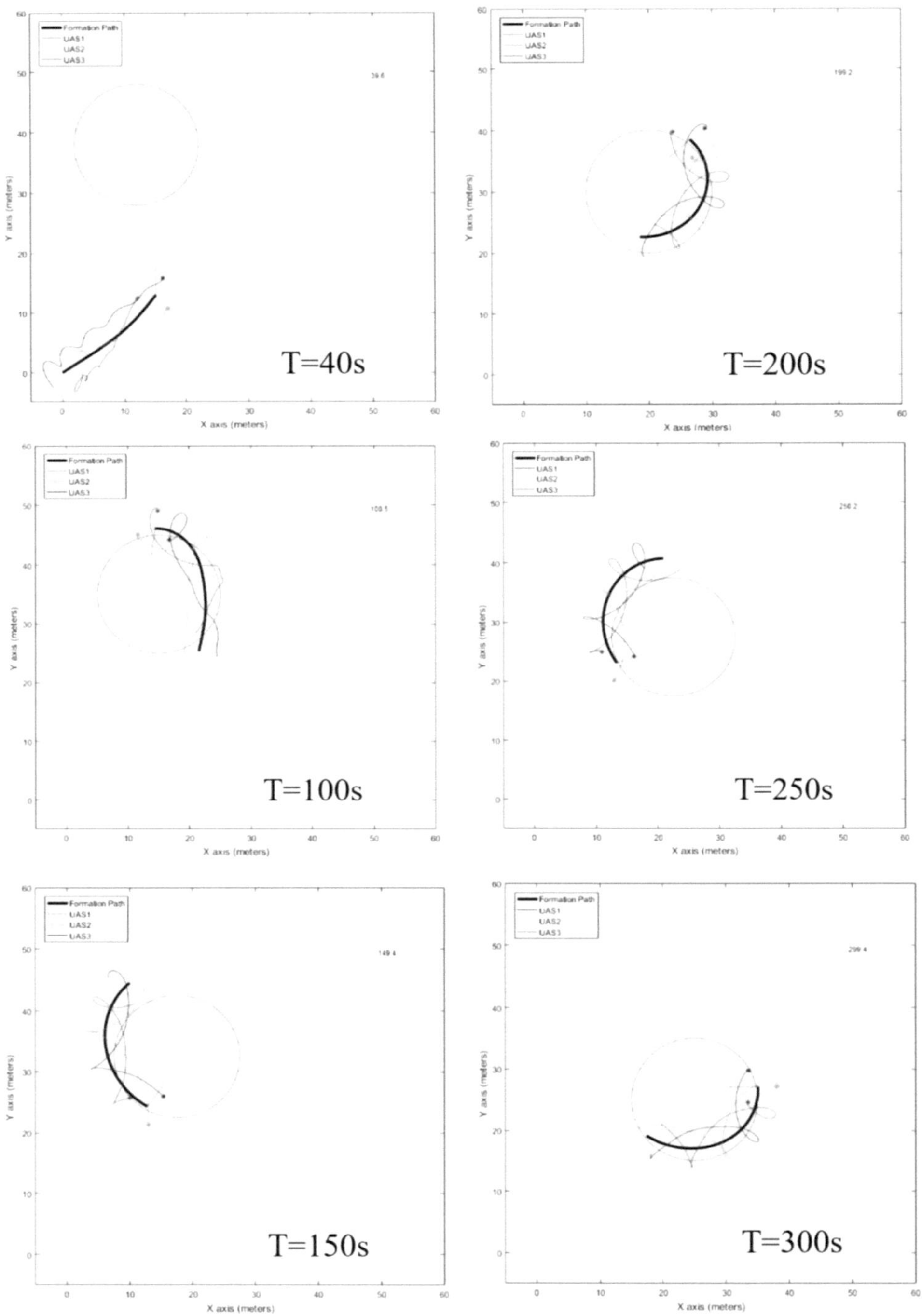

Figure 14.
Contour mapping simulation for a moving source.

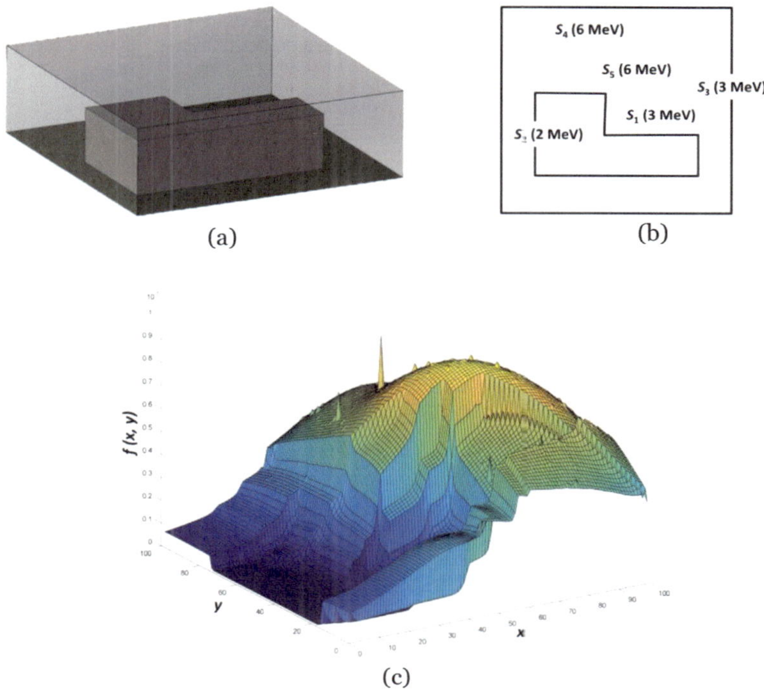

Figure 15.
Simulated radiation field: (a) model of the monitored volume; (b) radiation source locations; (c) simulated radiative map.

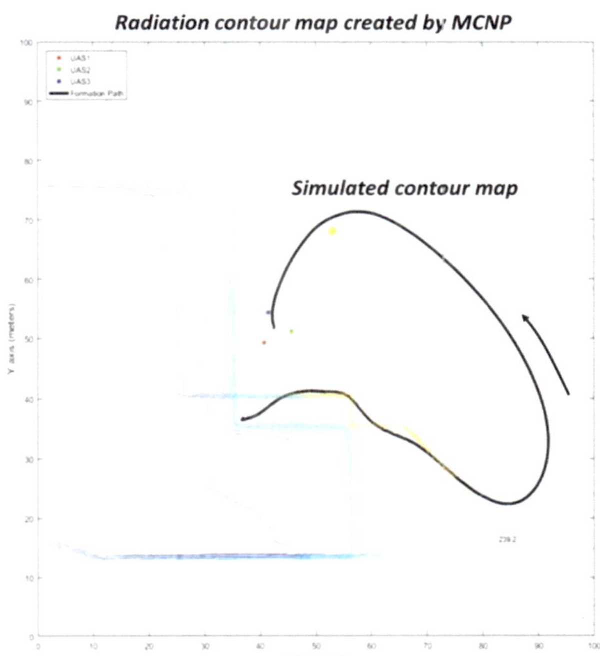

Figure 16.
Contour mapping in a radiation field computed using MCNP.

Mapping this source was possible in this particular case because the speed of the swarm was roughly seven times the speed of the source.

4.2 UAS swarm simulation in MCNP computed radiation field

A realistic gamma-ray flux distributions in a 3D volume with dimensions of 100 m × 100 m × 32 m that contained five photon sources with energy ranging from 3 to 6 MeV were computed using MCNP. A concrete building was used in the model to introduce a structure that attenuates and blocks radiation flux in certain areas of the volume monitored by onboard sensors of UAS. Therefore, complex gamma ray flux distributions were generated (**Figure 15**).

The reference contour along with performance of the contour mapping algorithm overlaid onto this radiation field is shown in **Figure 16**. A two dimensional radiation distribution at the height of 15 m was used in this simulation (the UAS platforms were moving in the same plane). In this simulation, the swarm's starting position was located inside the radiation field at a point with coordinates (35 m; 35 m). The algorithm allowed mapping the desired contour with a reasonable accuracy.

5. Experiments

Key algorithms of contour mapping and source seeking were validated in the indoor flight testbed. This testbed was outfitted with the OptiTrack motion capture technology which allows real-time feedback of the position and orientation of moving robots within the flight volume at 120 Hz rate.

The Crazyflie 2.0 by Birtcraze and the DJI Flamewheel 450 are two UAS platforms that were used; Crazyflie is an open source, easily modifiable, light weight quadcopter. Its small size enables validation of the contour mapping algorithm with a virtual source. The DJI Flamewheel 450 was used for validation of the source-seeking algorithm using a light source of the $1/R^2$ type. Each Flamewheel platform was equipped with an onboard omnidirectional light sensor. The Crazyflie and Flamewheel platforms with a single board computer mounted under the frame are shown in **Figure 17a** and **Figure 17b**, respectively.

5.1 Gradient estimation

To verify the algorithm of gradient estimation by three UAS in a circular formation, each platform was placed upside down roughly equiangular around the center

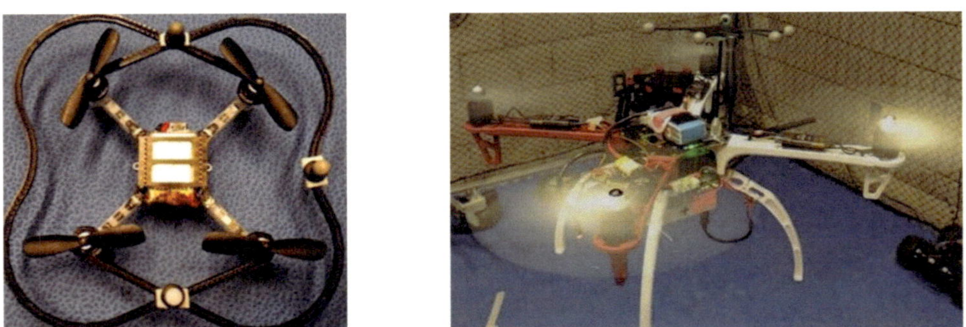

(a) (b)

Figure 17.
UAS platforms: (a) Crazyflie 2.0 and (b) DJI Flamewheel 450.

of the flight volume as shown in **Figure 18a**. Each UAS was placed at 0.5 m radius mark away from the center. The light source, acting as a radiation source analog, was moved around in a circle concentric with the swarm. The data stream captured through a wireless sensor network was input to the gradient estimation algorithm. Positions of a light source and three UAS were identified by the OptiTrack motion capture system, and used to compute true values of the gradient based on the $1/R^2$ assumption.

Figure 18b shows that the gradient direction estimation scheme agrees with the measured data reasonably well. It should be noted that the source distance of 0.7 m was used, which is rather too small for actual contour mapping application in the field. As expected, gradient direction estimation error reached almost 30 degrees which is rather too big for accurate mapping operation. However, it should be noted that experimental measurement data have a good agreement with the computed data.

5.2 Source-seeking behavior

Due to space constraints of the indoor flight volume, the source-seeking experiment was carried out using multiple Flamewheel platforms to test the gradient estimation algorithm for the heading angle rather than the contour mapping. A light source was placed on a movable dolly within the flight volume, and moved along the x-axis. It was shown that the gradient estimation algorithm was effective in determining the direction to the source using the light sensor's data from each UAS.

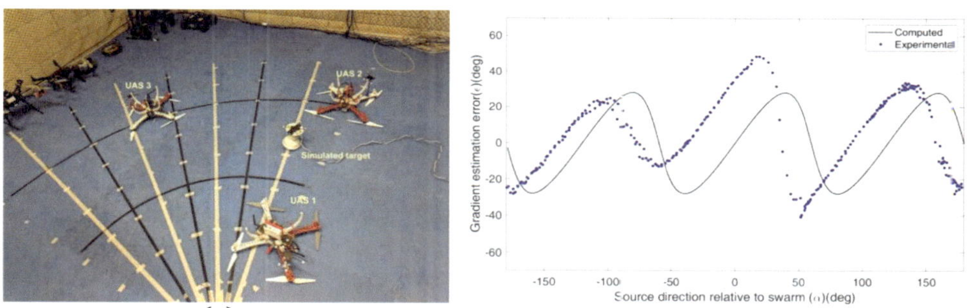

(a) (b)

Figure 18.
(a) Experimental setup for the gradient estimation and (b) plots of gradient estimation errors with respect to the source direction.

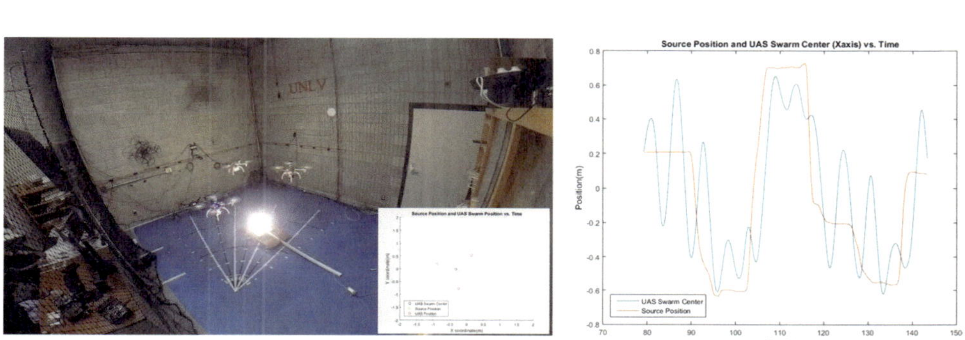

(a) (b)

Figure 19.
(a) Source-seeking behavior experiment using a light source and 3 lux sensors and (b) plot of source and swarm center positions.

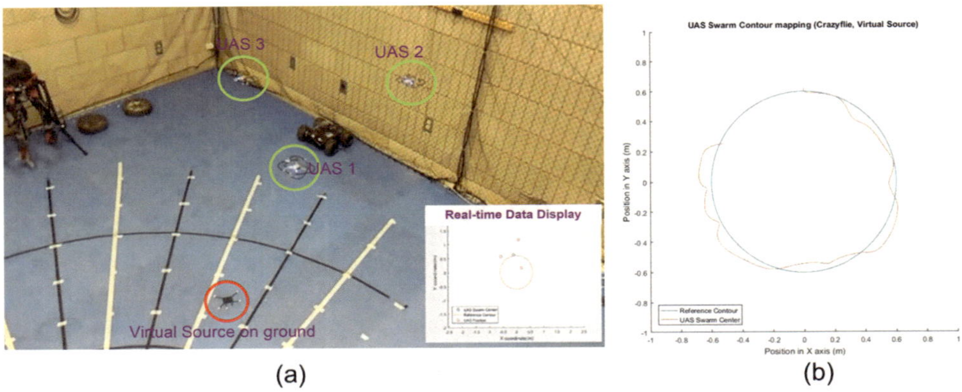

Figure 20.
(a) Contour mapping experiment with the swarm of three Crazyflie UAS (circled in green) and a virtual source (circled in red) along with a real-time data display window and (b) trajectory of the swarm's center.

The swarm was restricted to move along the x-axis and the reference position generation was bound to ±0.5 m, in order to minimize risk of the swarm crashing into the walls of the flight volume. The tracking of the light source and three UAS platforms was carried out using the OptiTrack. Note that an embedded window to display source-seeking performance in real time using the motion capture system in the flight volume is shown in **Figure 19a**. **Figure 19b** shows how the UAS swarm's center moves as the source is moved. As expected, the oscillatory motion of the swarm's center occurs due to gradient estimation errors using a finite number of light sensors.

5.3 Contour mapping behavior

To demonstrate the effectiveness of the contour mapping algorithm in the indoor flight volume, three Crazyflie platforms were used. A virtual source was used due to a limited payload and communication capability of the Crazyflie UAS. As shown in **Figure 20a**, a virtual source was located on the ground and the OptitTrack tracked the source and each UAS in the swarm. The 'source strength' needed for the gradient estimation algorithm was calculated using the $1/R^2$ model where R was obtained from the virtual source position data from each UAS. **Figure 20b** shows the experimental results on a plot of motion of the swarm's center following the reference contour defined with respect to the virtual source located on the floor. It maps the reference contour within ±0.1 m, which is less than 8% of the size of the contour.

6. Conclusion

Ambient temperature CZT and CLYC sensors were integrated onto the UAS platform using the plug-and-play approach. The CZT sensor was designed for high resolution gamma spectroscopy. The CLYC sensor enables gamma and neutron measurements with an excellent neutron-gamma pulse shape discrimination with the figure of merit 2.3. The automated spectral analysis code locating peaks and calculating their intensities was developed for both sensors.

USB hardware connections were used to link the sensors and the main controller with the UAS power source. ROS was used for the data communication and data fusion. To streamline the process of bridging disparate components into a cohesive network, the collection of libraries describing the publisher/subscriber communication of ROS nodes was developed for these sensors. The sensor's design supports hot swapping and does not require restarting the system.

The method of contour mapping and source seeking in the radiation field by the UAS swarm equipped with gamma-ray detectors was developed. The method is used for low altitude applications where fixed wing UAS platforms are not suitable. The source-seeking and contour mapping algorithms were validated using a realistic radiation field with scattering and attenuation of gamma flux computed with MCNP code. The algorithm was implemented to map the radiation contours for multiple radiation sources and also a moving source. Moreover, it showed an effective way of cutting down the flight trajectories of three flying platforms by adaptively updating a swarm's spin rate based on averaging the measured radiation data. The UAS swarm enables surveying an unknown, contaminated environment and mapping the area while locating the radiation sources, thus helping first responders to enhance situational awareness and to manage operations and safeguard personnel.

Acknowledgements

This work was supported by the Department of Energy Minority Serving Institution Partnership (MSIPP) managed by the Savannah River National Laboratory under SRNS contract MSIT00016.

Conflict of interest

The authors declare no conflicts of interest.

Author details

Monia Kazemeini, John Vargas, Alexander Barzilov* and Woosoon Yim
University of Nevada—Las Vegas, Las Vegas, NV, Unites States

*Address all correspondence to: alexander.barzilov@unlv.edu

References

[1] Till J, Grogan H, editors. Radiological Risk Assessment and Environmental Analysis. New York: Oxford University Press; 2008. 702 p. ISBN: 978-0-19-512727-0

[2] Moody K, Hutcheon I, Grant P. Nuclear Forensic Analysis. Boca Raton: CRC Press; 2005. 485 p. ISBN: 0-8493-1513-1

[3] Sen T, Moore L. An organizational decision support system for managing the DOE hazardous waste cleanup program. Decision Support Systems. 2000;**29**(1):89-109. DOI: 10.1016/S0167-9236(00)00066-X

[4] Runkle R, Bernstein A, Vanier P. Securing special nuclear material: Recent advances in neutron detection and their role in nonproliferation. Journal of Applied Physics. 2010;**108**:111101-1-111101-21. DOI: 10.1063/1.3503495

[5] Nagatani K et al. Emergency response to the nuclear accident at the Fukushima Daiichi nuclear power plants using mobile rescue robots. Journal of Field Robotics. 2013;**30**:44-63. DOI: 10.1002/rob.21439

[6] Hartman J, Barzilov A, Novikov I. Remote sensing of neutron and gamma radiation using aerial unmanned autonomous system. In: Proceedings of the IEEE Nuclear Science Symposium and Medical Imaging Conference (NSS/MIC'15); 31 October–7 November 2015; San Diego, CA, USA. IEEE; 2015. pp. 1-4. DOI: 10.1109/NSSMIC.2015.7581763

[7] Prokesch M, CdZnTe for gamma and x-ray applications. In: Awadalla S, editor. Solid-State Radiation Detectors: Technology and Applications. Boca Raton: CRC Press; 2015. pp. 17-48. ISBN: 9781482262209

[8] GR1 CZT gamma-ray detector spectrometer [Internet]. 2018. Available from: https://www.kromek.com/product/gamma-ray-detector-spectrometers-czt-based-gr-range/ [Accessed: 2018-11-12]

[9] Robot operating system: Powering the world's robots [Internet]. 2018. Available from: https://www.ros.org/ [Accessed: 2018-10-20]

[10] Kamuda M, Zhao J, Huff K. A comparison of machine learning methods for automated gamma-ray spectroscopy. Nuclear Instruments and Methods in Physics Research A. 2018. In press. https://doi.org/10.1016/j.nima.2018.10.063

[11] Medhat M. Artificial intelligence methods applied for quantitative analysis of natural radioactive sources. Annals of Nuclear Energy. 2012;**45**:73-79. DOI: 10.1016/j.anucene.2012.02.013

[12] Barzilov A, Kessler B, Womble P. Analysis of 14-MeV neutron induced gamma-ray spectra using multiwavelets. Radiation Measurements. 2015;**79**:43-49. DOI: 10.1016/j.radmeas.2015.06.007

[13] Barzilov A, Novikov I, Womble P. Material analysis using characteristic gamma rays induced by neutrons. In: Adrovic F, editor. Gamma Radiation. Rijeka, Croatia: InTechOpen; 2012. pp. 17-40. DOI: 10.5772/36372. ISBN: 978-953-51-0316-5. Available from: https://www.intechopen.com/books/gamma-radiation/material-analysis-using-characteristic-gamma-rays-induced-by-pulse-neutrons

[14] Barzilov A, Womble P. Study of Doppler broadening of gamma-ray spectra in 14-MeV neutron activation analysis. Journal of Radioanalytical and Nuclear Chemistry. 2014;**301**:811-819. DOI: 10.1007/s10967-014-3189-8

[15] Mariscotti M. A method for automatic identification of peaks in the presence of background and its application to spectrum analysis. Nuclear Instruments and Methods. 1967;**50**(2):309-320. DOI: 10.1016/0029-554X(67)90058-4

[16] D'Olympia N et al. Pulse-shape analysis of CLYC for thermal neutrons, fast neutrons, and gamma-rays. Nuclear instruments and methods in. Physics Research A. 2013;**714**:121-127. DOI: 10.1016/j.nima.2013.02.043

[17] Guckes A, Barzilov A. Development and deployment of the collimated directional radiation detection system. In: Proceedings of the SPIE Conference on Radiation Detectors in Medicine, Industry, and National Security XVIII; 7 September 2017; San Diego, CA, USA. SPIE; 2017. pp. 1039306-1-1039306-10. DOI: 10.1117/12.2273205

[18] Giaz A et al. The CLYC-6 and CLYC-7 response to gamma-rays, fast and thermal neutrons. Nuclear Instruments and Methods in Physics Research A. 2016;**810**:132-139. DOI: 10.1016/j.nima.2015.11.119

[19] Hartman J, Barzilov A, Peters E, Yates S. Measurements of response functions of EJ-299-33A plastic scintillator for fast neutrons. Nuclear Instruments and Methods in Physics Research A. 2015;**804**:137-143. DOI: 10.1016/j.nima.2015.09.068

[20] Uryasev S. Derivatives of probability functions and some applications. Annals of Operations Research. 1995;**56**(1):287-311. DOI: 10.1007/BF02031712

[21] Ogren P, Fiorelli E, Leonard N. Cooperative control of mobile sensor networks: Adaptive gradient climbing in a distributed environment. IEEE Transactions on Automatic Control. 2004;**49**(8):1292-1302. DOI: 10.1109/TAC.2004.832203

[22] Goorley T. Initial MCNP6 release overview. Nuclear Technology. 2012;**180**(3):298-315. DOI: 10.13182/NT11-135

[23] Barzilov A, Novikov I, Cooper B. Computational study of pulsed neutron induced activation analysis of cargo. Journal of Radioanalytical and Nuclear Chemistry. 2009;**282**:177-181. DOI: 10.1007/s10967-009-0298-x

[24] Fang Y, Xu C, Yao Y, Pelc N, Danielsson M, Badano A. Modeling charge transport in photon-counting detectors. Nuclear Instruments and Methods in Physics Research A. 2018;**899**:115-121. DOI: 10.1016/j.nima.2018.05.027

[25] Arranz L, Seuret A, De Wit C. Translation control of a fleet circular formation of AUVs under finite communication range. In: Proceedings of the IEEE Conference on Decision and Control (CDC); 15-18 December 2009; Shanghai, China. IEEE: 2009. pp. 8345-8350. DOI: 10.1109/CDC.2009.5400899

Section 3

Gamma Rays Applications

Chapter 7

Gamma Rays: Applications in Environmental Gamma Dosimetry and Determination Samples Gamma-Activities Induced by Neutrons

Hassane Erramli and Jaouad El Asri

Abstract

Gamma rays are high frequency electromagnetic radiation and therefore carry a lot of energy. They pass through most types of materials. Only an absorber such as a lead block or a thick concrete block can stop their transmission. In many alpha and beta transitions, the residual nucleus is formed in an excited state. The nucleus can lose its excitation energy and move to a "fundamental level" in several ways. (a) The most common transition is the emission of electromagnetic radiation, called gamma radiation. Very often the de-excitation occurs not directly between the highest level of the nucleus and its basic level, but by "cascades" corresponding to intermediate energies. (b) The gamma emission can be accompanied or replaced by the electron emission so-called "internal conversion", where the energy excess is transmitted to an electron in the K, L or M shell. (c) Finally, if the available energy is greater than $2m_ec^2$ = 1022 keV, the excited nucleus can create a pair (e^+, e^-). The excess energy appears as a kinetic form. This internal materialization process is very rare. In this chapter we are presenting two applications of gamma rays: On the one hand, TL dosimeters and field gamma dosimetry are studied, a careful study of the correcting factors linked to the environmental and experimental conditions is performed. On the other hand, we are presenting a calculation method for controlling neutron activation analysis (NAA) experiments. This method consists of simulating the process of interaction of gamma rays induced by irradiation of various samples.

Keywords: gamma-ray, gamma attenuation, gamma spectroscopy, Monte Carlo simulation, medicine, industry, TL dosimetry

1. Introduction

The gamma rays are emitted from a nucleus or from the annihilation of positron with electrons. The most intense sources of gamma rays are radioactive sources. The photons resulting from de-excitation of nuclei have energies ranging from less than 1 to about 20 MeV.

The photons resulting from annihilations event can have much larger energy: the neutral pion (π^0), for example, produce two photons of about 70 MeV.

Gamma rays are ionizing electromagnetic radiation, obtained by the decay of an atomic nucleus. Gamma rays are more penetrating, in matter, and can damage living cells to a great extent. Gamma rays are used in medicine (radiotherapy), industry (sterilization and disinfection) and the nuclear industry. Shielding against gamma rays is essential because they can cause diseases to skin or blood, eye disorders and cancers.

The interaction of gamma rays with matter may be divided into three main categories depending on the energy of the photon. These three mechanisms are the photoelectric effect, Compton scattering and pair production. All results in the energy of the photon being transferred to electrons which subsequently lose energy by further interactions.

2. Origin of the gamma rays

The gamma radiations are constituted by photons, characterized by their energy, inversely proportional to their wavelength. The gamma rays come from the nuclei during the nuclear reactions, it is monoenergetics for a given characteristic reaction. So, the β-decay of the ^{137}Cs to ^{137}Ba produces a gamma radiation of 660 keV. The β-decay of the ^{60}Co to ^{60}Ni produces a double radiation emission of 1.17 and 1.33 MeV.

Natural gamma radiation sources can be easily divided into three groups according to their origin. The first group includes potassium (^{40}K) with a half-life of 1.3×10^9 years, uranium-238 (^{238}U) with a half-life of 4.4×10^9 years, uranium-235 (^{238}U) with a half-life of 7.1×10^8 years and thorium (^{232}Th) with a half-life of 1.4×10^{10} years.

The second group includes radioactive isotopes from the first group. Those have half-lives ranging from small fractions of a second to 10^4 to 10^5 years. The third group will include it isotopes created by external causes, such as the interaction of cosmic rays with the earth and its atmosphere.

2.1 Gamma radiation production mechanism

The Gamma rays are produced in number of ways:

- **Thermal radiation:** Only an extremely hot medium ($T = 10^8$ K) is likely to produce gamma radiation. Such media are extremely rare, and this process is not fundamental to the production of this radiation.

- **Inverse Compton effect:** During a collision with low energy photon, a relativistic electron can transfer to it a significant part of its energy, the photon can then have an energy of 100 MeV.

- **The Synchrotron radiation:** A relativistic electron spiraling through the force lines of a magnetic field radiates electromagnetic energy. Thus, energy electrons 3×10^8 MeV in a magnetic field of 3×10^{-6} Gauss will create a gamma radiation. But, by radiating the electrons lose energy: their lives are therefore limited.

- **Bremsstrahlung or braking radiation:** An electron passing near a nucleus is influenced by its Coulombian field. The deceleration of the electron is accompanied by a loss of energy in the form of gamma radiation when the electron has a relativistic speed.

- **Production of meson π^0:** A meson can be produced during reactions of several types:
  - Collision between two protons or a proton and alpha (helium nucleus).
  - Collision between a proton and an antiproton; if the antimatter exists in the Universe, the observation of the gamma radiation produced is a method for detecting it. The meson (π^0) then disintegrates into two photons.
- **Nucleus de-excitation:** Just like an atom or a molecule, a nucleus has energy levels whose transitions between the least excited levels give rise to gamma radiation. Such nucleus de-excitation may occur either during the interaction of a nucleus with neutrinos or during certain thermonuclear reactions.

3. Gamma attenuation

Unlike charged particles that gradually loses up their energy to matter, when gamma rays traverse matter, some are absorbed, some pass through without interaction, and some are scattered as lower energy photons. Electromagnetic radiation vanishes brutally as a result of interaction. We can no longer talk about a slowdown. We have to introduce the attenuation notion.

Although a large number of possible interaction mechanisms are known, when monoenergetic gamma rays are attenuated in the matter, only four major effects are important: photoelectric effect, Compton effect, elastic or Rayleigh scattering and pair production (with a threshold energy of 1022 keV).

The probability which has a photon of energy given to undergo an interaction during the penetrating in a given material is represented by the attenuation coefficient for this material absorber. The more range of gamma rays in the absorber is long, the more the interaction probability increases. The attenuation coefficient is the sum of coefficients of the various interaction modes (Compton, photoelectric, pair production), the proportion of each of these effects varying with the radiation energy and nature of the absorber.

3.1 Attenuation law

When a narrow beam of N_0 (number of photons/unit area) monoenergetic photons with energy E_0 passes through an homogeneous absorber of thickness x, the number of photons that reach a depth (in cm) without having an interaction is given by:

$$N = N_0 e^{-\mu(\rho,Z,E)x} \tag{1}$$

where μ (in cm) is the linear attenuation coefficient for material of physical density (in g/cm^3) and atomic number Z.

The total linear attenuation coefficient can be decomposed into the contributions from each above described mode of photon interaction as:

$$\mu_{tot} = \mu_{ph} + \mu_C + \mu_{coh} + \mu_p \tag{2}$$

where μ_{ph}, μ_C, μ_{coh} *and* μ_p are the attenuation coefficient for photoelectric effect, Compton scattering, Rayleigh scattering and pair production respectively. Although linear attenuation coefficients are convenient for engineering applications and shielding calculations, they are proportional to the density of the absorber, which depends on the physical state of the material.

The relationship (1) can be written as:

$$N = N_0 e^{-\left(\frac{\mu(Z,\rho)}{\rho}\right)\rho x} \tag{3}$$

where μ/ρ (in cm^2/g) is the mass attenuation coefficient.

If the absorber is a chemical compound or mixture, its mass attenuation coefficient μ/ρ can be approximately evaluated form the coefficients for the constituent elements according to the weights sum

$$\left(\frac{\mu}{\rho}\right)_{compound} = \sum_i \left(\frac{\mu}{\rho}\right)_i w_i \tag{4}$$

where w_i is the proportion by weight of the ith constituent the material.

The mass attenuation coefficient of a compound or a mixture can be, therefore, calculated from the mass attenuation coefficient of the components [1].

The total linear attenuation coefficient, $\mu_{\text{compound or mixture}}$ of the compound or mixture can then be simply found by multiplying the total mass attenuation coefficient, $(\mu/\rho)_{\text{compound}}$ with its density, ρ. Thus,

$$\mu_{compound} = \left(\frac{\mu}{\rho}\right)_{compound} \times \rho \tag{5}$$

Figure 1 shows the linear attenuation of solid sodium iodide, a common material used in gamma-ray detectors.

3.2 Attenuation coefficients versus atomic number and physical density

The total mass attenuation coefficient μ/ρ is also proportional to the total cross section per atom σ^2_{tot}. His relation is:

$$\frac{\mu}{\rho} = \sigma^2_{tot}\left(\frac{cm^2}{atom}\right)\cdot\left(\frac{N_A}{A}\right)\left(\frac{atoms}{g}\right) \tag{6}$$

where $N_A = 6.03 \times 10^{23}$ is Avogadro's number and A is the atomic mass of the absorber (in g).

Since there are Z electrons per atom:

$$\frac{\mu}{\rho} = \sigma^2_{tot}\cdot\left(\frac{ZN_A}{A}\right) = \sigma^2_{tot}\cdot\delta_e$$

where σ^2_{tot} (cm^2/electron) is the total cross section per electron and the term $\left(\frac{ZN_A}{A}\right)$ represents the electron density (number of electrons per g).

For all elements except hydrogen, δ_e approximately equals $N_A/2 = 3 \times 10^{23}$, because Z/A = ½. For hydrogen, δ_e is equal to N_A, and is therefore twice the « normal » value.

This indicates that atomic composition dependence of $\frac{\mu}{\rho}$ is uncorrelated to the term δ_e, and is related to the term σ^2_{tot}. The linear attenuation coefficient μ is, therfore, approximately depending on Z in the physical density term ρ and expressly depending on Z in term σ^2_{tot}.

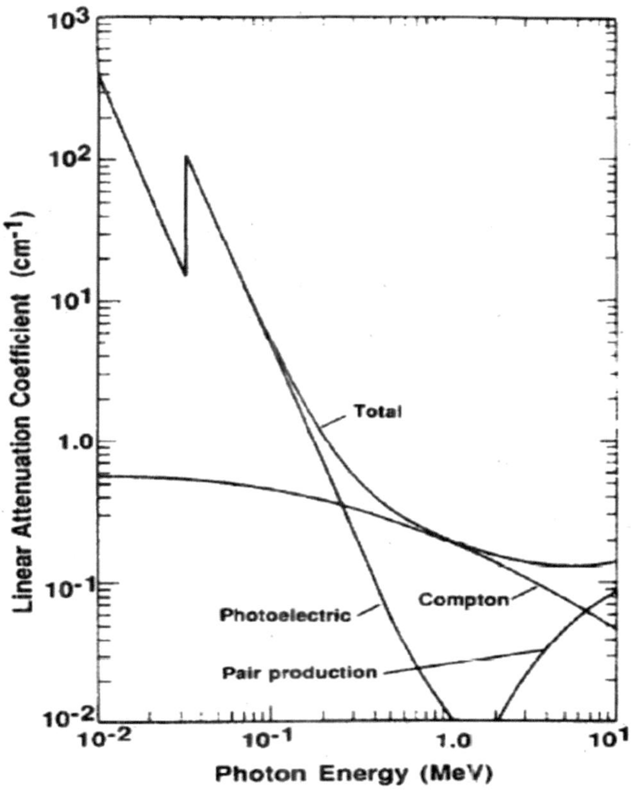

Figure 1.
Linear attenuation coefficient of NaI showing contributions from photoelectric absorption, Compton scattering, and pair production [2].

3.3 Calculation of HVL, TVL and relaxation length (λ)

The linear attenuation coefficient is inversely proportional to a quantity called a half-value layer (HVL), which is the material thickness needed to attenuate the intensity of the incident photon beam to half of its original value. This means that when × = HVL, $\frac{N_0}{N} = \frac{1}{2}$.

From Eq. (3) it can be shown that:

$$\frac{N_0}{2} = N_0 e^{-\mu HVL}$$
$$HVL = \frac{ln\,2}{\mu} = \frac{0.693}{\mu} \tag{7}$$

The HVL of a given material thus characterizes the quality (penetrance or hardness) of a gamma beam.

Figure 2 shows the relationship between the linear attenuation coefficient and the HVL for a soft tissue [3].

3.4 Calculation of relaxation length (λ)

The average distance between two successive interactions is called the relaxation length (λ) or the photon mean free path which is determined by the equation:

$$\lambda = \frac{\int_0^{\infty} xe^{-\mu x}dx}{\int_0^{\infty} e^{-\mu x}dx} = \frac{1}{\mu} \tag{8}$$

where μ is the linear attenuation coefficient and x is the absorber thickness.

N.B: The relaxation length is the thickness of a shield for which the photon intensity in a narrow beam is reduced to 1/e (or 0.37) of its original value.

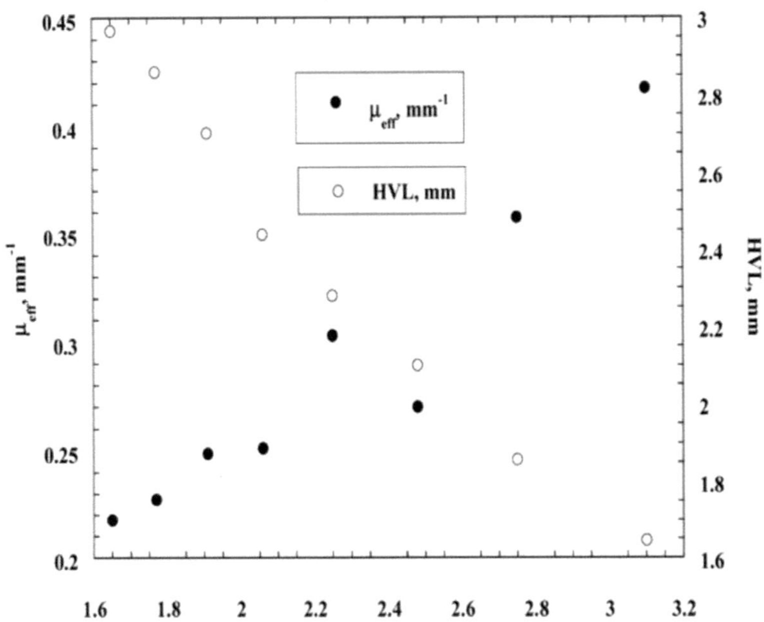

Figure 2.
Relationship between the linear attenuation coefficient and the HVL for a soft tissue.

4. Application field

Apart from the use of nuclear energy for the supply of electricity, the applications of radioactivity are numerous in many areas: medical physics, earth sciences, industry and preservation of cultural heritage.

The properties used for these various applications are:

- Time decline of radioactivity
- Radiation emission
- Sensitivity of detection

4.1 Medical physics applications

The major application of radiation in medicine is radiotherapy and/or treatment by ionizing radiation. A few months after the discovery of X-rays, there is over a century, it has become clear that biological action radiation could be used in the treatment of cancers. Cancers cells divided more quickly are more sensitive, than normal cells to ionizing radiation. By sending these cells a certain dose of radiation, it is possible to kill them and eliminate the tumor.

4.2 Sterilization of objects by gamma radiation

- **Irradiation of surgical and food material:** Irradiation is a privileged means to destroy micro-organisms (fungi, bacteria, virus...). As a result, many applications radiation exists for sterilization of objects. For example, most medical-surgical equipment (disposable syringes, etc.) is today radio-sterilized by specialized industrialists. Similarly, the treatment by irradiation of food ingredients allows improve food hygiene: sterilization spices, elimination of salmonella from shrimp and frog legs. This technics is also known as food ionization.

- **Irradiation of art objects:** Treatment with gamma rays helps to eliminate larvae, insects or bacteria lived inside objects, to protect them from degradation. This technics is used in the treatment of conservation and restoration of arts objects, ethnology and archaeology. It is applicable to different types of materials: wood, stone, leather, etc.

4.3 Industrial applications

- **Elaboration of materials:** Irradiation causes, under certain conditions, chemical reactions that allow the development of more resistant materials, more lightweight, capable of superior performance.

4.4 Gamma rays spectrometry

The development of γ-spectrometry began with the development of nuclear sciences and technology to meet the needs for the control, characterization and analysis of radioactive materials. This measurement technics exploits a fundamental property observed for unstable nuclei: the emission of radiation from the process of nuclear decay. It is thus known as non-destructive because it respects the integrity of the object to be analyzed [4].

The interest in γ spectrometry has continued to grow over the years, both from a point of view metrological and a point of view applications. This development was made possible by a better understanding of the process of photon interaction with matter, and especially by the appearance of semiconductor detectors in the 1960s. Spectrometry γ then became a powerful tool for studying decay patterns. It is now used in a wide variety of sectors (for example: dating, climatology, astrophysics, medicine) and in virtually all stages of the fuel cycle.

Photon spectrometry is a commonly used nuclear measurement technique to identify and quantify gamma emitting radionuclides in a sample. It is non-destructive and does not require specific sample preparation. Conventional spectrometers are designed around semiconductor detectors, usually with high purity germanium (hyper-pure germanium).

The radionuclides measured by this method emit gamma photons of specific energies and their interactions with the detector depend on several variables (geometry or conditioning: physical shape of the object, density, measured quantity, container type, emission energy, size, shape, nature of the detector, etc.).

5. Example of gamma rays applications

5.1 Environmental gamma dosimetry

The thermoluminescence dating method (TL) requires a very accurate knowledge of the annual radiation doses deposited, in the minerals that are used, by the alpha, beta, gamma and cosmic rays [5, 6].

In the annual radiation doses, we can distinguish two components:

- internal dose: deposited by the radiations with a short range (alpha and beta) coming out of the bulk of the sample.
- external dose: deposited by the long range radiations (gamma rays and cosmic) coming out of the surroundings of the sample.

Gamma-ray dose-rate may be measured by a TL dosimeter. But as this dose is not valid for the dosimeter itself, corrections must be made to know the one corresponding to the soil.

These corrections are related to the complexity of the energy spectrum gamma radiation incidents; their origins are:

- The difference in composition between TL dosimeter and the soil, resulting in absorption and therefore deposited different doses, especially for an energies less than 100 keV.
- The capsule enclosing the TL dosimeter induce absorptions of low energy gamma rays.

The correction factors had already been investigated theoretically [7] and experimentally [8, 9]. Below, we describe a theoretical evaluation method, for these factors, which does not require excessive computer time and so can be easily extended to a wide variety of site conditions.

5.1.1 Environmental conditions

The calculations presented here refer to $CaSO_4$: Dy as dosimeter, and two encapsulating materials, polyethylene and copper, of various thicknesses (**Figure 3**). However, the resulting computer programs can be easily extended to other materials.

Several kinds of soils were considered. For this case, we retained two extreme cases of real soils:

- a siliceous soil of low <Z> (<Z> = 11.26), hereafter referred to as 'S soil';
- a very calcareous soil of high <Z> (<Z> = 14.07) referred to as 'C soil'.

The relative energies and intensities of the lines taken into consideration are given in **Table 1** [10]. For the uranium series, the contribution of uranium-235 and its descendants were taken into account.

The bulk of the calculation therefore comes down to a few dozen successive numerical integrations, with about 220 steps each time. It is therefore a much

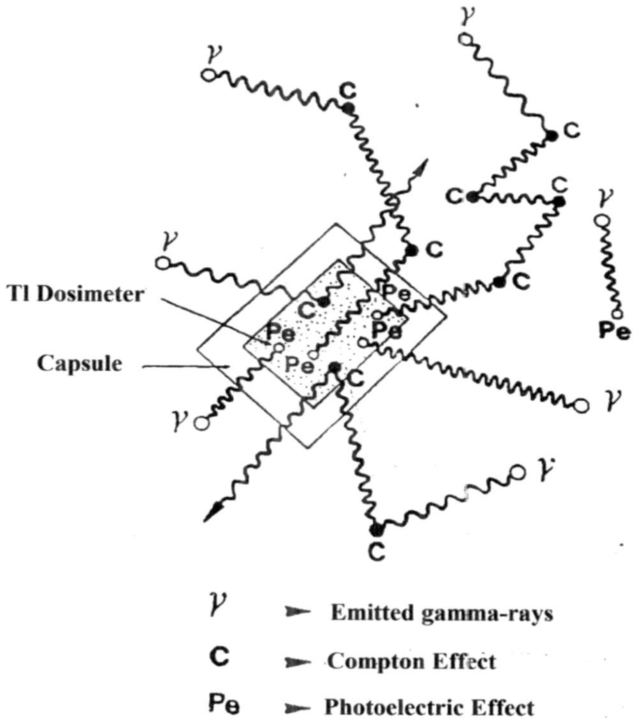

Figure 3.
TL dosimeter irradiated by an incident γ-ray flux.

lighter and less demanding procedure in computer time than the Monte-Carlo method for a similar definition.

5.1.2 Outline of the calculations of the dosimeter/soil dose ratio

The ratio ε of the dose deposited in an encapsulated dosimeter by an incident gamma flux to that deposited in soil by the same flux can be evaluated using the expression:

$$\varepsilon = \frac{\int \phi(E) E\mu_{a1}.e^{-\mu_{a2}t} dE}{\int \phi(E) E\mu_{a3} dE} \tag{9}$$

where E is the photon energy: $\phi(E)$ is the incident γ-ray flux density (γ per unit area and unit time); μ_{a1}, μ_{a2}, μ_{a3} are the total mass absorption coefficients for γ-rays, respectively in dosimeter, capsule walls, and soil; t is the capsule thickness.

In fact, formula Eq. (9) is only approximate and a more exact formulation is:

$$\varepsilon = \frac{\int E\mu_{a3}\{(1-e^{-\mu_{02}t})\phi(E) + \int e^{-\mu_{02}t}.\left(\frac{\sigma_{02}}{\mu_{02}}\right) F(E,E')\phi(E')dE'\}dE}{\int \phi(E) E\mu_{a3} dE} \tag{10}$$

where μ_{02} and σ_{02} are, respectively the total and Compton mass attenuation coefficients for the encapsulating material; *F(E, E')* is the energy distribution of secondary γ-rays (energy E') from a Compton interaction induced by a photon of

Energy (keV)	Intensity	Energy (keV)	Intensity
$^{238}U + ^{235}U$			
13	90	768	9
49	30	934	5
75	22	1120	17.5
92	14	1238	8
186	7	1377	8.6
242	10	1509	4.4
295	21	1764	23.3
352	40	2204	8.2
609	49		
^{232}Th			
12	79	511	8.6
77	30	583	32.6
87	16	727	7.2
129	7	795	8.5
209	6	911	30
239	48	969	25.4
277	10	1588	11.6
338	16.4	2615	35.8
463	7		

Table 1.
U and Th series γ spectra.

energy E. $F(E, E')$ can be obtained from the well-established Klein and Nishina cross-section for Compton effect [11].

For the use of these formulas, it is assumed that there is an electronic equilibrium, that is, the dimensions of the dosimeter are equal to or greater than the secondary electron range (of the order of a few mm). Murray [9] showed that this is verified if the dosimeter mass is greater than 100 mg.

In applying formula Eq. (9), the main difficulty lies in the calculation of the incident photon flux density $\phi(E)$. $\phi(E)$ includes not only the discrete primary emission spectrum of the considered γ source, but also the continuum spectrum of γ rays from successive Compton interactions.

5.1.3 Infinite medium γ-ray spectrum

Here one assumes an infinite environment and homogeneous, in which the radioelements are evenly distributed. The interactions of gamma rays with the material taken into consideration are the photoelectric effect and the Compton effect; the pair production is considered negligible to the considered energies (<2.6 MeV).

Let $N_0(E)$ be the primary spectrum (number of photons per unit time and unit mass of soil). After one interaction, the energy spectrum of secondary γ-rays will be:

$$N_1(E') = \int N_0(E)\frac{\sigma_{01}(E)}{\mu_{01}(E)}F(E,E')dE \quad (11)$$

And the ratio of secondary to primary γ-rays is:

$$Q_1 = \frac{\int N_0(E).\left(\frac{\sigma_{01}}{\mu_{01}}\right).dE}{\int N_0(E).dE} \tag{12}$$

In the same way, the spectrum of third generation γ-rays can be deduced from $N_2(E)$, and in a more general manner the following recurrence relations hold:

$$N_i(E') = \int N_{i-1}(E).\frac{\sigma_{01}(E)}{\mu_{01}(E)}F(E,E')dE \tag{13}$$

And the total spectrum will be:

$$N(E) = \sum_{i=1}^{\infty} N_i(E) \tag{14}$$

Of course, there is a rapid decrease with i of the mean γ energy and of the total energy carried.

Figure 4 an example of various order spectra for ^{4C}K.

Practically, the calculation will be stopped at a generation order i such that $Q_i \ll 1$. The condition $Q_i < 10^{-3}$ was imposed, corresponding to i = 15–32 according to the medium involved.

$N(E)$ is the photon spectrum corresponding to the number of photons created per unit mass; to get ϕ(E), which is the number of incident photons per unit area, the mean relaxation length λ(E) of photons must be taken into account:

$$\phi(E) = \lambda(E).N(E) = \sum \frac{N_i(E)}{\mu_{01}} \tag{15}$$

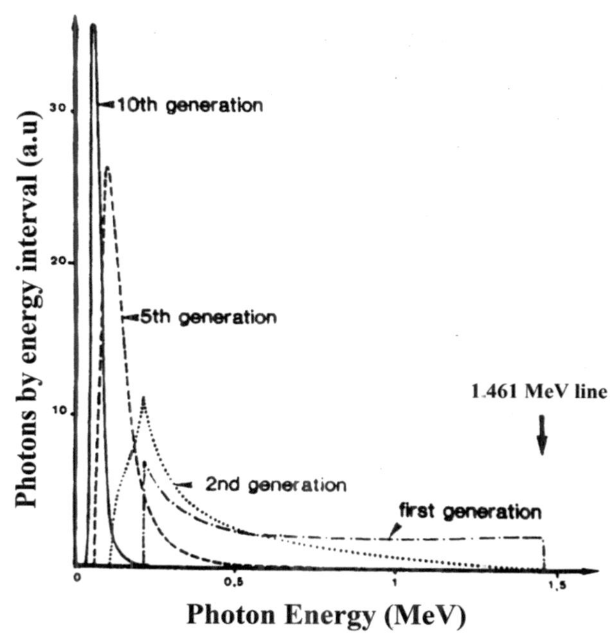

Figure 4.
Computed spectra of secondary γ-rays from ^{40}K after 1, 2, 5, 10 Compton interactions [5].

$$(\text{With}) : \lambda(E) = \frac{1}{\mu_{01}}$$

The value of absorption or attenuation coefficients was obtained for the Compton effect from the Klein-Nishina formula.

For the photoelectric effect, they were deduced from the data of Hubbell [12].

Some typical spectra are shown in **Figure 5** for ^{40}K. The low energy cut-off is determined by photoelectric effect and occurs at higher energy for high-Z media. And for ^{232}Th in pure water (**Figure 6**).

In the case of water, a “light” medium and a strong low energy ray contribution (below 100 keV) can be noticed. Similar features are also noticed for the more complex case of U and Th series.

These spectra show the overlapping of the primary ray spectrum and the degraded continuous spectrum, which carry most of the energy (60%).

There is a similarity in the degraded spectrum, even for two very different gamma sources (for example, thorium and potassim-40). On the other hand, the degraded spectrum is higher in energy as the <Z> of the medium is high: this is due to the increasing influence of the photoelectric effect which “cuts” the low gamma ray energy.

Our results are in good agreement with those obtained by G. Valladas [7] by the Monte-Carlo method for potassium-40 in siliceous medium.

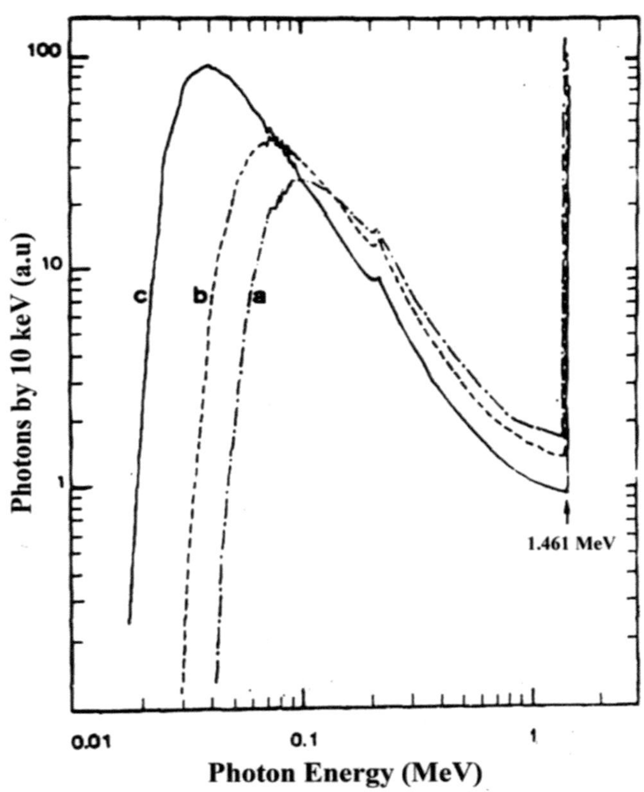

Figure 5.
Computed energy spectra for ^{40}K in (a) dry C soil, (b) wet S soil, and (c) pure water [5].

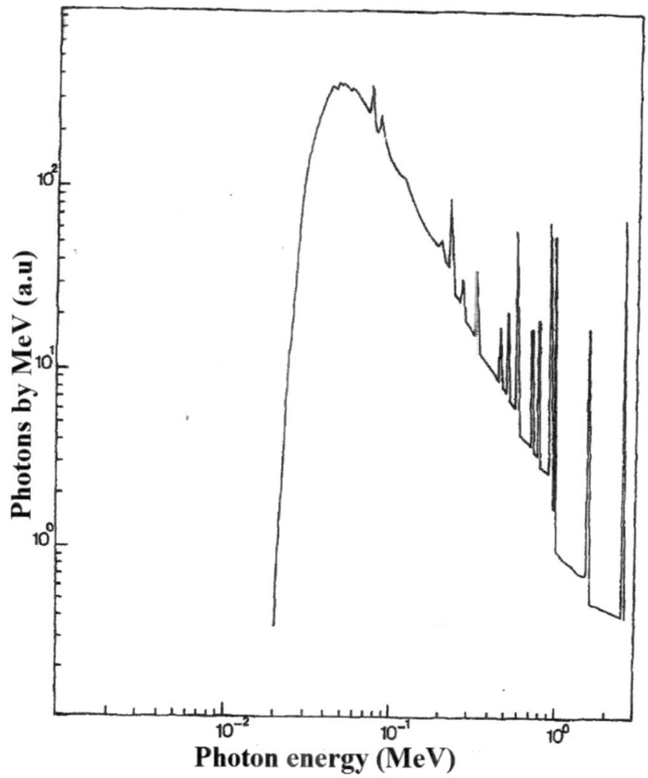

Figure 6.
Computed energy spectra for ^{232}Th in pure water [5].

5.1.4 Dosimeter to soil dose ratio for infinite homogeneous soils

The calculation of ε was done for the list of floors already cited (from the gamma ray energy spectra already calculated), for the three radiation sources and for two types of 1 g/cm^2 thick absorbers, sufficient to stop beta radiation.

- Copper, $<Z>$ medium, frequently used
- Polyethylene, $<Z>$ low

The results are given in **Table 2**. Calculations are performed here with, as a TL dosimeter, $CaSO_4$: Dy, but are readily feasible for any combination of known dosimeters and compositions.

It can be found that ε it may be less than or greater than 1; two antagonistic effects act: on the one hand, the absorption by the walls, which tends to decrease ε on the other hand, the difference in nature between soil and TL dosimeter which has the opposite effect, the coefficient of absorption of the dosimeter being the highest.

The dispersal of the value of ε is smaller between the different soil for a copper capsule than for a polyethylene capsule. This can be explained by the fact that copper absorbs the lower part of the spectrum of energy for which the difference dose between the dosimeter and the soil is significant.

Surrounding media	^{40}K		Uranium		Thorium	
	Encapsulating		Encapsulating		Encapsulating	
	Polyethylene	Copper	Polyethylene	Copper	Polyethylene	Copper
C soil	0.96	0.95	0.94	0.89	0.95	0.91
C soil + 15% water	0.97	0.95	0.96	0.90	0.96	0.91
S soil	1.03	0.97	1.10	0.95	1.08	0.96
S soil + 15% water	1.05	0.98	1.13	0.96	1.11	0.97
S soil + 30% water	1.06	0.98	1.16	0.97	1.14	0.98
S soil + 50% water	1.07	0.96	1.20	0.95	1.17	0.96
Pure water	1.35	1.01	1.82	1.03	1.70	1.03

Table 2.
Values of ε for infinite uniform medium with 1 g m^{-2} capsule wall thickness [5].

For a copper capsule, the difference of ε between dry a soil and moderately wet (e.g., less than 15% water) is negligible (less than 1%); therefore, it is not necessary in this case to know exactly the soil humidity to perform the correction, which is an advantage.

In practice, the limestone soil and the siliceous soil are two extreme cases, which make only a difference of 5% for copper on the value of ε, so only a coarse knowledge of the cashing medium is necessary for the calculation of e with an high accuracy.

The variations of ε for ^{40}K with the absorber or medium are less important than for uranium or thorium. This is due to a smaller contribution of low energies to the ^{40}K spectrum (no low energy lines).

An important consequence of the foregoing is that, for the experimenter, copper (or a close material) is well used when the composition (especially moisture) of the soil is not well known. The error made a priori on the value of ε will thus be reduced.

On the other hand, in most cases, the dispersion on ε related to the relative intensity of the three gamma radiation sources is insignificant. Indeed, the potassium-40, uranium and thorium series generally intervene for approximately 1/3 each, and significant deviations from this proportion are rare.

In general, it is difficult to determine the respective contribution of the potassium-40, uranium and thorium series to the total dose rate. However, in most cases, the contribution of these three radioelements does not exceed 60% and approximate values on the whole with acceptable dispersion may be proposed.

For example,

- $\varepsilon = 0.97 \pm 0.01$ for S soil + 15% water
- $\varepsilon = 0.92 \pm 0.01$ for dry C soil.

The results obtained with our program under the same conditions regarding the dosimeter, the surrounding environment, the capsules and the sources of the gamma rays are slightly superior to the theoretical results of G. Valladas [7]. This small difference (in the order of 3%) can be explained by the fact that the nominal values for capsule thickness have been used instead of the mean values which take into account geometry, also to a lesser extent by the fact that self-absorption has not been taken into account in our calculation. The effect of this last correction was estimated to be less than 1%.

5.2 Calculation method adapted to the experimental conditions for determining samples γ-activities induced by 14 MeV neutrons

5.2.1 Introduction

The gamma radiation self-absorption coefficient is of great interest in activation analysis. Since it is difficult to measure this coefficient, various calculation methods have been developed.

Measuring the self-absorption coefficient is not a simple thing. The physicists who have faced this problem, for a long time, have always used methods of statistical or non-statistical computation: Parallel beam methods, Monte-Carlo method and many other methods. For our part, we have developed an original technique calculating the self-absorption coefficient of multienergetic γ-radiations [13].

In this chapter, we are presenting a method that we have developed, this which allows us control and calibrate the activation analysis experiments [13]. This method consists of simulating the interaction processes of gamma rays induced by neutron activation of various samples by using the Monte Carlo method adapted to experimental conditions.

5.2.2 Samples and standards induced gamma-activities

Different disk shaped red beet samples and standards were irradiated with 14 MeV neutrons. Standards were prepared by mixing pure graphite and high purity chemical compounds powders (NaCl, Na_2HPO_4 and K_2CO_3) [14]. The induced gamma activities on the sodium, potassium, chlorine and phosphorus elements have been experimentally measured by means of hyper-pure germanium spectrometer.

The analyzed beet samples and standards have a 23 mm diameter and a 6 mm thickness.

The different parameters of the nuclear reactions used (cross section, isotopic abundance, etc.) are summarized in **Table 3**.

After the irradiation (irradiation time = t_{irr}) and cooling times (t_d), the number of produced radionuclides is given by:

$$N(t_{irr}+t_d)=\frac{TN_0\sigma\phi}{\ln 2}\theta I_\gamma\left(1-e^{-\left(\frac{ln2}{T}\right)t_{irr}}\right).e^{-\left(\frac{ln2}{T}\right)t_d} \tag{16}$$

where σ is the nuclear reaction cross section, N_0 is the number of target nuclei, ***T*** is the half-life of the produced radionuclide, θ is the isotopic abundance of the studied element, I_γ, is the emitted gamma rays intensity and ϕ the neutron flux. To take into account the activity measuring time, relation Eq. (16) should be multiplied by the term $\left(1-e^{-\left(\frac{ln2}{T}\right)t_m}\right)$.

Element	θ (%)	Nuclear reaction	σ (mb)	T	E_γ (keV) (I_γ %)	t_{irr} (s)	t_d (s)	t_m (s)
Na	100	$^{23}Na(n, p)^{23}Ne$	43 ± 5	38.0 s	440 (33%)	200	30	200
Cl	75.5	$^{35}Cl(n, 2n)^{34m}Cl$	43 ± 5	32.4 min	146.5 (45%)	600	30	600
P	92.2	$^{31}P(p, \alpha)^{28}Al$	43 ± 5	2.30 min	1779 (100%)	600	30	600
K	93.1	$^{39}K(n, 2n)^{38g}K$	43 ± 5	7.70 min	2167 (100%)	600	30	600

Table 3.
The produced nuclear reactions by irradiating the samples (standards) with 14 MeV neutrons.

5.2.3 Calculation method

The transmission probability of a gamma photon generated from point P_1 (**Figure 7**) after crossing a path length $l_1 = P_1S_1$ in a homogeneous disk shaped sample (standard) of radius R_t in cm, depth D in cm, and density ρ_1 in $g.cm^{-3}$ is given by:

$$P(l_1) = e^{-\mu_1 l_1}$$

where μ_1 is the total attenuation coefficient of the gamma rays in the irradiated sample.

The interaction probability of a gamma photon after crossing a path $l_2 = Q_2S_2$ (**Figure 7**) in a disk shaped detector of radius R_2 in cm, depth e in cm and density ρ_2 in $g.cm^{-3}$ is given by:

$$P(I_2) = \left(1 - e^{-\mu_2 l_2}\right) \quad (17)$$

where μ_2 is the total attenuation coefficient of the gamma photon in the detector.

The absorption probability of a gamma photon in the detector is given by:

$$P = e^{-\mu_1 l_1}\left(1 - e^{-\mu_2 l_2}\right) \quad (18)$$

Consequently the detection rate for N photons is:

$$\overline{\omega} = \frac{1}{N}\sum_{i=1}^{N} e^{-\mu_1 l_{1i}}\left(1 - e^{-\mu_2 l_{2i}}\right) \quad (19)$$

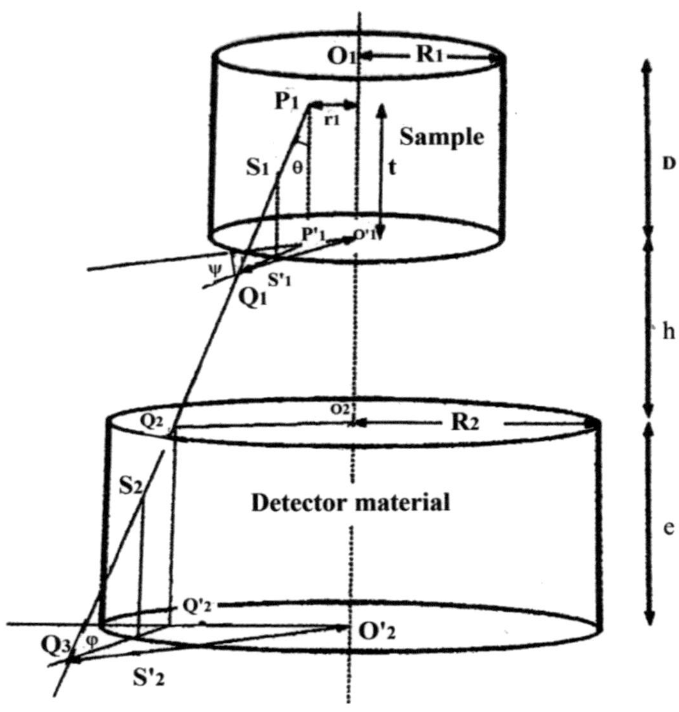

Figure 7.
The irradiated sample (standard) to γ-detector arrangement used in the activities calculation.

By combining relations Eq. (16) and Eq. (19) we get the number of the detected gamma rays.

$$N_c = \frac{TN_0\sigma\phi}{\ln 2}\ \theta I_\gamma \varpi \left(1 - e^{-\left(\frac{ln2}{T}\right)t_{irr}}\right).\ e^{-\left(\frac{ln2}{T}\right)t_a}.\left(1 - e^{-\left(\frac{ln2}{T}\right)t_m}\right) \tag{20}$$

The total attenuation coefficients are calculated by using respectively the Klein-Nishina theory formula [12], Allen Brodsky's approximation [15], and Max Born's approximation [16], for the Compton, photoelectric and pair production effects.

The calculation of the paths lengths l_1 and l_2 consists firstly on generating random numbers by using a programme based on a congruentia 1 method.

The path length l_1 is given by [17]:

$$l_1 = \begin{cases} \dfrac{t}{cos\theta}\ if\ R_1 \geq \overline{O_1'O_1} \\ l(r,\theta,\psi)\ if\ R_1 < \overline{O_1'O_1} \end{cases}$$

where:

$$\overline{O_1'O_1}^2 = r_1^2 + t^2.tan^2(\theta) + 2r_1.t.tang(\theta).cos(\psi)$$

$$\text{And} : l(r,\theta,\psi) = \frac{\sqrt{R_1^2 - r_1^2 sin^2(\psi)} - r_1 cos(\psi)}{sin(\theta)}$$

The uniform random sampling of the emission point P_1 and emission direction is achieved by computing the distance from the center r_1, the depth t, cos(θ) and ψ, with four uniform random numbers [17]:

$$r_1 = R_1\sqrt{\xi_1}$$
$$t = D\xi_2$$
$$cos(\theta) = \xi_3$$
$$\psi = 2\pi\xi_4$$
$$(\text{With})\ 0 \leq \xi_i \leq 1$$

We have developed the following theory to calculate the path length l_2 which is given by:

$$\begin{cases} 0\ \ if\ R_2 \leq \overline{O_2Q_2} \\ \dfrac{e}{cos\theta}\ if\ R_2 > \overline{O_2Q_2}\ \ and\ \ R_2 \geq \overline{O_2'Q_3} \\ X_2\ if\ \overline{O_2Q_2} < R_2 < \overline{O_2'Q_3} \end{cases}$$

where

$$\overline{O_2Q_2}^2 = r_1^2 + (t+h)^2.tan^2(\theta) + 2r_1.(t+h).tang(\theta).cos(\psi)$$
$$\overline{O_2'Q_3}^2 = r_2^2 + e^2.tan^2(\theta) + 2r_2.e.tang(\theta)\ cos(\psi)$$

h is the distance between the sample and the detector (**Figure 7**). r_2 is the distance from the axis of the detector and:

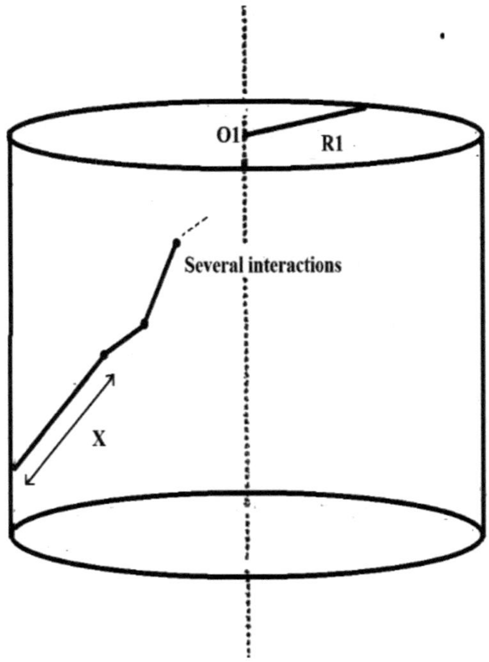

Figure 8.
Scheme shows changes in gamma-ray direction in the case of multiple interactions.

$$X_2 = \begin{cases} 0 \ if \ tan\ \theta \geq tan\ \theta_{max} \\ \dfrac{\sqrt{R_2^2 - r_2^2 sin^2\varphi} - r_2 cos\varphi}{sin\theta} \ if \ tan\theta < tan\ \theta_{max} \end{cases}$$

$$(\text{With})\ tan\ \theta_{max} = \frac{R_2 - R_1}{t + h}$$

To complete this study, we have developed another program based on the EGC method [18]. This program results in determining the energy loss predominant phenomenon that occurs when gamma rays interact with the absorber.

For this, we compare the range x to the relaxation length (λ) of the γ-rays in the material (**Figure 8**).

5.3 Results and discussions

The measured N_m and calculated N_c activities of some different irradiated standards containing Na, K, Cl, and P are shown in **Table 4**.

Element	Element mass (mg)	ϕ	N_m	N_c	$R = \frac{N_c}{N_m}$
Na	0.51	1.0×10^7	50 ± 7	58.0 ± 0.2	1.16
K	2.32	2.3×10^8	17 ± 4	14 ± 0.1	1
Cl	0.92	5.0×10^8	447 ± 21	432 ± 1	0.96
P	1.61	2.0×10^9	2533 ± 50	2700 ± 5	1.06

Table 4.
Data obtained for different irradiated standards with a 14 MeV neutron flux by experimental N_m and calculation N_c methods.

We notice that the results obtained by the two methods (experimental and calculation) are in good agreement with each other. The calculation method has the advantage of being accurate (error is smaller than 3%) and rapid (the calculation time is of about 2 min).

6. Conclusion

In the first part of the chapter, a careful study of the correcting factors linked to the environmental and experimental conditions is performed.

In the second part, the calculation method was developed. It is very accurate, rapid, adapted to the experimental conditions, it does not necessitate the use of a very expensive detection chain, and can be used to determine the trace element concentrations in materials. This technique is a good test for neutron activation analysis experiments. It allows these experiments to be calibrated in cases where it is difficult to achieve standards.

Author details

Hassane Erramli[1]* and Jaouad El Asri[2]

1 Faculty of Sciences Semlalia, University Cadi Ayyad, Marrakech, Morocco

2 Nuclear Reactor and Nuclear Security, Faculty of Sciences, University Mohammed V, Rabat, Morocco

*Address all correspondence to: hassane@uca.ac.ma

References

[1] McCullough EC. Photon attenuation in computed tomography, Medical Physics. 1975;**6**:307

[2] Hubbell JH. Photon cross sections, attenuation coefficients, and energy absorption coefficients from 10 keV to 100 GeV. National Bureau of Standads report NSRDS-NSB 29; August 1969

[3] AL-Akhras M-A, Aljarrah K, Omari AH, Khateeb HM, Albiss BA, Azez K, et al. Influence of extremely low energy radiation on artificial tissue: Effects on image quality and superficial dose. Spectroscopy. 2008;**22**(5):419-428

[4] Lyoussi A. Mesure nucléaire non destructive dans le cycle du combustible. Techniques de l'Ingénieur, Réf. BN3405; Janvier 2005

[5] Erramli H. Développement de la dosimétrie appliquées à la datation par thermoluminescence [PhD thesis]. France: Université Blaise Pacal; 1986

[6] Fain J, Erramli H, Miallier D, Montret M, Sanzelle S. Environmental gamma dosimetry using TL dosimeters: Efficiency and absorption calculations. Nuclear Tracks. 1985;**10**(4–6):639-646

[7] Valladas G. Mesure de la dose γ annuelle de l'environnement d'un site arch~ologique par un dosimetre TL. Photodynamic Antimicrobial Chemotherapy. 1982;**6**:77-85

[8] Mejdahl V. Measurement of environmental radiation at archaeological sites by means of TL dosimeters. Photodynamic Antimicrobial Chemotherapy. 1978;**2**:70

[9] Murray AS. Environmental radiation studies relevant to thermoluminescence dating [PhD thesis]. Oxford; 1981

[10] Ewbank WB. Status of transactinium nuclear structure data. In: The Evaluated Nuclear Structure Data File. IAEA-TECDOC-232. Vienna: International Atomic Energy Agency; 1980. pp. 109-141

[11] Evans RD. Gamma rays. In: American Institute of Physics Handbook. New York: McGraw-Hill; 1972. pp. 8-190-8-218

[12] Hubbell JH. Photon cross-sections attenuation coefficients and energy absorption coefficients for 10keV to 100GeV photons, NSRD-NBS-29. Washington; VS Government Printing Office; 1969

[13] Rzama A, Erramli H, Misdaq MA. A new calculation method adapted to the experimental conditions for determining samples y-activities induced by 14 MeV neutrons. Nuclear Instruments and Methods. 1994;**B93**: 464-468

[14] Berrada M, Misdaq MA, Thallouarn P. Determination of potassium in beet by 14 MeV neutron activation analysis. Journal of Radioanalytical Chemistry. 1979;**54**:361

[15] Brodsky A. Handbook of Radiation Measurement and Protection. Section A. Vol. 1. West Palm Beach, FL: CRC Press; 1978

[16] Heitler W. The Quantum Theory of Radiation. 3rd ed. London: Oxford University Press; 1954

[17] Gotoh H. Calculation of the self-absorption of gamma rays in a disc shaped source. Nuclear Instruments and Methods. 1973;**107**:199

[18] Nelson WR, Hirayama H, Rogers DW. SLAC-report-265; December 1985

Chapter 8

Application of Radiation and Genetic Engineering Techniques to Improve Biocontrol Agent Performance: A Short Review

Seyed Mahyar Mirmajlessi, Hossein Ahari Mostafavi, Evelin Loit, Neda Najdabbasi and Marika Mänd

Abstract

Biological control is a potential nonchemical method to manage plant pathogens by beneficial microorganisms. To improve antagonistic potential of biocontrol agents, mutation by radiations, chemicals, and genetic manipulations has been used. Genetic techniques and ionizing radiation containing direct or indirect emissions play the greatest role for selection of useful microorganisms to enhance the efficiency of biological systems. Indeed, genetic engineering has a main role in increasing antimicrobial metabolites, host colonization ability, and endurance in micro-ecosystem. Genetic improvement can be achieved by protoplast fusion, genetic modification (GM), and chemical (genotoxic agents) and physical mutations. However, ultraviolet light and ionizing radiations can induce modifications in the genome of an organism. Irradiation, particularly gamma rays, is also applied for controlling postharvest diseases. Indeed, irradiation cannot completely eliminate pathogens, but it might result in cell injury and directly damage the chromosomal DNA of a living cell. This technology has been used for many reasons including disinfestation of foods, reducing foodborne pathogens, and extending shelf life many fruits, vegetables, and nuts. In the current review, we discuss advances in the radiation and molecular genetic techniques with the aim to improve antagonistic potential of microorganisms as it is applied to the suppression of plant pathogens.

Keywords: biocontrol agents, genetic engineering, ionizing radiations

1. Introduction

The risks associated with chemical residues on the leaves and fruits have highlighted the need for more useful and safer alternative control treatments. Biological control is a potential nonchemical mean to manage plant pathogens including fungi, bacteria, nematodes, or weeds by beneficial microorganisms such as fungi, yeast, or bacteria [1]. The virulence of pathogens probably changes if genetic mutations occur in the genes related to pathogenicity of microorganisms upon mutagenic treatments [2]. It is quite obvious that the application of ionizing radiations such as gamma rays and X-rays and genetic methods plays the greatest role in the development of techniques for the selection of useful microorganisms

[3]. Indeed, irradiation is the process of exposing an amount of energy in the form of speed particles or rays for improving food safety and reducing microorganisms that destroy agricultural crops. For instance, studies on transcriptional changes of *Salmonella typhimurium* using gamma irradiation showed that the expression of the virulence genes in irradiated mutants was reduced in comparison with non-irradiated controls [4]. It has been also shown that gamma irradiation alone or in combination with other methods can improve the microbiological safety [5]. Likewise, the wholesomeness (lack of teratogenicity, mutagenicity, and toxicity) of irradiated products has been studied extensively, in which food irradiation has not provided any evidence of increased threat of mycotoxin formation in irradiated food [6]. There is an increasing interest in commercial application of genetically modified microorganisms with improved biocontrol properties toward plant pathogens [7]. To that end, this chapter is an advanced survey reviewing the radiation and molecular genetic techniques with the aim to improve antagonistic potential of microorganisms as it is applied to the suppression of plant diseases.

2. Genetic improvement of microorganisms

The biological and molecular characterizations of biocontrol agents and bioactive compound producers are very important for the modern agriculture [8]. Since environmental conditions are subject to change, the biocontrol agent requires genetic improvement for effective performance [9]. To improve the efficiency and productivity of biological systems, genetic engineering has a main role in increasing antifungal and antibacterial metabolites, host colonization ability, and endurance in micro-ecosystem. Genetic improvement can be achieved by chemical and physical mutations, protoplast fusion, and transformation [10].

2.1 Use of protoplast fusion

Protoplast fusion is an important technique for gene manipulation. It breaks down the barriers to genetic exchange and is a relatively new flexible technique to induce or promote genetic recombination in a variety of prokaryotic and eukaryotic cells [11]. By protoplast fusion (**Figure 1**), interspecific or even intergeneric hybrids can be produced, and it is feasible to transfer useful genes, for attributes (such as disease resistance, enzyme and phytotoxin production, rapid growth rate, nitrogen fixation, protein quality, and drought resistance), from one species to another [12].

To breed new strains with improved production of the spore and phytotoxin, protoplast fusion between the *Helminthosporium gramineum* subsp. *echinochloa* and *Curvularia lunata* was carried out [10]. Prabhavathi et al. [13] showed that the isolated protoplast from *Trichoderma reesei* strain PTr2 significantly increased enzyme activities in two fusants SFTr2 and SFTr3 as compared to the parental strain PTr2. Bakhtiari et al. [14] produced a new recombinant of *Tolypocladium inflatum* with cyclosporine 2.8 times more than parental strain by protoplast fusion between different strains.

2.2 Genetic transformation

Nowadays, some wild-type fungal and bacterial species are used as biocontrol agents. But, there are limitations that restrict their efficacies. Poor survival of the inoculant in particular soils and low production of required metabolites are the most important limitations [15]. The technology of genetic modification (GM) has the capacity to create new strains in which these problems are overcome [16]. Genetic manipulation requires the development of vector-mediated transformation

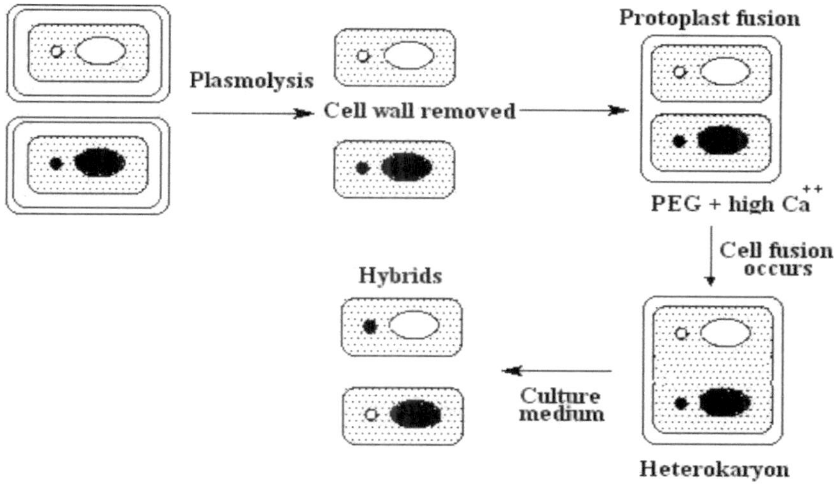

Figure 1.
Schematic illustration of haploid cell construction by protoplast fusion.

systems that include, first, infusion of exogenous DNA into receiver cells; second, expression of genes present on the incoming DNA; and, finally, stable preservation and replication of the inserted DNA, leading to the expression of the desired phenotypic trait (**Figure 2**).

The development of an effective strain that can tolerate the environmental adversities and securing of the essential regulatory approval are primary obligations for the successful use of a GM inoculant as biocontrol agents [17]. Resca et al. [18] genetically modified the strains of *P. fluorescens* F113Rif (pCU8.3) and *P. fluorescens* F113Rif (pCUP9) to increase phenazine1 carboxylate (Phl) production for biocontrol efficacy against *Polymyxa betae* on sugar beet. An endochitinase-encoding gene was cloned from the *T. viride*. This gene was introduced into *Chaetomium globosum* CG10 after ligating it with the promoter and terminator of Trp synthetase from *Aspergillus nidulans*. Endochitinase activity significantly increased in 30% of the transformants for improvement of the biocontrol activity of *Chaetomium globosum* [19].

2.3 Improvement of the bioagents via mutagenesis

Breeding of antagonists is directed to achieve effective strains for biocontrol of plant pathogens under a wide range of environmental conditions. Mutation techniques have been used to improve antagonistic potential of biocontrol agents to manage phytopathogens. Physical and chemical mutagens have been applied by many researchers to generate new biotypes [3]. Ultraviolet light (UV), ionizing radiation, and chemicals (as genotoxic agents) can randomly induce modifications in organism's genome. About 25 years ago, X-ray, as an active mutagen, was used to produce the first mutant strain of *Penicillium chrysogenum*. The next mutagenic technique used was the application of ultraviolet light on *P. chrysogenum*. The produced mutant displayed three times more activity (with respect to penicillin contains) than the original strain induced by X-irradiation.

Later on, fast neutrons and gamma irradiation were used successfully. The effectiveness of gamma rays significantly surpassed that of UV and X-rays. Also, mutagenic effects of chemicals in combination with UV light on *P. chrysogenum* were investigated. The mutant strains were about 1500 times as productive as the wild type [19]. Mess et al. [20] demonstrated that gamma irradiation at 130 Gy using 137Cs generated an avirulent mutant (avr-mutant) of *F. oxysporum*

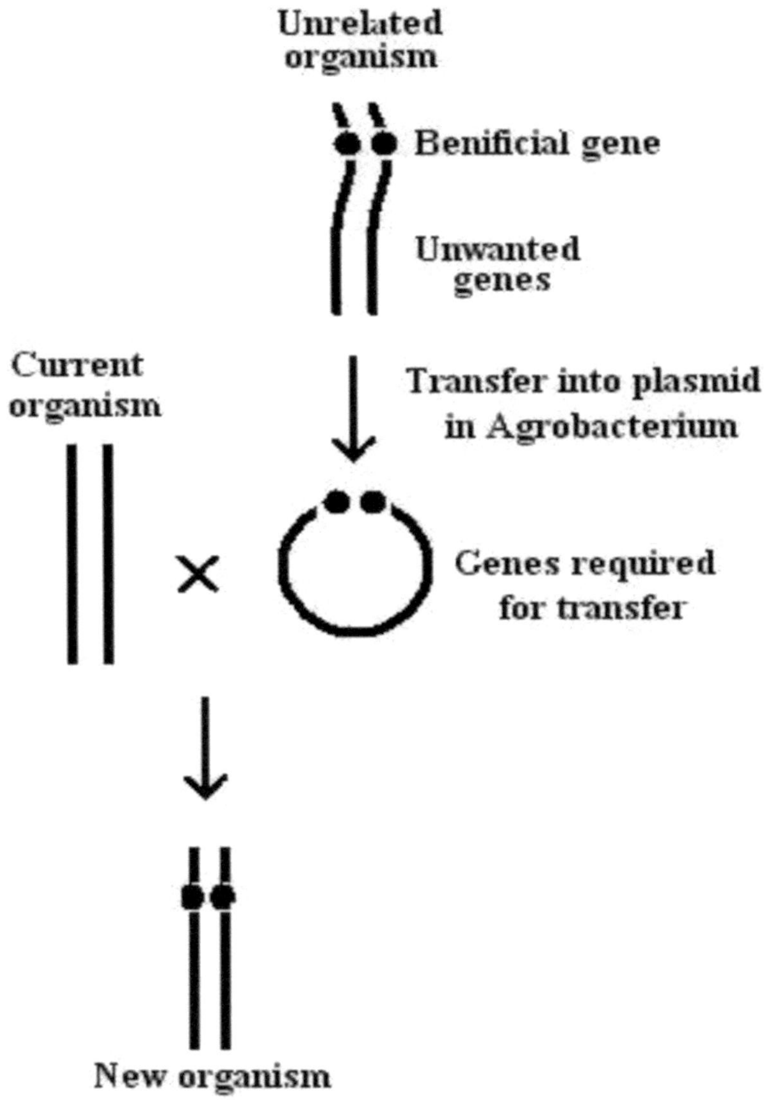

Figure 2.
Schematic illustration of plasmid-mediated transformation system.

f. sp. *lycopersici*. One mutant showed the expected loss of fusarium root rot on tomato plants. Some successful studies have been made to improve the biocontrol potential of *Trichoderma* and *Gliocladium* species by exposing them to physical mutagens such as gamma ray. Mutagenesis of *Trichoderma atroviride* by gamma irradiation greatly improved their capabilities to produce antibiotics [21]. Rugthaworn et al. [22] investigated the improvement of Actinomycete strains by gamma irradiation for biocontrol of *Fusarium sporotrichioides*, *Rhizoctonia solani*, and *Sclerotium rolfsii*. The induced mutants were detected by observing the ability of chitinase production and showed higher inhibitory effect on three phytopathogenic fungi than the wild type. Mohamed and Haggag [23] showed that gamma irradiation generated a *T. harzianum* mutant against *F. oxysporum* through increasing of antifungal metabolites (hydrolytic enzymes, antibiotics, and total phenols) under salt stress conditions.

Irradiation of *T. harzianum* produced salt-tolerant mutants that greatly were improved in sporulation, growth rate, and biocontrol ability against *Fusarium oxysporum*, the causal agent of tomato wilt disease. Ahari-mostafavi et al. [24] showed the possibility of biological control of bean root rot disease, using of avirulent mutants of *Fusarium solani* f. sp. *phaseoli* isolate produced by gamma irradiation. Furthermore, stable gamma-irradiated mutants of *G. virens* have immense potentiality as biocontrol agents [25]. Haggag [26] improved the production of *P. fluorescens* antibiotics (including phenazine, pyrrolnitrin, and phloroglucinol) and siderophore against damping-off pathogens (*F. solani*, *F. oxysporum*, and *R. solani*) by UV light. Papavizas et al. [27] developed mutant isolates of *G. virens* by means of irradiation and a chemical mutagen. They showed evidence that it is feasible to induce benomyl-tolerant mutants of *G. virens* by long and repeated exposure to UV irradiation combined with ethyl methanesulfonate. Mohamed et al. [7] applied the UV mutagenesis technique to enhance three hydrolytic enzymes effective in the biocontrol ability of *T. viride* against two of the important plant fungal pathogens, *Sclerotia rolfesii* and *Sclerotinia sclerotiorum*. In biological control experiments against root rot and white rot diseases caused by *S. rolfesii* and *S. sclerotiorum*, respectively, in bean plants under artificial and natural infested soil, complete control of the disease was achieved. Treatment of the bean seeds with *T. viride* mutants resulted in reducing colonization of *S. rolfesii* and *S. sclerotiorum* in bean rhizosphere compared with treatment with their parental wild type and increased plant yield.

Moreover, Balasubramanian et al. [28] tested biocontrol efficacy of the UV mutants and wild strain of *T. harzianum* against phytopathogens such as *F. oxysporum*, *Bipolaris oryzae*, *R. solani*, and *Alternaria* sp. A mutant strain showed maximum inhibition of the above pathogens through more extracellular chitinase and protein production, when compared to the wild strain. Also, UV mutagenesis increased 2.29-fold in phytase activity of *T. lanuginosus* over that of their parental strain [29]. Chitinase production was improved by sixfold in *Ophiostoma floccosum* using UV mutagenesis [30]. Furthermore, the ethyl methanesulfonate (EMS) mutant with an enzyme activity of 25.56 U ml^{-1} was obtained by further exposure to UV radiation and yielded an activity of 34.12 U ml^{-1} [31].

2.3.1 Mechanism of ionizing radiation effects on bioagents

Irradiation can have direct or indirect effects on organisms. In the case of a direct hit, electromagnetic radiance and high-energy particles break chemical bonds or reactive oxygen-centered (•OH) radicals originated from the radiolysis of water. Hydroxyl radicals are very reactive and are known to interfere with the bonds between nucleic acids within a single strand or between opposite strands. An indirect effect occurs on organisms when radiation ionizes a neighboring molecule, which in turn reacts with the genetic material [32]. Ionizing radiation effects (on microorganisms) are primarily the result of disruption of the DNA or RNA (**Figure 3**).

Since the DNA is much larger than other molecular structures in a cell, the induced biological and chemical changes by either primary ionizing or through secondary free radical attack can prevent replication and destroy cells [33] or generate some genetic traits of microorganisms to have higher antagonistic potential [2]. Therefore, the induced biological and chemical changes are related to the absorbed radiation dose [33]. The pathogenicity of infectious organisms is diminished by irradiation. Lim et al. [4] showed that the expression of the virulence genes in *S. typhimurium* gamma-irradiated mutants was reduced and expression of toxin genes of *Vibrio* spp.-irradiated mutants did not increase, compared with non-irradiated wild types.

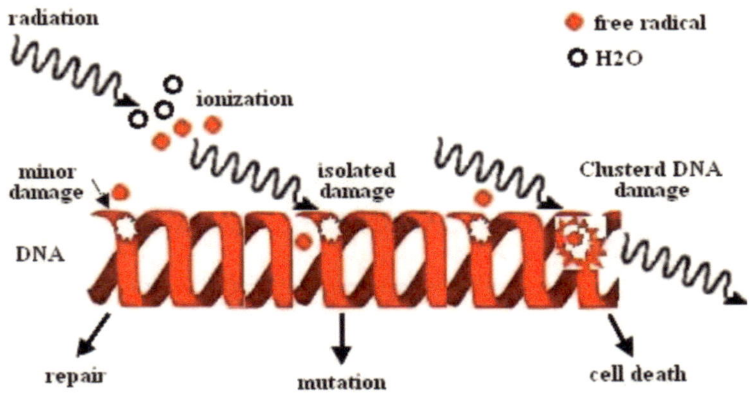

Figure 3.
Schematic illustration of DNA mutation by ionizing mutations drawn by anonymous.

2.3.2 Irradiation sources

Ionizing radiation is mainly known to high-energy photons of radio nuclides (gamma rays) and X-rays from machine sources with energies up to 5 MeV and accelerated electrons with energies up to 10 MeV generated by electron accelerating machines [34]. Gamma ray occurs in the short wavelength, with high-energy region of the spectrum, and has the greatest penetrating power. Gamma rays come from spontaneous breakdown of radionuclides including the radionuclide's cobalt-60 (60Co) with a half-life of 5.3 years or cesium-137 (137Cs) with a half-life of 30 years [35]. X-ray is similar to gamma radiation based on radioactive isotopic sources. Although their effects on materials are generally similar, these kinds of radiations differ in their energy spectra, angular distributions, and absorbed dose rates [36, 37]. Electrons can be produced from machines capable of accelerating electrons as light speed by means of a linear accelerator. In comparison with gamma ray and X-rays, electrons cannot penetrate very far into materials and do not have deep penetrating power. So, electrons as beta particles are usually chosen to treat the surface of materials [32, 34]. Generally, the strength of the source and the length of time a material is exposed to the ionizing energy determine the irradiation dose, measured in grays (Gy) or kilo grays (1 kGy = 1000 Gy). One gray is equal with one joule of energy absorbed in a mass of one kilogram [38]. Based on available evidence, the safety of the irradiation technology in food industry was considered and judged acceptable. This has developed in international bodies such as the World Health Organization (WHO), the Food and Agriculture Organization (FAO), the International Atomic Energy Agency (IAEA), the and Codex Alimentarius Commission [35, 39].

3. Conclusion

Plant pathogens are a worldwide problem. A variety of approaches may be applied to control plant diseases. Beyond suitable agronomic performances, farmers frequently rely on chemicals. Environmental pollution and carcinogenic effects of pesticides are limiting factors in the success of their application. Nowadays, there are severe regulations and political pressure to remove the most dangerous chemicals from the market. So, biological control could be the best alternative against plant pathogens, and development of mutants is an important technique in strain improvement toward plant pathogen suppression, which yields reliable strains for biocontrol. Since strains bred by mutagenesis can get registration (from environmental protection agencies) more easily than strains produced by protoplast fusion and transformation or via gene cloning for field use, more attention should be paid to the mutagenic methods. In an agricultural environment, mutants are interacting and competing with various communities of microorganisms that can have intense effects on the survival and performance of the introduced mutants. Hence, before commercial application of such inoculants in an open environment, their behaviors and potential impact on ecosystems should be investigated as part of the risk assessment. Additionally, integrated treatment of irradiation and biocontrol agent has the potential as an alternative means for postharvest disease control. In fact, there is a limited dose rate for its application on postharvest diseases of fruits and vegetables. Thus, the combination of irradiation and biocontrol agent increases applied range of irradiation for postharvest control by decreasing of dose rate to which the product has been exposed. From our point of view, considering different aspects of irradiation may provide useful information for managing harmful microorganisms on crops, so that in the near future, this technique will be used as one of the most important research tools for biotechnologists, plant pathologists, and molecular biologists.

Acknowledgements

Authors want to thank Prof. Vahe Minassian from Tarbiat Modares University for critical reading of the manuscript.

Author details

Seyed Mahyar Mirmajlessi[1,2]*, Hossein Ahari Mostafavi[3], Evelin Loit[4], Neda Najdabbasi[1,2] and Marika Mänd[1]

1 Department of Plant Health, Estonian University of Life Sciences, Estonia

2 Department of Plants and Crops, Faculty of Bioscience Engineering, Ghent University, Ghent, Belgium

3 Department of Plant Protection, Agricultural, Medical and Industrial Research School, Nuclear Science and Technology Research Institute, Karaj, Iran

4 Department of Crop Science and Plant Biology, Estonian University of Life Sciences, Estonia

*Address all correspondence to: m.mirmajlessi@gmail.com

References

[1] Agrios GN. Plant Pathology. New York: Elsevier Academic Press; 2005. 952 p

[2] Barkai-Golan R. Postharvest Diseases of Fruits and Vegetables: Development and Control. Hardbound. Academic Press; 2001. 442 p

[3] Rey M, Delgado JJ, Rincon AM, Limon CM, Benitez T, Perez EA, et al. Improvement of Trichoderma strains for biocontrol. Micologia Industrial Y Micopatotogia. 2000;**17**:531-536

[4] Lim S, Jung J, Kim D. The effect of gamma radiation on the expression of the virulence genes of *Salmonella typhimurium* and Vibrio spp. Radiation Physics and Chemistry. 2007;**76**(11):1763-1766

[5] IAEA (International Atomic Energy Agency). Irradiation to Ensure the Safety and Quality of Prepared Meals, Vienna, Austria; 2009. p. 375

[6] Kottapalli B, Wolf-Hall CE, Schwarz P. Effect of electron-beam irradiation on the safety and quality of Fusarium-infected malting barley. International Journal of Food Microbiology. 2006;**110**(3):224-231

[7] Mohamed HAA, Haggag WM, Attallah AG. Genetic enhancement of *Trichoderma viride* to overproduce different hydrolytic enzymes and their biocontrol potentiality against root rot and white mold diseases in bean plants. Agriculture and Biology Journal of North America. 2010;**1**(3):273-284

[8] Spadaro D, Gullino ML. Improving the efficacy of biocontrol agents against soilborne pathogens. Crop Protection. 2005;**24**:601-613

[9] Hornok L. Genetically modified microorganisms in biological control. Novenyvedelem. 2000;**36**:229-237

[10] Zhang ZB, Burgos NR, Zhang JP, Yu LO. Biological control agent for rice weeds from protoplast fusion between *Curvularia lunata* and *Helminthosporium gramineum*. Weed Science. 2007;**55**:599-605

[11] Bhojwani SS, Powar JB, Cocking EL. Isolation, culture and division of protoplast. Plant Science Letters. 1977;**8**:85-89

[12] Murlidhar RV, Panda T. Fungal protoplast fusion: A revisit. Bioprocess and Biosystems Engineering. 2000;**22**:429-431

[13] Prabhavathi VR, Mathivanan N, Sagadevan E, Murugesan K, Lalithakumari D. Intra-strain protoplast fusion enhances carboxymethylcellulase activity in *Trichoderma reesei*. Enzyme and Microbial Technology. 2006;**38**:719-723

[14] Bakhtiari MR, Fallahpour M, Foruzanfakhr P, Moazami N. Protoplast fusion technique in *Tolypocladium inflatum* for increasing cyclosporine production. Journal of Biotechnology. 2007;**131**(2):135-136

[15] Van-veen JA, Van-overbeek LS, Van-elsas JD. Fate and activity of microorganisms introduced into soil. Microbiology and Molecular Biology Reviews. 1997;**61**:121-135

[16] Morrissey JP, Walsh UF, O'Donnell A, Moënne-Loccoz Y, O'Gara F. Exploitation of genetically modified inoculants for industrial ecology applications. Antonie Van Leeuwenhoek. 2002;**81**(4):599-606

[17] Ye X, Al-Babili S, Kloti A, Zhang J, Lucca P, Beyer P, et al. Engineering the provitamin A (beta-carotene) biosynthetic pathway into (carotenoid-free) rice endosperm. Science. 2000;**287**:303-305

[18] Resca R, Basaglia M, Poggiolini S, Vian P, Bardin S, Walsh UF, et al. An integrated approach for the evaluation of biological control of the complex *Polymyxa betae*/Beet Necrotic Yellow Vein Virus, by means of seed inoculants. Plant and Soil. 2001;**12**:215-226

[19] Haggag WM, Mohamed HA. Enhancement of antifungal metabolites production from gamma-ray induced mutants of some Trichoderma species for control onion white rot disease. Plant Pathology Bulletin. 2002;**11**:45-56

[20] Mess J, Wit R, Christa ST, Francis DG, Michel AH, Ben JC. Loss of avirulence and reduced pathogenecity of a gamma-irradiated mutant of *Fusarium oxysporum* f. sp. *Lycopersici*. Phytopathology. 1999;**89**:1131-1137

[21] Brunner K, Susanne Z, Woo L, Lorito M, Christian P, Kubicek P, et al. Improvement of the fungal biocontrol agent *Trichoderma atroviride* to enhance both antagonism and induction of plant systemic disease resistance. Applied Environmental Microbiology. 2005;**71**(7):3959-3965

[22] Rugthaworn P, Dilokkunanant U, Sangchote S, Piadang N, Kitpreechavanich V. A search and improvement of actinomycete strains for biological control of plant pathogens. Kasetsart Journal: Natural Science. 2007;**41**:248-254

[23] Mohamed HAA, Haggag WM. Biocontrol potential of salinity tolerant mutants of *Trichoderma harzianum* against *Fusarium oxysporum* causing tomato wilt disease. Arab Journal of Biotechnology. 2005;**8**(1):35-48

[24] Ahari-Mostafavi H, Safaie N, Naserian B, Fathollahi H, Dorri H, Lak M, et al. Possibility of biological control of bean root rot disease, using of avirulent mutants of *Fusarium solani* f. sp. *phaseoli* isolate. Journal of Plant Protection. 2009;**16**(3):123-134

[25] Jash S, Khalko S, Bose S, Roy M, Pan S. Morphological and physiological characterization of some mutant isolates of *Gliocladium virens*, a potential mycoparasite of sclerotial plant pathogens. Indian Journal of Agricultural Research. 2006;**40**(2):114-118

[26] Haggag WM. Enhancement of suppressive metabolites from *Pseudomonas fluorescence* against tomato damping-off pathogens. Arab Journal of Biotechnology. 1999;**2**:1-14

[27] Papavizas GC, Roberts DP, Kim KM. Development of mutants of Gliocladium virens tolerant to benomyl. Canadian Journal of Microbiology. 1990;**36**:484-489

[28] Balasubramanian N, Thamil-Priya V, Gomathinayagam S, Shanmugaiah V, Jashnie J, Lalithakumari D. Effect of chitin adapted and ultra violet induced mutant of *Trichoderma harzianum* enhancing biocontrol and chitinase activity. Australian Journal of Basic Applied Science. 2010;**4**(10):4701-4709

[29] Gulati HK, Chadha BS, Saini HS. Production, purification and characterization of thermostable phytase from thermophilic fungus *Thermomyces lanuginosus* TL-7. Acta Microbiologica et Immunologica Hungarica. 2007;**54**(2):121-138

[30] Wu C, Teo VS, Farrell RL, Bergquist PL, Nevalainen KM. Improvement of the secretion of extracellular proteins and isolation and characterization of the amylase I (amy1) gene from *Ophiostoma floccosum*. Gene. 2006;**384**:96-103

[31] Ul-Haq I, Ali S, Aslam A, Qadeer MA. Characterization of a *Saccharomyces cerevisiae* mutant with enhanced production of beta-D

fructofuranosidase. Bioresource Technology. 2008;**99**(1):7-12

[32] Ahari-Mostafavi H, Mirmajlessi SM, Fathollahi H. The potential of food irradiation: Benefits and limitations. In: Trends in Vital Food and Control Engineering. Croatia, Rijeka: Intech Open Access Publisher; 2012. pp. 43-68

[33] Scott-smith J, Suresh P. Irradiation and food safety. Food Technology. 2004;**58**(11):48-55

[34] Hvizdzak AL, Beamer S, Jaczynski J, Matak KE. Use of electron beam radiation for the reduction of *Salmonella enterica* serovars Typhimurium and Tennessee in peanut butter. Journal of Food Protection. 2010;**73**:353-357

[35] Ahari-Mostafavi H, Fathollahi H, Motamedi F, Mirmajlessi SM. Food irradiation: Applications, public acceptance and global trade. African Journal of Biotechnology. 2010;**9**(20):2826-2833

[36] Johnson M, Estes-Reynolds A. Consumer acceptance of electron-beam irradiated ready-to-eat poultry meats. Journal of Food Process Preservation. 2004;**28**:302-319

[37] Suresh P, Leslie A, Braby L. Electron Beam Technology for Food Irradiation, The International Review of Food Science and Technology (Winter 2004/2005). An Official Publication of the International Union of Food Science and Technology (IUFoST); 2005

[38] EFSA (European Food Safety Authority). Statement summarizing the conclusions and recommendations from the opinions on the safety of irradiation of food adopted by the BIOHAZ and CEF panels. The EFSA Journal. 2011;**9**(4):2107

[39] Landgraf M, Gaularte L, Martins C, Cestari A, Nunes T, Aragon L, Destro M, Behrens J, Vizeu D, Hutzler B. Use of Irradiation to Improve the Microbiological Safety of Minimally Processed Fruits and Vegetables. The IAEA Technical Documents (IAEA-TECDOC) 1530; 2006. pp. 41-59

Chapter 9

Study of Bio-Based Foams Prepared from PBAT/PLA Reinforced with Bio-Calcium Carbonate and Compatibilized with Gamma Radiation

Elizabeth C.L. Cardoso, Duclerc F. Parra, Sandra R. Scagliusi, Ricardo M. Sales, Fernando Caviquioli and Ademar B. Lugão

Abstract

Foamed polymers are future materials, considered "green materials" due to their properties with very low consumption of raw materials; they can be used to ameliorate appearance of structures besides contributing for thermal and acoustic insulation. Nevertheless, waste disposal has generated about 20–30% of total of solid volume in landfills besides prejudicing flora and fauna by uncontrolled disposal. The development of biodegradable polymers aims to solve this problem, considering that in 2012, bio-plastics market was evaluated in 1.4 million tons produced and in 2017 attained 6.2 million tons. Biodegradable polymers as poly(lactic acid) (PLA) and poly(butylene adipate-*co*-terephthalate) (PBAT) are thermoplastics which can be processed using the most conventional polymer processing methods. PLA is high in strength and modulus but brittle, while PBAT is flexible and tough. In order to reduce interfacial tension exhibited by PLA/PBAT blends, it was used as compatibilizing agent 5 phr of PLA previously gamma-radiated at 150 kGy. Ionizing radiation induces compatibilization by free radicals, improving the dispersion and adhesion of blend phases, without using chemical additives and at room temperature. As a reinforcement agent, calcium carbonate from avian eggshell waste was used, at 10 ph of micro particles, 125 μm. Admixtures were further processed in a single-screw extruder, using CO_2 as physical blowing agent (PBA). Property investigations were performed by DSC, TGA, XRD, SEM, FTIR, and mechanical essays.

Keywords: PBAT/PLA foams, eggshells, PBA, gamma radiation, compatibilization

1. Introduction

Natural polymers, biopolymers, and synthetic polymers based on annually renewable resources are the basis for the twenty-first-century portfolio of sustainable, eco-efficient plastics. The interest on these polymers is considerable, due to a decrease of world resources in oil; in addition, there is a concern to limit the plastics' contribution to waste disposal. The degree of concern has been [1–8] raised along with the development in urbanization. The development of biodegradable polymers generally catches the attention of researchers due to environmental problems associated with the disposal of petroleum-based polymers.

The depletion of petroleum resource led to considerable research efforts on the development of biodegradable polymeric materials. Biodegradable polymers offer a great variety of advantages to environmental conservation; based on their non-harmful effects, they can be classified into two major categories: natural polymers and synthetic polymers; polymers obtained basically from renewable sources are a new generation of material capable to significantly reduce the environmental impact in order to achieve certain technical requirements besides being fully biodegradable. In addition, natural polymer-based materials offer a feasible alternative to the traditional polymeric materials when recycling of synthetic polymer is not cost-effective or technically impossible [9–15].

Poly(butylene adipate-*co*-terephthalate) (PBAT) is an aliphatic-aromatic random co-polyester, fully biodegradable, and prepared from 1,4–butanediol, adipic acid, and terephthalic acid: a synthetic polymer based on fossil resources, 100% biodegradable, with high elongation at break, and very flexible [16]. PBAT is an elastomeric polymer intended to improve mechanical properties; it can be used in several applications, such as, packaging materials, hygiene products, biomedical fields, and industrial composting, among others [17–21]; nevertheless, PBAT has poor thermal and mechanical properties, which can be overcome through the addition of fillers; in addition, it is a versatile polymer that allows the manufacturing from films up to shaped devices, and it can be used in food and dairy industries as well in hygiene packing [22, 23].

Polylactide or poly(lactic acid) (PLA) is the front-runner in the emerging bioplastics market with the best availability and the most attractive cost structure: PLA is a linear, aliphatic thermoplastic polyester, used for different applications ranging from medical to packaging, resorbable, and biodegradable under industrial composting conditions [24]. Therefore, its rheological properties, especially its shear viscosity, have important effects on thermal processes. Despite all its advantages, some properties of PLA such as inherent brittleness, low toughness, slow crystallization, poor melt strength, narrow processing window, and low thermal stability, besides high cost, pose considerable scientific challenges that limit their large-scale applications (film blowing, injection molding, and foaming) [25–27].

So, combining the high toughness of PBAT and the relatively low price of PLA can result in a novel blend. PLA was blended with PBAT flexible polymer, considering its high toughness and biodegradability. Poly(lactic acid) (PLA) and poly(butylene adipate-*co*-terephthalate) (PBAT), both biodegradable aliphatic polyesters, semicrystalline, and thermoplastic, can be processed by conventional methods. Their resulting blends provide interesting materials for industrial and hygiene packaging applications, due to their increased ductility in function of PBAT content.

PLA and PBAT binary blends exhibited improved properties concerned with higher elongation at break but lower tensile strength and modulus than pure PLA. Therefore, the addition of filler to PLA/PBAT blends led to a modulus approaching that of pure PLA.

In this paper bio-calcium carbonate from avian egg shells was used. Daily, tons of chicken eggshells are discarded, generating commercially devalued waste from restaurants, food industry, and homes. Currently, egg production throughout the world is 65.5 million metric tons per year, with Asia as a key contributor to global egg output growth [28]. The eggshell is rich in calcium carbonate, a natural bioceramic composite with a unique chemical composition of high inorganic (95% of calcium carbonate in the form of calcite) and 5% of organic (type X collagen, sulfated polysaccharides) components; this eggshell characteristic structure combined with substantial availability makes eggshells a potential source of bio-fillers that can be efficiently used for polymer composites [29].

Unfortunately, PLA/PBAT blends filled with calcium carbonate ($CaCO_3$) have poor mechanical properties due to the poor interfacial adhesion. Many polymers are immiscible and form heterogeneous systems when blended. In order to cope with this problem, irradiation was used to improve the compatibility between immiscible polymers in a blend. In comparison with other methods of compatibilization based on the reactivity of functional groups grafted on the polymer backbone, the changes are not limited to the interface. Irradiation leads to changes not only in the interphase but also in the bulk of both polymers (chain scission, crosslinking, etc.). Therefore, it is very difficult to determine whether macroscopic properties change due to the compatibilizing effect of irradiation or due to modification of the polymers in bulk. A great number of authors working on irradiated immiscible polymer blends claim in their articles to have increased the compatibility between the two polymers just considering mechanical properties. Compatibilization is essential in order to decrease interfacial tension exhibited by PLA/PBAT blends: herein it was used as compatibilizing agent of PLA previously gamma-radiated at 150 kGy, air environment, 10.5 kGy h^{-1}. Güven and collaborators have proposed the use of ionizing radiation in replacing chemical compatibilizing agents for thermoplastic materials with enhanced properties [30–38].

Foam technology has been developing since 1930, using blowing agents in polymer processing. Polymer foams consist of two phases: a polymeric matrix and entrapped, well-dispersed cells generated by blowing agents. Foams have several advantages: low density, insulating capability, energy absorption, etc. These make foams a desired product in many applications such as packaging, floating materials, paddings, shields for reducing noise, shoes, etc. Foam density varies across a wide range from several kg/m^3 to near thousands kg/m^3 [39]. Carbon dioxide was used as a physical blowing agent (PBA): it has a regular solubility and is considered as an eco-friendly gas. A PBA is capable to produce a cellular structure via foaming process, and it is typically applied when blown material is in liquid stage. Cellular structure in a matrix reduces density, increasing thermal and acoustic insulation [40, 41].

The proposal of the present work is the development of biodegradable foams from PBAT/PLA blends, reinforced with bio-calcium carbonate from avian eggshells, 125 μm particle size, compatibilized with PLA gamma-radiated at 150 kGy, and further assessed for DSC, TGA, XRD, SEM, FTIR, and mechanical essays.

2. Experimental section

2.1 Materials

PLA and PBAT polymers, with main characteristics described in **Table 1**.

Both PLA and PBAT were dried at 70°C for 12 h before processing.

PLA, irradiated in a Cobalt-60 source, 150 kGy, 10.5 kGY h^{-1} dose ratio, at multipurpose reactor, in CTR/IPEN, Instituto de Pesquisas Energéticas e Nucleares, São Paulo.

Carbon dioxide (CO_2): physical blowing agent, selected according to good diffusion in PLA foaming [42].

Calcium carbonate ($CaCO_3$) from avian eggshells: white chicken eggshells were subjected to a thorough cleaning using tap water for removing of internal membranes. Afterward, clean eggshells were kept for 4 h in a 100°C water bath

Characteristics of PLA	Characteristics of PBAT
Grade: ingeo biopolymer 3251 D	Commercial name: Ecoflex FS
Supplier: nature works	Supplier: BASF
Melting point: 168°C	Melting point: 110–120°C
Glass transition temperature: 62°C	Glass transition temperature: −30°C
Average molecular weight: 100,000 g mol^{-1}	Average molecular weight: 40,000 g mol^{-1}

Table 1.
Main characteristics of used polymers.

Designation	PBAT (wt%)	PLA (wt%)	$CaCO_3$ (phr)	PLA 150 kGy (phr)
PBAT	100	—	—	—
PBAT50	50	50	—	—
PBAT65	65	35	—	—
PBAT82	82	18	—	—
PBAT50CI	50	50	10	5
PBAT65CI	65	35	10	5
PBAT82CI	82	18	10	5
PLA	—	100	—	—

Table 2.
Material designation and composition for PBAT/PLA/$CaCO_3$/PLA 150 kGy gamma-irradiated.

and finally dried at 100 ± 2°C for 2 h in an air-circulating oven. Eggshells were size reduced to fine powder, particle size equal or lower than 125 μm, by using ball mills and granulometric sieve, respectively. Then they were dried again at 100 ± 2°C, for 24 h, in order to reduce its moisture content to less than 2%.

2.2 Preparation and processing

Composite materials were prepared according to **Table 2**; they were first homogenized by melting extrusion process, using a corotating twin-screw extruder (HAAKE Rheomex 332p, 3.1 L/D, 19/33 compression ratio), by using a 120–145°C temperature profile and 50 rpm.

Homogenized samples (pellets) were further subjected to extrusion under pressure, by expansion physical method using carbon dioxide (CO_2) as blowing agent, at 10 bar (approximately 10 kgf cm^{-2}). A mono-screw specific for foaming was used, maintaining the same temperature profile: 130–145°C.

3. Characterization

3.1 Differential scanning calorimetric analyses (DSC)

Thermal behavior was examined in a DSC Mettler Toledo apparatus, according to ASTM D3418-08. A set of heating/cooling ramps was carried out following a three-step process; the samples were firstly heated to 200°C and kept in the molten state for 10 min to erase the thermal history of the material. They were then cooled down to 30°C at 10°C min^{-1} to evaluate the ability of PLA, PBAT, and their compositions listed in **Table 2** to crystallize upon cooling. After cooling treatment, the samples were heated back to 200°C at 10°C min^{-1}. The percent crystallinity of each one was calculated separately, upon the second heating by using Eq. 1 [43]:

$$x_c\ (\%\text{Crystallinity}) = \frac{\Delta H_m}{\Delta H_m^o} \times \frac{100}{w}. \quad (1)$$

where ΔH_m is the measured heat of fusion, w is the weight fraction of PLA or PBAT in the blend, and $\Delta H^o{}_m$ is the enthalpy of fusion for a crystal having infinite crystal thickness (93 J g^{-1} for PLA and 114 J g^{-1} for PBAT).

3.2 Thermogravimetric analyses (TG)

Thermogravimetric analyses provide complimentary and supplementary characterization information to DSC, by measuring the amount and rate (velocity) of change in the mass of a sample as a function of temperature or time in a controlled atmosphere. Measurements are used primarily to determine the thermal and/ or oxidative stabilities of materials as well as their compositional properties. The technique can analyze materials that exhibit either mass loss or gain due to decomposition, oxidation, or loss of volatiles (such as moisture). TGA were performed using a DSC Mettler Toledo apparatus, according to ASTM E1641-07, by using 5–9 mg of foam sample, within a 25–600°C program, at 10°C min^{-1}, in a nitrogen flow of 50 ml min^{-1}.

3.3 X-ray diffraction analysis (XRD)

X-ray diffraction is a technique used for determining anatomic structure: it consists in a constructive interference of a wave from X-ray incident beam in relation to a uniform atomic spacing.

In this technique Bragg's law is applied, defined by $n\lambda = 2dsen\theta$, where $n\lambda$ is an entire value for wavelength generated by a specific target according to a given electronic transition and $sen\theta$ is the angle where the constructive interference occurs; therefore, it is possible to determine interplanar distances (d) for each crystalline plane. The identification of crystalline phase of a material is given from a database defined by the Joint Committee on Powder Diffraction Standards (JCPDS) that compares position of obtained peaks with intensity relationship.

It was employed herein a X-Ray diffractometer, RigakuMultiflex, graphite monochromator, 40 kV, 20 mA, X-rays tube, copper anode $\lambda Cu_{k\alpha} = 1,5418\,\mathring{A}$, scanning 2θ within 3°–60°, speed 0.06°/4 s, fixed time. It provides, among others, information on sample crystallinity, via diffractograms, distinguishing between amorphous and crystalline states.

3.4 Scanning electron microscopy (SEM)

Electronic microscopy technique is a major tool for the study of material structure and morphology; it allows the visualization of details in a micrometric scale of changes in the material.

Morphology investigations were accomplished in a FEG-SEM equipment, model F-50, capable to read up to 20 nanometers, in various magnification micrographs. Samples were freeze-fractured in liquid nitrogen and gold coated in a Balzers SCD 050 sputtering before accomplishment of analyses.

3.5 Attenuated total reflection Fourier-transform infrared spectroscopy (ATR-FTIR)

FTIR is a sensible method for identifying chemical modification in a material and, so, is capable to detect chemical modifications in a polymeric material. This method detects vibrational movements imparted from chemical bonds for the material that is being analyzed. As each chemical group absorbs vibrational energy at a given value, it is possible to differentiate them via infrared spectrum. Spectra were obtained from a PerkinElmer, universal ATR sampling accessory spectrum 100 FTIR spectrometer. Setup collection sample was adjusted for 64 scans, within a 4000–650 cm^{-1} range.

3.6 Tensile and elongation at break

Tensile and elongation at break essay are relevant instruments for evaluating loss of properties and evolution of degradative process of the polymer. Parameters that contribute for mechanical behavior of polymers are chemical structure, crystallinity degree, molar mass, moisture, and reinforcing agent present, among others. All these properties are modified during degradation processes. In case of reinforcing agents, the concentration is not changed; nevertheless, their interaction can be modified in consequence of chemical modifications suffered by the polymer. Tensile and elongation at break tests were accomplished at 25 ± 5°C, in an EMIC model DL 300 universal essay machine, 20 kN load cell, 50 mm min^{-1}, in accordance with ASTM D 638-14. Specimens were conditioned at 25 ± 5°C and 50 ± 5% relative humidity, for 24 h, prior to testing.

4. Results and discussion

4.1 Differential scanning calorimetric analyses (DSC)

DSC heating curves of PLA, PBAT, and PLA/PBAT (50/50) blends, after crystallizing from melt, are shown in **Figure 1**.

PLA was primarily amorphous when it was cooled from melt, and this result suggests that PLA was not able to crystallize within the cooling time frame.

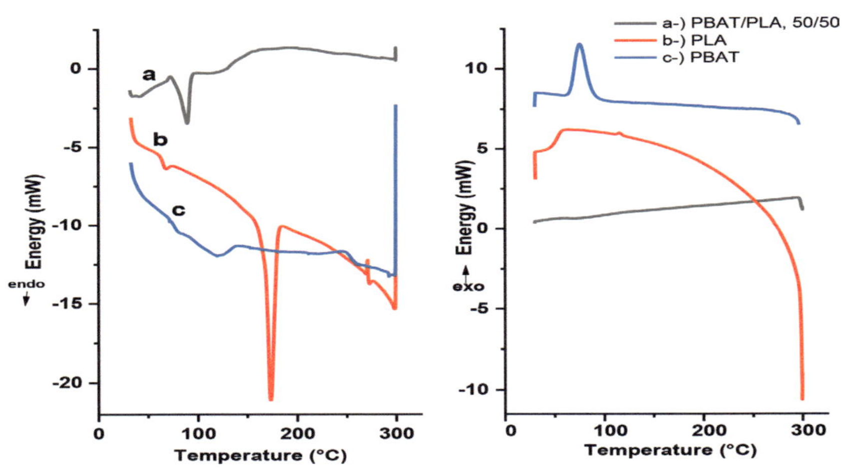

Figure 1.
Melting and crystallization curves for PLA, PBAT, and PLA/PBAT (50/50).

In **Table 3** a brief summary of thermal properties of PLA, PBAT, and PBAT/PLA (50/50) is depicted:

4.2 Thermogravimetric analyses (TG)

TG was carried out to investigate the effect of processing on the thermal decomposition of PLA and PBAT under nitrogen atmosphere; in **Figure 2** behavior of samples studied is shown.

The onset temperature of the decomposition of PLA slightly decreases with the extrusion process; nevertheless, its blend with PBAT and PBAT purely showed a higher onset temperature. This change in PLA could be originated from the degradation of the polymer, leading to the presence of shorter polymer chains and an increase in the number of chain ends per mass. Chain ends then promote a dominant degradation pathway at the temperature range of 270–360°C.

4.3 X-ray diffraction analysis (XRD)

X-ray diffraction patterns of all studied samples are shown in **Figure 3**.

In order to provide a more effective visualization of involved samples, as well as their behavior in the present study, components were separated into individual graphs, according to **Figures 4-7**, as follows.

Designation	Tg (°C)	Tm (°C)	Xc (%)
PBAT	−29.7 [44]	120.0	14.8
PLA	60.4	158.0	61.1
PBAT/PLA, 50/50	—	86.9	29.9

Tg = glass transition temperature; Tm = melting temperature, second fusion; Xc = crystallinity.

Table 3.
Thermal properties of materials studied.

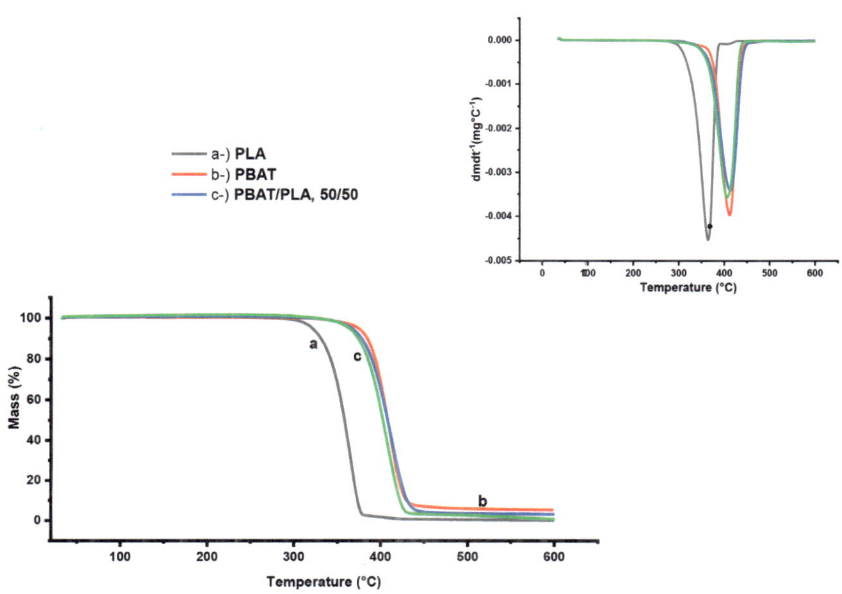

Figure 2.
TG and DTG curves for PLA, PBAT, and PBAT/PLA (50/50).

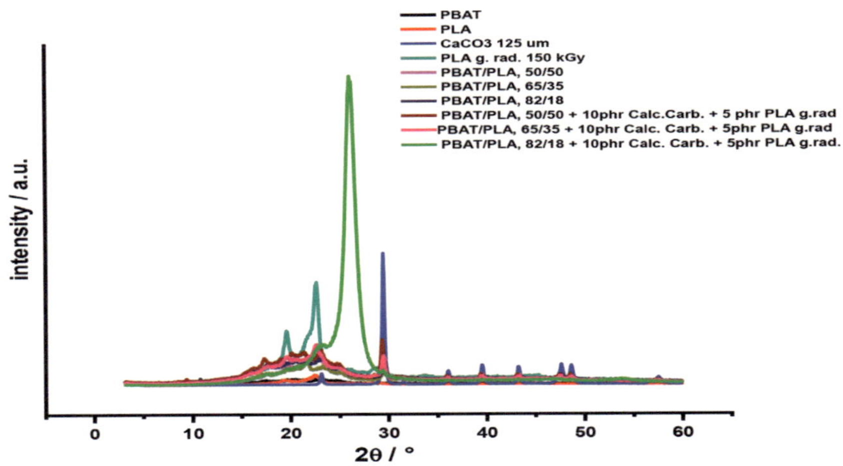

Figure 3.
DRX diffractograms of all studied samples.

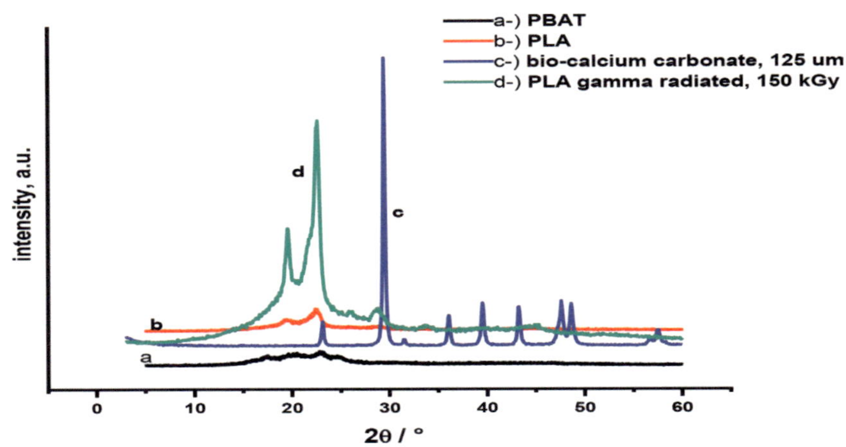

Figure 4.
DRX diffractograms of basic components: PBAT, PLA, $CaCO_3$, and PLA gamma-radiated at 150 kGy.

In general, the sample is composed by crystals and amorphous phases: the sharp peaks are related to crystallite diffraction, and larger peaks are related to amorphous phases. In **Figure 4**, pure PBAT and PLA exhibited four peaks in 17.5, 20.5, 22.5, and 24.5°, in which 22.5° 2θ was the most intense. Bio-$CaCO_3$ exhibited the most intense peak at 30.0 2θ, among other crystalline ones. PLA gamma-irradiated at 150 kGy exhibited two peaks at 20.0 and 22.0, 2θ, proving the efficacy of gamma irradiation treatment.

PBAT/PLA blends, 82/18 and 65/35, corresponding to **Figures 5** and **6**, respectively, as well as their composites, exhibited two intense peaks at 22.5 and 30.0 2θ, emphasizing that composites showed a higher intensity for peaks than based blends.

PBAT/PLA blend (50/50) and its composite showed just one intense peak at 30.0 2θ, much more intense for corresponding composite.

4.4 Scanning electron microscopy (SEM)

The cell morphology of all formulations processed using SEM is shown in **Figures 8** and **9**. Images were taken in a 100 × magnification, confirming structural foam nature [45].

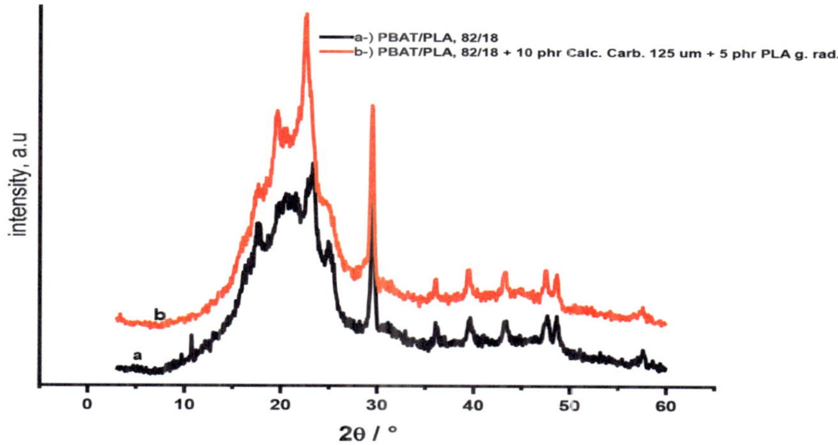

Figure 5.
DRX diffractograms of PBAT/PLA (82/18) and their compositions with 10 phr of $CaCO_3$ and 5 phr of PLA gamma-radiated at 150 kGy.

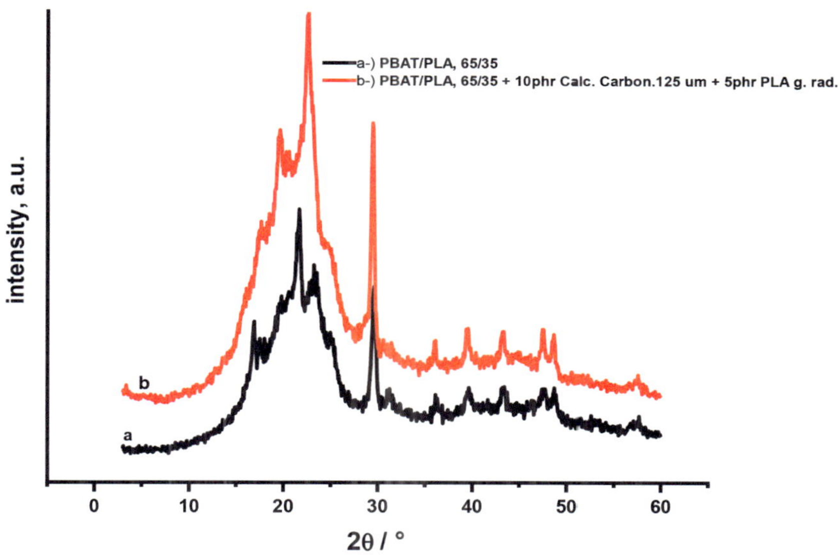

Figure 6.
DRX diffractograms of PBAT/PLA (65/35) and their compositions with 10 phr of $CaCO_3$ and 5 phr of PLA gamma-radiated at 150 kGy.

The higher PBAT concentration in PLA/PBAT blends, the easier will be the miscibility between both PBAT and PLA, as shown in **Figure 8**; in **Figure 9** pure PLA and PBAT micrographs are presented.

PLA shows an irregular dispersion and PBAT a continuous phase in blends; that is, PLA has a typical and irregular morphology *island-phase* type and the PBAT a *sea-phase* type morphology, as can be observed in **Figure 9a** and **b**.

Addition of PLA gamma-irradiated at 150 kGy contributed for an effective distribution of bio-calcium carbonate 125 μm reinforcement in PBAT/PLA compositions and buildup of structural foams, as can be seen in **Figure 10**.

In **Figure 11** foamed samples obtained from 4 mm die extruder and final specimens are shown.

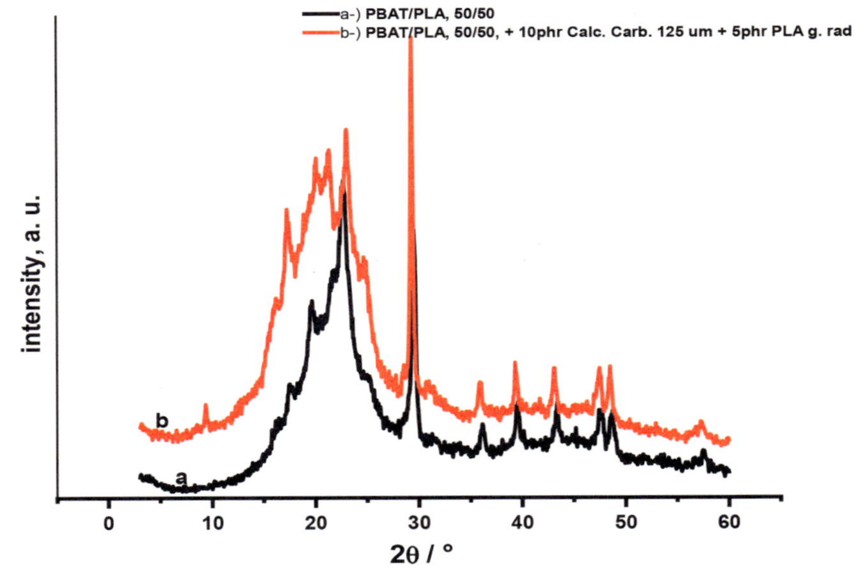

Figure 7.
DRX diffractograms of PBAT/PLA (50/50) and their compositions with 10 phr of $CaCO_3$ and 5 phr of PLA gamma-radiated at 150 kGy.

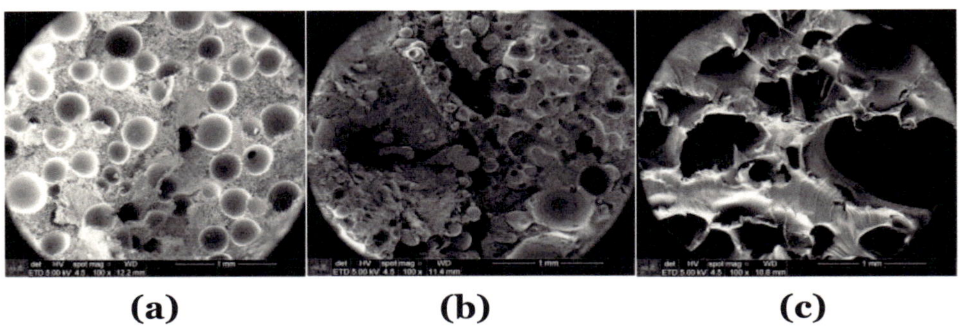

Figure 8.
SEM micrographs of PLA/PBAT blends, 100 X magnification: (a) PBAT/PLA, 82/18; (b) PBAT/PLA, 65/35; (c) PBAT/PLA, 50/50.

4.5 Attenuated total reflection Fourier-transform infrared spectroscopy (ATR-FTIR)

FTIR spectra of PBAT/PLA blends and PBAT/PLA blends with 10phr of bio-$CaCO_3$ and 5 phr of PLA gamma-radiated at 150 kGy are shown, respectively, in **Figures 12** and **13**.

For PLA, the peak at around 752 cm^{-1} associated with the rocking vibration of α-methyl; peak at around 864 cm^{-1} associated with the ester (O-CH-CH_3); the peak at around 1042, , and 1180 cm^{-1} associated with the stretching vibration of C-O-C; the peak at 1381 cm^{-1} associated with the CH symmetric bending vibration; the peak at around 1450 cm^{-1} associated with the CH_3 antisymmetric; the peak at 1748 cm^{-1} associated with the carbonyl C=O stretching vibration; and the symmetric and antisymmetric stretching vibration of CH3 of saturated hydrocarbons were found at 2943 and 2997 cm^{-1}, respectively [46, 47].

For PBAT, the peak at 725 cm^{-1} associated with the bending vibration of CH-plane of benzene ring; the symmetric stretching vibration of trans-C-O was found at 937 cm^{-1}; the peak at 1018 cm^{-1} associated with the bending vibration at the surface of adjacent hydrogen atoms on the phenyl ring; the peak at 1103 cm^{-1} associated with the left-right symmetric stretching vibration of C-O; the peak at 1265 cm^{-1} associated with the C-O symmetric stretching vibration; the peak at 1408 cm^{-1} associated with the trans-CH_2-plane bending vibration; the peak at 1504 cm^{-1} associated with the skeleton vibration of the benzene ring; the peak at 1713 cm^{-1} associated with the C-O stretching vibration; and the peak at 2959 cm^{-1} associated with the CH_2 asymmetric stretching vibration [46, 47].

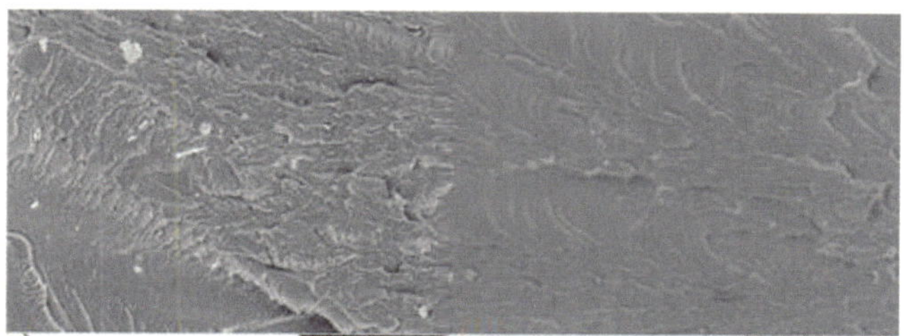

a) b)

Figure 9.
SEM micrographs, 500 X magnification, for pure PLA (a) and pure PBAT (b).

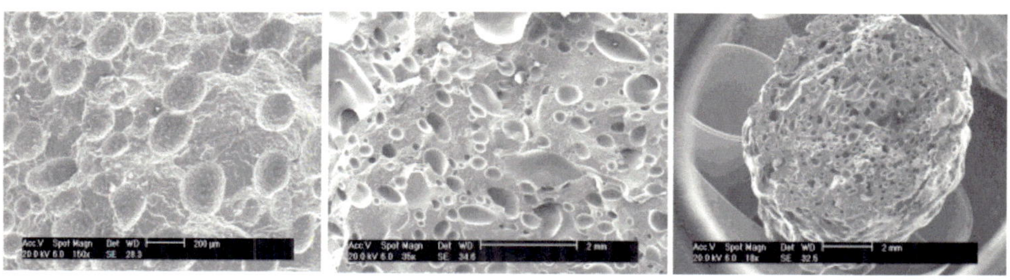

(a) (b) (c)

Figure 10.
SEM micrographs of foams, randomly chosen, with different magnifications: 150 (a), 35 (b), and 18 (c) X, respectively.

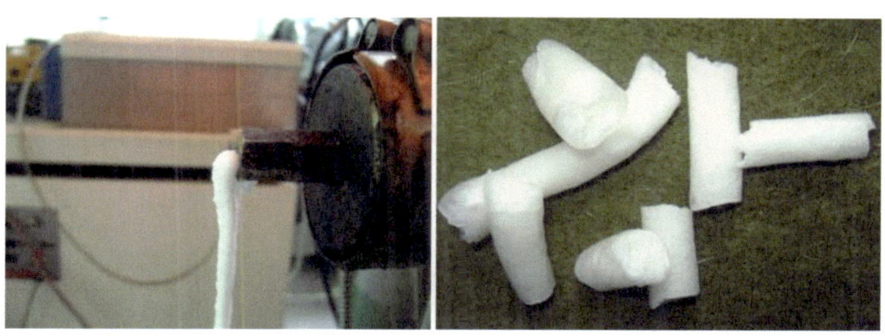

(a) (b)

Figure 11.
Structural foams: (a) extruded foams from a 4 mm die extruder; (b) cylinder structural foams, of approximately 400 kg m^{-3} density.

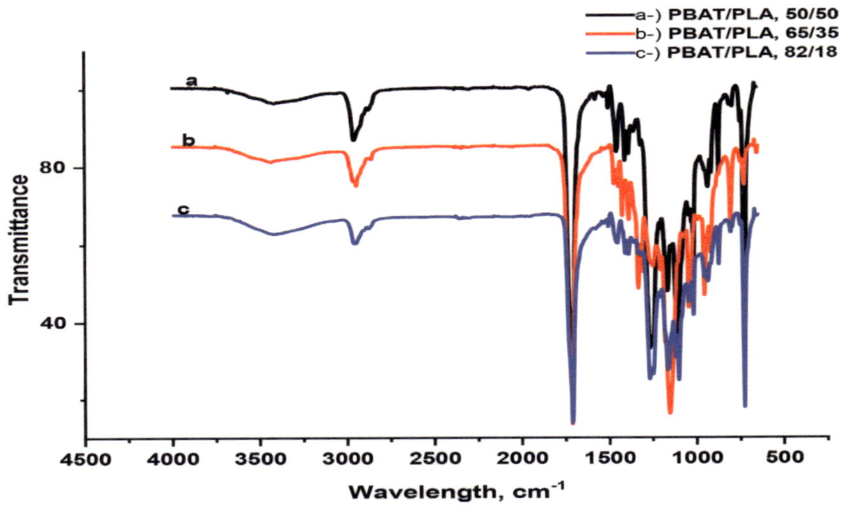

Figure 12.
FTIR spectra of PBAT/PLA blends.

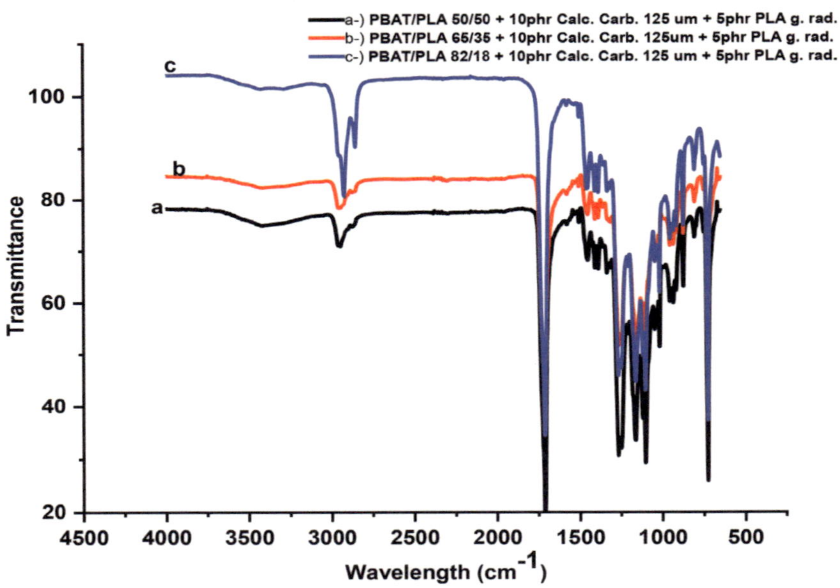

Figure 13.
FTIR of PBAT/PLA blends incorporated with c (10 phr of bio-$CaCO_3$) and I (5 phr of PLA gamma-radiated at 150 kGy).

Absorption spectral of PLA/PBAT blends showed the upshift of CH-plane of the benzene ring vibration from 725 to 729 cm^{-1}. Ester vibration peak in PLA shifted from 864 to 872 cm^{-1} [7]. There was however no clear evidence of interaction between PLA and PBAT in the blends.

4.6 Tensile and elongation at break

Tensile mechanical properties of PBAT/PLA blends and PBAT/PLA blends with 10 phr of bio-calcium carbonate and 5 phr of PLA gamma-radiated at 150 kGy are presented in **Figures 14** and **15**.

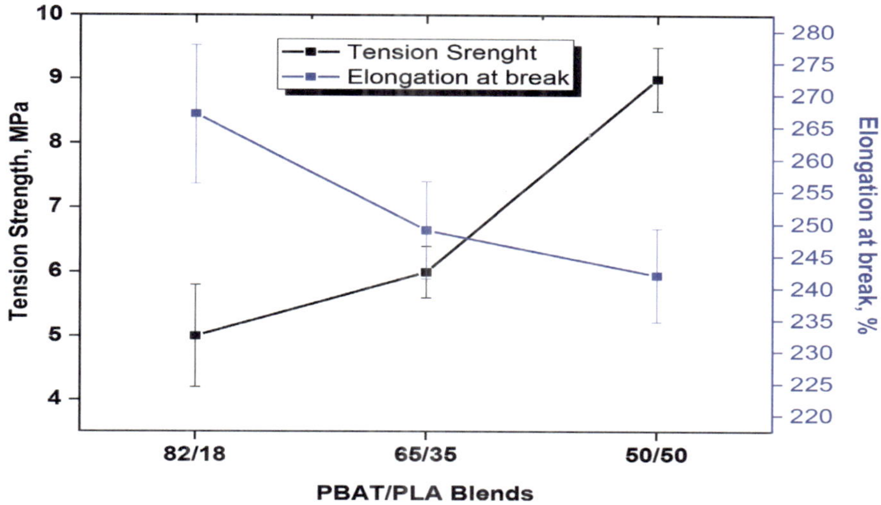

Figure 14.
Tensile mechanical properties for PBAT/PLA blends: 82/18, 65/35, and 50/50.

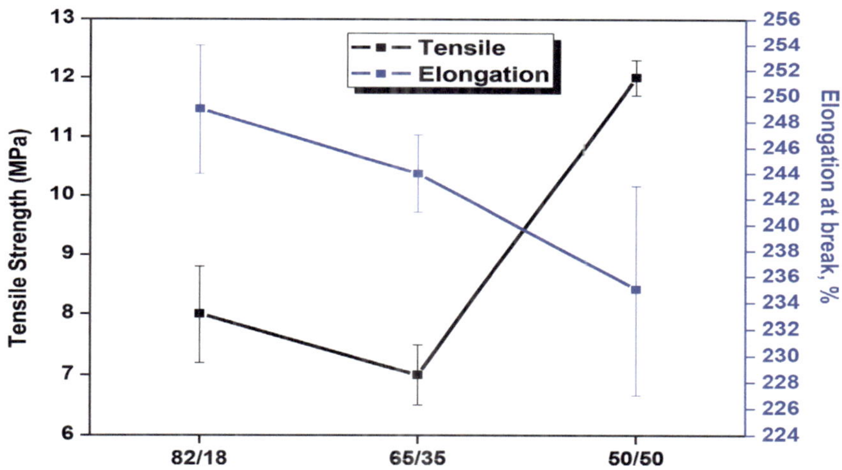

Figure 15.
PBAT/PLA blends, 82/18, 65/35, and 50/50 with 10 phr of bio-$CaCO_3$ and 5 phr of PLA gamma-radiated at 150 kGy.

Blends and composites	Tensile strength (MPa)	Elongation at break (%)	Elasticity modulus (MPa)
PBAT50	9.0	242.0	112.6
PBAT65	6.0	250.0	93.1
PBAT82	5.0	265.0	215.0
PBAT50CI	12.0	235.01	116.2
PBAT65CI	7.0	244.0	126.9
PBAT82CI	8.0	249.0	138.5

Table 4.
Tensile properties of PBAT/PLA blends and their composites.

In **Table 4** tensile properties presented in **Figures 14** and **15** are summarized.

From **Table 4**, PBAT/PLA 50/50 presented a higher value for tensile strength and PBAT/PLA 82/18 a higher value for elongation at break and elasticity modulus. PBAT/PLA compositions with bio-calcium carbonate and PLA gamma-radiated at 150 kGy presented results slightly higher than base compositions, following the same tendency.

5. Conclusions

Interaction between PLA and PBAT, registered from thermal analyses, proved to be fundamental for accomplishment of investigations. Addition of calcium carbonate from avian eggshells proved to be effective for reinforcement of PBAT/PLA blends, according to mechanical tests. PLA gamma-radiated at 150 kGy, used as compatibilizing agent, provided a higher crystallinity in assessed samples, as it can be seen from DRX analyses, exhaustively shown in separate graphs: in summary, it contributed for the effective interaction between components and further good performance in mechanical essays. Spectra obtained from infrared determinations were typical for PLA, PBAT, and their blends; nevertheless, insertion of bio-$CaCO_3$ and PLA gamma-radiated at 150 kGy contributed for more defined peaks, within 2750 and 3200 cm^{-1}. SEM analyses pointed toward the acquisition of structural closed-cell foams, with no interference of naturally immiscible PLA and PBAT; this efficacy can be attributed to PLA gamma-radiated at 150 kGy, capable to provide a complete and expected interaction between bio-$CaCO_3$ and PBAT/PLA blends.

Acknowledgements

The authors gratefully acknowledge NatureWorks and BASF for raw materials, IPEN/SP for radiation, and CNEN/RJ for financial support.

Author details

Elizabeth C.L. Cardoso*, Duclerc F. Parra, Sandra R. Scagliusi, Ricardo M. Sales, Fernando Caviquioli and Ademar B. Lugão
Instituto de Pesquisas Energéticas e Nucleares, São Paulo, SP, Brazil

*Address all correspondence to: eclcardo@ipen.br

References

[1] Jia W, Gong RH, Hogg PJ. Poly(lactic acid) fibre reinforced biodegradable composites. Composites. Part B, Engineering. 2014;**62**:104-112

[2] JalaliDil E, Carreau PJ, Favis BD. Poly(butylene adipate-co-terephthalate) blends. Polymer. 2015;**68**:202-212

[3] Pivsa-Art W, Chaiyasat A, Pivsa-Art S, Yamane H, Ohara H. Preparation of polymer blends between Poly(lactic acid) and Poly(butylene adipate-co-terephthalate) and biodegradable polymers as compatibilizers. Energy Procedia. 2013;**34**:549-554

[4] Yu T, Li Y. Influence of poly(butylene adipate-co-terephthalate) on the properties of the biodegradable composites based on ramie/ poly(lactic acid), Composites. Part A, Applied Science and Manufacturing. 2014;**58**:24-29

[5] Zhao P, Liu W, Wu Q, Ren J. Preparation, mechanical and thermal properties of biodegradable polyesters/ poly(lactic acid) blends. Journal of Nanomaterials. 2010;**2010**:287082

[6] Chaishome J, Brown KA, Brooks R, Clifford MJ. Thermal degradation of flax fibres as potential reinforcement in thermoplastic composites. Advanced Materials Research. 2014;**894**:32-36

[7] Chaishome J, Rattanapaskorn S. International conference on mining: Material and metallurgical engineering. 2nd. Materials Science and Engineering. 2017. p. 191

[8] Gupta A, Kumar V. New emerging trends in synthetic biodegradable polymers-polylactide: A critique. European Polymer Journal. 2007;**43**(10):4053e74

[9] Klemm D, Heublein B, Fink HP, Bohn A. Cellulose: fascinating biopolymer and sustainable raw material edition. Angewandte Chemie, International Edition. 2005;**44**:3358

[10] Habibi Y, Lucia LA, Rojas OJ. Cellulose nanocrystals: Chemistry, self-assembly and applications. Chemical Reviews. 2010;**110**:3479

[11] Dufresne A. Nanocellulose: A new ageless bionanomaterial. Materials Today. 2013;**16**:220

[12] Azizi Samir MAS, Alloin F, Dufresne A. Review of recent research into cellulosic whiskers, their properties and their application in nanocomposite field. Biomacromolecules. 2005;**6**:612

[13] Nonato RV, Mantelatto PE, Rossell CEV. Production of biodegradable plastic (PHB), sugar and ethanol in a sugar mill. Applied Microbiology and Biotechnology. 2001;**57**:1

[14] Ravi Kumar MN. A review of chitin and chitosan applications. Reactive and Functional Polymers. 2000;**46**:1

[15] Blackburn RS. Natural polysaccharides and their interactions with dye molecules: Applications in effluent treatment. Environmental Science & Technology. 2004;**38**:4905

[16] Rodrigues BVM, Silva AS, Melo GFS, Vasconscellos LMR, Marciano FR, Lobo AO. Influence of low contents of superhydrophilic MWCNT on the properties and cell viability of electrospun poly(butylene adipate-co-terephthalate) fibers. Materials Science and Engineering. 2016;**59**:782

[17] Van de Velde K, Kiekens P. Biopolymers: Overview of several properties and consequences on their applications. Polymer Testing. 2002;**21**:433

[18] Gross RA, Kalra B. Influencing factors and process on in situ

degradation of Poly (Butylene succinate) film by strain biomectria ochroleuca, BFM-X1 in soil. Science. 2002;**297**:803

[19] Pereira da Silva JS, Farias da Silva JM, Soares BG, Livi S. Fully biodegradable composites based on poly (butylene adipate-coterephthalate)/ peach palm trees fiber. Composites: Part B. 2017;**129**:117

[20] Santana-Melo GF, Rodrigues BVM, da Silva E, Ricci R, Marciano FR, Webster TJ, et al. Electrospun ultraphin PBAT/nHap fibers influenced the in vitro and improved the mechanical properties of neoformed bone. Colloids and Surfaces, B: Biointerfaces. 2017;**155**:544

[21] de Castro JG, Rodrigues BVM, Ricci R, Costa MM, Ribeiro AFC, Marciano FR, et al. Designing a novel nanocomposite for bone tissue engineering using electrospun conductive PBAT/polypyrrole as a scaffold to direct nanohydroxyapatite electrodeposition. RSC Advances. 2016;**6**:32615

[22] Fukushima K, Wu MH, Bocchini S, Rasyda A, Yang MC. PBAT based nanocomposites for medical and industrial applications. Materials Science and Engineering: C. 2012;**32**:1331-1351

[23] Chen JH, Chen CC, Yang MC. Characterization of nanocomposites of poly (butylene adipate-co-terephthalate) blending with organoclay. Journal of Polymer Research. 2011;**18**:2151-2159

[24] Alexandre M, Dubois P. Polymer-layered silicate nanocomposites: Preparation, properties and uses of a new class of materials. Materials Science and Engineering. 2000;**28**(1e2):1e63

[25] Maazouz A, Lamnawar K, Mallet B. Improvement of thermal, stability, rheological and mechanical properties of PLA, PBAT and their blends by reactive extrusion with functionalized epoxy, polymer, degradation and stability. International Journal of Engineering Science. 2012;**XXX92012**:1-17

[26] Lamnawar K, Maazouz A, Mallet B. Patent; 2010. International patent C08J5/

[27] Gu SY, Zhang K, Ren J, Zhan H. Melt rheology of polylactide/ poly(butylene adipate-co-terephthalate) blends. Carbohydrate Polymers. 2008;**74**(1):79e85

[28] Global poultry trends: "Asia is a key to global egg output growth". The Global Poultry Site. 2013. Available from: http://www.thepoultrysite.com/articles/2735/global-poultry-trends-asia-is-key-to-global-egg-output-growth

[29] Hassan SB, Aigbodion VS, Patrick SN. Development of polyester egg shell particulate composites. Tribology in Industry. 2012;**34**(4):217-225

[30] Guven O et al. Polymer recycling: Potential application of radiation technology. Radiation Physics and Chemistry. 2002;**64**(1):41-51

[31] Sun Y, Chmielewski AG. Applications of Ionizing Radiation in Materials Processing. Institute of Nuclear Chemistry and Technology. Warszawa, Erasmus; Vol. 2. 2017

[32] Rangari VK et al. Value added biopolymer nanocomposites from waste eggshell-based $CaCO_3$ nano particles as fillers. ACS Sustainable Chemistry & Engineering. 2014;**2**(4):706-717

[33] Utracki LA. Polymer blends, rapra review report, 11, Report 123. 2000

[34] Sonnier R, Rouif S, Taguet A. Modification of polymer blends by E-beam and gamma irradiation. 2012.

Available from: www.researcggate.net/publication/285296722

[35] Makuuchi K, Cheng S. Radiation Processing of Polymer Materials and its Industrial Applications. Nova York: John Wiley & Sons Inc; 2012. p. 1

[36] Chmielewski A, Haji-Saeid M. Radiation technologies: Past, present and future. Radiation Physics and Chemistry. 2004;**71**:17

[37] Zhao W, Pan X. Technology of Radiation Processing and its Applications. Beijing, China: Weapon Industry; 2003

[38] Telnov AV, Zavyalov NV, Khokhlov YA, Sitnikov NP, Smetanin ML, Tarantasov VP, et al. Radiation degradation of spent butyl rubbers. Radiation Physics and Chemistry. 2002;**63**:245

[39] Zhang Y. Polyolefin formulations for improved foaming: effect of molecular structure and material properties [unpublished doctoral dissertation]. Ontario, Canada: Queen's University Kingston; 2013

[40] de Vries DVWM. Characterization of polymeric foams. MT 09.22-0611747. 2009

[41] Borealis AG. Daploy™ High Melt Strength PP. 2004

[42] Gendron R, Mihai M. Extrusion foaming of polylactide. In: Polymeric Foams: Innovations in Processes, Technologies and Products. 2016. Available from: https://www.researchgate.net/publication/310646111

[43] Garlotta D. A literature review of Poly(lactic acid). Journal of Polymers and the Environment. 2002;**9**(2):63

[44] Sikorska W et al. Forensic engineering of advanced polymeric materials—Part V: Prediction studies of aliphatic—Aromatic copolyester and polylactide commercial blends in view of potential applications as compostable cosmetic packages. Polymers. 2017;**9**:257

[45] Cardoso ECL. Desenvolvimento de espumas parcialmente biodegradáveis a partir de blendas de PP/HMSPP com polímeros naturais e sintéticos [tese]. São Paulo: Instituto de Pesquisas Energéticas e Nucleares (IPen)/Universidade de São Paulo (USP); 2014

[46] Weng YX, Jin YJ, Meng QY, Wang L, Zhang M, Wang YZ. Biodegradation behavior of PBAT, PLA and their blend under soil conditions. Polymer Testing. 2013;**32**(5):918-926

[47] Xiuyu M, Yufeng W, Jianqing W, Yaning X. In: MATEC Web of Conferences; 88, 02009. 2017

Chapter 10

Gamma Radiation in the Vicinity of the Entrance to Linac Radiotherapy Room

Kinga Polaczek-Grelik, Aneta Kawa-Iwanicka, Marek Rygielski and Łukasz Michalecki

Abstract

Radiotherapy using high-energy photon beams (10–20 MV) is accompanied by the production of secondary neutron radiation via (γ/X,n) reactions. These interactions as well as subsequent neutron capture are the source of induced gamma radioactivity. When studied with standard range of spectrometric systems, only decay gamma radiation is usually registered, whereas a significant part of radiation—prompt gammas—is omitted, what might result in a significant underestimation of occupational risk for therapists in the vicinity of the door to the treatment room during therapeutic beam emission. Presented study has shown the main components of gamma radiation field in this localization investigated with the use of high-purity germanium spectrometry. Among them, prompt gamma radiation in light elements of concrete and in metal construction of the door, as well as 477.6 and 2224.6 keV photons emitted by neutron absorbing layers, contributes the most. Effective dose values depend on thickness of the door as well as on neutron production by particular linac and are within the range of 1.8–56.2 μSv/h. Standard environmental radiometry could underestimate these values by about 60% due to low efficiency for high-energy photon counting.

Keywords: induced radioactivity, HPGe spectrometry, prompt gamma radiation, photon radiotherapy, radiation protection, occupational hazard

1. Introduction

Radiotherapy is one of the commonly used anticancer treatment methods, either applied alone or (what is more widely used) in conjunction with surgery and/or chemotherapy, as induction, supplement, or sensitizing agent. Therefore, it is administered as neoadjuvant, concurrent, or sequential (adjuvant) therapy [1–3]. In the first case, irradiation of a solid tumor may cause its shrinkage, making the subsequent surgery less extensive. In the second case, it is about to kill the cancer cell clusters too small to be seen and removed by the surgeon, limiting the risk of local recurrence or lymph node metastasis. The last issue is to gain the success of systemic chemotherapy, even in reducing treatment toxicity [1–3].

Generally, the distinction according to the localization of medically used radiation source relative to the patient's body divides radiotherapy into teletherapy—externally located radiation source, and internally located either sealed (brachytherapy) or unsealed (nuclear medicine) radiation source in the form of radioactive nuclide. The use of external beams in radiotherapy (EBT) requires increased penetration of radiation to reach deeply located tumors spearing healthy tissues at the same time. Therefore, EBT generally employs higher energies of radiation, whereas brachytherapy benefits from limited range of ionizing radiation, which helps to spear health tissues neighboring the tumor region.

In the connection with the above, external beam radiotherapy is of higher concern in terms of radiological protection and safety work around ionizing radiation devices. Special attention is paid to the construction of the treatment room, what will be discussed later.

Every ionizing radiation type could be used for the purpose of radiotherapy and each of them has its advantages and disadvantages, making their availability and applicability common or restricted to a specific cases. Limited range of electrons in the tissue makes their use practical for shallow location of tumors, especially when healthy radiosensitive tissues are located close below/behind. On the other hand, forming uniform dose distribution with high target conformity is hardly to achieve using electron beams. Neutron beam production and guidance is a task difficult enough to limit the usage of fast neutron EBT, despite of lower oxygen enhancement ratio (OER—dependence of tumor cell sensitivity on its oxygenation) and entrance/surface dose in comparison with electron beams [4]. Photon beams in the form of Bremsstrahlung radiation produce similar shape of depth dose distribution as electron and neutron beams, showing regions of build-up, dose maximum, and quasiexponential decrease; however, skin spearing effect is more pronounced and radiation is more susceptible on shaping with the use of collimation/absorption blocks. The major advantage of using photons with energies of the order of MeV is the high penetration property. Therefore, using multiple directions of incident beam, it is possible to reach with therapeutic dose the tumor surrounded by healthy tissues from the outside. However, exit dose is not negligible and region of low doses due to scattered radiation is extended. In contrary, heavy charged particles give quite different dose distribution, follow the Bragg peak shape. It means that energy transfer in case of hadrons increases with decreasing their energy, just the opposite to neutral particles and electrons [4]. Moreover, high degree of tissue spearing is achieved at localizations beneath the tumor, since physical dose behind the peak tends to be negligible, what is of high beneficial when therapy concerns structures near radiation-sensitive organs. Among the biggest disadvantages, expensiveness of accelerator technology, unsure radiobiological and physical interactions of high-energy heavy charged particles in tissues are usually mentioned.

Despite the diverse advantages of every radiation type, radiotherapy with the use of electron and photon beams remains the most widespread and widely available technology. Modern techniques for accelerating electrons are both miniaturized and efficient, what enables to generate beams with a wide range of energies, from several keV to tens of MeV. However, the trend in radiotherapy is to replace electron beams with photon (X-ray) beams when employing dynamic, intensity, and volumetric modulated delivery of therapeutic dose distribution. Electrons are more easily controlled and accelerated in short sections than heavy charged particles and as such, linear electron accelerators for medical purpose are the most common once. Even nowadays, the development of devices producing X-ray beams for radiotherapeutic use is taking place, although rather in terms of increasing the number of degree of freedom in beam delivery than in terms of new method of beam producing and controlling. This development has led X-ray machines in radiotherapy from orthovoltage Roentgen machines through classical linacs to devices such as Tomotherapy® or CyberKnife®. Among them, only linear accelerators in their classical form are designed to produce X-rays at accelerating potentials from a quite wide range (4–25 MV). This feature enables to clinically use highly penetrating X-ray beams with mean energies of 1–6 MeV with long tail of high-energy photons up to the end-point values of 4–25 MeV, respectively. Extensive comparison of linac beam spectra could be found in [5–10].

From physics point of view, variety of interactions of therapeutic beams with accelerator elements, room equipment, air and human body should be taken into account for two general reasons:

- precise planning of dose distribution in patient's body requires advanced dosimetric modeling of treatment unit and taking into account majority of ionizing radiation interaction types in terms of dose deposition, scattering, and density corrections [4];

- there is usually a need to limit the ionizing radiation field for radiation protection purposes, which should take into account secondary radiation generated by highly penetrative ionizing radiation.

The first one is crucial for radiotherapy beneficial outcome, whereas the second one is important as radiation safety issue. This second aspect is subjected in the present work studied with the use of gamma radiation spectrometry.

The most frequently used radiotherapy machine set in oncological center includes medical electron linear accelerator (linac), brachytherapy unit (usually high dose rate (HDR) type), and CT scanner with virtual simulation option for therapy positioning purposes. Usually, several linacs are installed to secure non-disturbance of radiotherapy process. From a radiation protection point of view, shielding vault should be designed with respect to the most penetrative radiation type among those used in particular room.

Several organizations have been releasing recommendations regarding shielding design and radiation safety issues since 1970s:

- National Council on Radiation Protection and Measurements (NCRP) [11–14];

- American Association of Physicists in Medicine (AAPM) [15, 16];

- International Atomic Energy Agency (IAEA) [17];

- Institute of Physics and Engineering in Medicine (IPEM) [18, 19];

- International Organization of Standards (ISO) [20];

- International Electrotechnical Commission (IEC) [21, 22].

German Industrial Norms (DIN) [23–26] also serve as practical reference worldwide. The relevant reports are constantly updating to include most recent development of radiotherapy machines as well as treatment techniques, e.g., the usage of photon energies above 10 MV, dual or even triple photon energy machines, intensity modulated radiation therapy (IMRT), stereotactic body radiation therapy (SBRT), or total body irradiation (TBI), which in comparison with static simple geometry fields require more radiation in terms of linac monitor units (MUs) to be emitted to deposit therapeutic dose in planning target volume (PTV) or usage of nonstandard field dimensions. New modalities in photon radiation therapy include also more complicated irradiation geometry, i.e., more incident beam directions and thus, more complicated scattered radiation patterns (see: CyberKnife, Gamma Knife, or Tomotherapy).

Shielding considerations of medical linear accelerator room distinguish the following radiation groups, schematically presented in **Figure 1a**:

- primary beam—therapeutically useful radiation always directed to the linac isocenter (point of gantry rotation), what means when using rotational technique, this beam might incident on four out of six walls of treatment room;

- leakage radiation—ionizing radiation of the treatment beam type leaving the linac head through unattended ways—directed from the source (e.g., Bremsstrahlung conversion target) to outside; at the linac construction stage, it should be limited to a maximum of 0.2% and an average of 0.1% of the maximum absorbed dose [22] in $10 \times 10\ cm^2$ radiation field by, e.g., using beam stoppers or lead shielding;

- scattered radiation originating from the patient body, room walls, and equipment, directed in full solid angle, with wide range of energies.

The characteristic design of the therapeutic room, which is schematically presented in **Figure 1b**, meets the protective requirements against all of the above-mentioned radiation groups and contains:

- primary barrier, the thickest or made of concrete enriched with heavy material, e.g., barite;

- secondary barrier against leakage and scattered radiation, which is usually of lower intensity and/or energy than primary beam; therefore, the barrier is thinner than the primary barrier;

- maze, as a construction protecting the entrance from direct incidence of unattenuated beam, with the purpose of lengthening the scattering radiation path.

The use of high-energy radiotherapeutic beams (E > 10 MeV) is additionally accompanied by the aspect of generation of secondary radiation, which will be widely addressed below.

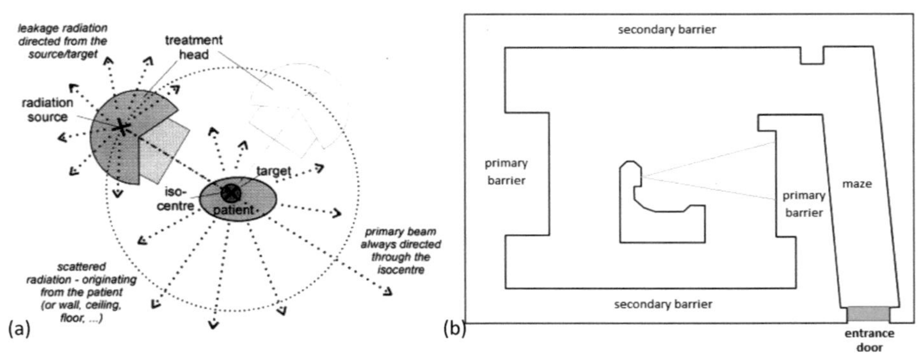

Figure 1.
(a) Schematic presentation of different radiation groups the linac room shielding must face with, available from [27]; (b) scheme of typical radiotherapy treatment vault. Medical accelerator head is able to rotationally move on 360° around the axis perpendicular to the beam (gray-line cone) central axis. The entrance door as the localization relevant for the presented study is also marked.

Radiotherapeutic photon beams in the form of Bremsstrahlung radiation generated on conversion target by electrons accelerated at potentials of the order of MV have wide-energy spectrum up to energy determined by the nominal accelerating potential used. Therefore, in high-energy therapeutic beam (10–20 MV) even up to about 20% of photons could have energies above 8 MeV, which is the approximate threshold energy for photonuclear reactions. However, for tungsten (the main component of linac collimation system)—nuclear photo effect starts approximately from 5 MeV. Generally, the higher mass number of nuclide, the lower threshold energy, and for defined energy—the higher cross section for photon absorption by the nucleus is observed, what is presented in **Figure 2**. Among the products of such reactions are secondary gamma rays (when inelastic scattering occurs) or nucleons (protons, neutrons, or their groups: deuterons, alphas), when a reaction through a stage of compound nucleus has occurred. The final nucleus either nucleon-deficient (photonuclear reaction product) or nucleon-excess (neutron absorption product) could be unstable and undergo radioactive decay. From occupational radiation protection point of view, high-penetrative radiation, i.e., prompt and decay gamma rays as well as neutrons are of importance, since (1) they are able to reach entrance to the treatment room, (2) they form a significant part of radiation leakage from the treatment room through the door.

The energy range of photons and electrons used in linac radiotherapy is sufficient to trigger a nuclear reaction via (γ/X,n), (γ/X,p), and (e,e'n) mechanisms and to observe subsequent nuclear reactions of secondary generated particles, among

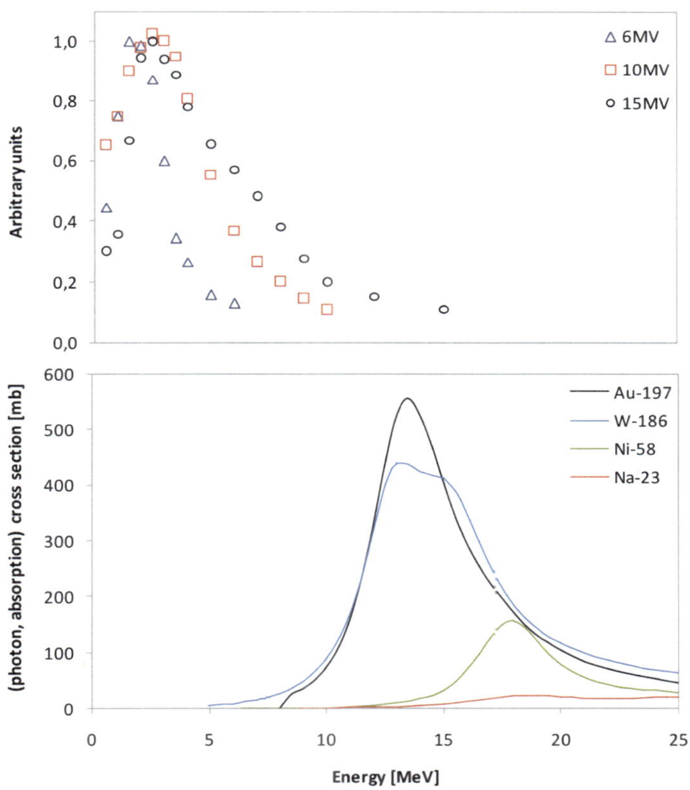

Figure 2.
The comparison of photon beam spectra (modeled by us in commercial dose verification system) with cross sections of photon absorption nuclear reactions [28] for commonly observed activation target nuclides, in terms of energy overlapping regions.

which (n,n'γ) and (n,γ) are the most commonly observed, for fast and thermalized neutrons, respectively. Every mechanism mentioned above could activate radionuclides. Nevertheless, the majority of induced radioactivity is found in construction materials of the accelerator head, mostly in heavy elements of collimation and beam shaping system. The contribution of particular elements of linac head in overall induced radioactivity is studied mostly with Monte Carlo simulations, as in: [29, 30]. However, gamma radiation spectrometry is a good tool for identification of particular radionuclides and their contribution in this phenomenon, for example, see: [31]. The apparent linac radioactivity depends on the localization of measuring point; therefore, the radiation hazard due to this phenomenon is different for patients and for the staff, with the dominant contribution of tungsten collimator or head casing, respectively [32]. Induced radioactivity has been also observed and investigated in tissues [33–36], air [37], treatment couch [38], and treatment accessories stored inside the linac room [39]. Moreover, the dependence of induced activity on the therapeutic dose rate could be observed in some cases, i.e., when half-life of radioisotope is comparable with the time of beam emission, and is more pronounced for higher nominal accelerating potentials [35].

Among the mechanisms of radionuclide activation outside the field of irradiation, neutron capture contributes the most. Linacs used nowadays are not routinely equipped with shielding constructions dedicated for neutrons; therefore, neutron fluence all over the treatment room is reported [32, 40–44] in the amount sufficient for inducing radioactivity at measurable level. Therefore, medical linear accelerators are often characterized in terms of neutron source strength Q [14, 44], which depends on beam nominal potential, as presented in **Figure 3**.

The spectrum of neutron flux undergoes changes via scattering mechanisms. Leaving the linac head, the mean energy of neutrons is of the order of 1 MeV, on treatment couch, an additional peak at thermal energies is already observed and neutrons impinging the door have an average energy of ~0.2 MeV [45, 46].

Neutron radiation weighting factor for effective dose calculation strongly depends on energy, having maximal values around 1 MeV [47]. The cross section of (n,γ) nuclear reaction follows the 1/E dependence with some resonance peaks at intermediate energies [28]. Therefore, high-energy neutrons contribute mostly to the dose, whereas slow neutrons to the phenomenon of induced radioactivity.

It is of high importance to be aware of the physical mechanisms of radiation absorption and removal from the beam. These are in principle different for various radiation types. Nevertheless, similar mechanisms might be observed for various radiation types but occurring with different efficiency.

The readily used in diagnostic radiology heavy metal shielding is no longer valid in high-energy radiotherapy rooms due to the generation of secondary

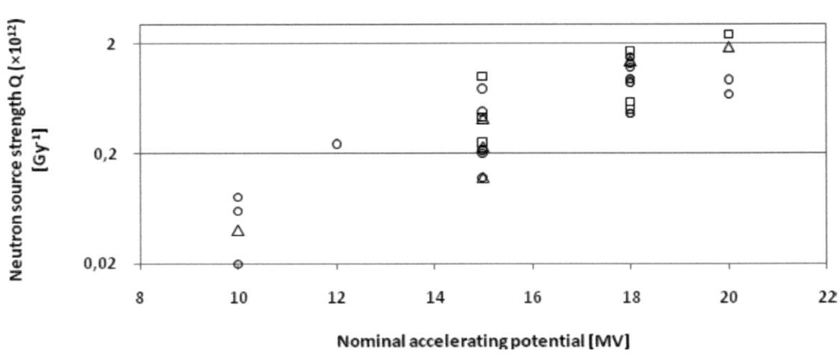

Figure 3.
The comparison of neutron source strength values reported in [14] (○), [44] (□), and obtained by us (Δ).

radiation via nuclear reactions, as discussed above. Therefore, concrete, barite-concrete, earth bricks, and similar materials are preferred during solid shielding construction. These, built of mostly light elements, for which neutron production threshold energies are relatively high, i.e., tens of MeV, should not gain the production of secondary radiation. Additional lamination of maze walls as well as using multibend geometries are the solutions advised for increasing the neutron absorption before reaching the entrance. These solutions help to slim down the room door or even built the door-less entrance, minimizing secondary radiation at the entrance [48]. Typical door construction contains the most inner layer of neutron-absorption material (polyethylene, paraffin, or borax), enclosed with heavy photon-absorption layer (lead, tungsten) coated with industrial material, typically of stainless steel or wood. Unfortunately, to maintain an acceptable mechanics/kinetics of the door, the weakness of this radiation barrier must be accepted. Therefore, from radiation protection point of view, the vicinity of entrance to the treatment room is not an advised place for staying during radiotherapy beam emission as a location with increased occupational radiation hazard. The standard radiometric methods used in such case could seriously underestimate the radiation indications since they are calibrated on ^{60}Co or ^{137}Cs sources. Although average energy of leakage/scattered radiation reaching the entrance door is close to the energies of these radionuclide sources, prompt gamma rays produced during neutron capture are much more energetic (over a dozen of MeV), therefore, detected with very low efficiency by these devices. Moreover, standard spectrometric range of detected energies is aimed at measuring decay gamma rays up to about 3 MeV and therefore omits significant range of prompt gammas. That is the reason which makes gamma spectrometry with extended energy range to be adequate for more precise investigation of the occupational radiation hazard near the entrance door to the high-energy medical linac room.

2. Semiconductor spectrometry and its application for radiation characterization

Semiconductor high-purity germanium (HPGe) detectors are most suitable for the investigation of gamma radiation spectra of unknown origin since the excellent energy resolution enables the exact identification of any radionuclide, which contributes to the radiation field in measured localization. Their use is, however, limited due to the need of liquid nitrogen cooling; therefore, scintillation or room-temperature semiconductor detectors, both with limited resolution, are used instead.

The spectrometric system used in this study, as shown in **Figure 4**, consists of:

- coaxial HPGe detector with reversed electrodes (ReGe), manufactured by Canberra Inc., having 40% relative efficiency and characterized by the resolution of 2.1 keV FWHM @ 1332 keV; the use of a standard spectrometric gain of 5.0 enables for spectra registration up to 3.2 MeV;

- InSpector™ 2000 MultiChannel Analyzer (MCA) with 8194 channels;

- Genie™ 2000 v.3.2.1 Gamma Acquisition and Analysis Software (Canberra Inc.).

The carbon-composite entrance window enables the registration of low-energy photons (above 7 keV). The end-point energy of measured spectra has been set

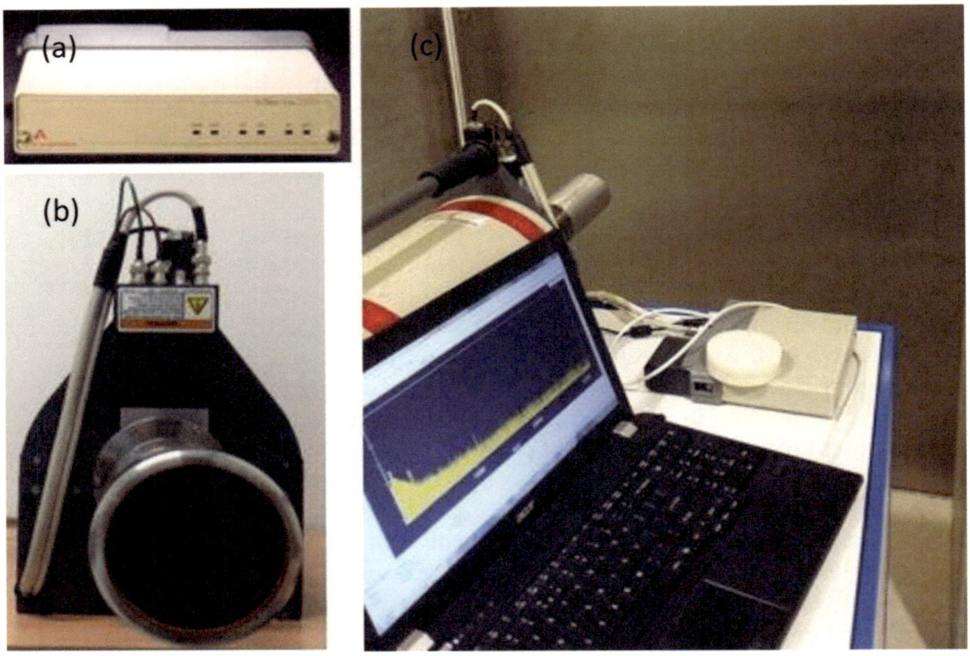

Figure 4.
ReGe gamma spectrometer used in present study: (a) multichannel analyzer, (b) front view of high-purity germanium detector with carbon-composite entrance window, and (c) measurement configuration 50 cm from the entrance door to the linac radiotherapy room.

by adjusting the gain value of MCA. This enables for spectra registration up to the energy of 3–10 MeV.

Energy calibration of spectrometric system has been performed for standard MCA gain (5.0) with the use of radioisotope sealed sources (1 cm Ø) having activity of the order of 10 kBq. Subsequently, it was checked that the calibration scales linearly inversely with the spectrometric gain.

Spectrometric efficiency has been modeled in In Situ Object Counting Software (ISOCS™, Canberra Inc.), applying full factory characterization of a given detector performed with the use of NIST-traceable sources and MCNP Monte Carlo modeling code, supplied by the manufacturer. The geometry of rectangular complex plane for calculating the detection efficiency has been chosen from the ISOCS predefined templates as best matching to the experimental conditions, giving the possibility to include the multilayer design of the entrance door. The energy-dependent photon detection efficiency (ε) of spectrometric system has been finally described with the following function:

$\ln(\varepsilon) = a + b\cdot\ln(E) + c\cdot\ln(E)^2 + d\cdot\ln(E)^3 + e\cdot\ln(E)^4 + f\cdot\ln(E)^5$, with the fitting parameters of a–f.

The analysis of registered spectra (photopeaks' identification and net areas counting) has been performed using unidentified second differential and nonlinear LSQ fit in Genie™ 2000 software. The sources of gamma radiation (activated nuclides) have been identified on the base of photopeaks' energies, whereas areas under these photopeaks were used for photon flux density (Φ) assessment on the basis of Eq. (1), for a defined detector front surface (S_{Ge}) and life time (LT) of each measurement.

$$\Phi(E)\ \left[cm^{-2}s^{-1}\right] = \frac{Peak\_net\_area(E)}{\varepsilon(E)\cdot S_{Ge}\left[cm^2\right]\cdot LT[s]} \qquad (1)$$

The recommended quantity for the purpose of limitation of ionizing radiation exposure is the effective dose (E_d). However, radiation protection and operational quantities are distinguished according to the relevance to radiation health effects and possibility to be measured, respectively. These are presented in reports 26, 60, and 103 of International Commission on Radiological Protection (ICRP) and in reports 66 and 85 of International Commission on Radiation Units and Measurements (ICRU) [47, 49–52]. Moreover, conversion coefficients of radiation fluence to organ absorbed doses as well as equivalent doses and effective dose are also supplied in ICRP 74 and 116 as well as in ICRU 57 reports [53–55]. This enables the risk estimation on the basis of different measuring quantities.

3. Results and discussion

Gamma radiation spectrometry system was placed 50 cm in front of linac treatment room entrance door, at the height of 1 m above the floor (see **Figure 4**), at a localization representative for dose rate assessment for the staff waiting for the end of patient irradiation.

The spectra were registered during emission of 6–18 MV photon beams with a dose rate of 450 MU/min. The gantry angle of 90 or 270° and irradiation field size of 40 × 40 cm^2 were set to achieve maximal intensity of radiation reaching the entrance door, to study the worst scenario of occupational hazard.

3.1 Comparison of spectra for various beam energies

The detailed characteristics of gamma ray spectra acquired in standard and extended energy range near the linac room door are presented in **Table 1**. The energies of photons in 6 MV beam are too low to trigger nuclear reactions; therefore, induced radionuclides are not observed on the spectrum outside the door in this case, as shown in **Figure 5**. Nevertheless, the increase in low-energy continuous part of the registered spectrum in comparison with natural background radiation indicates that part of scattered radiation from therapeutic beam penetrates the door.

Spectra registered during high-energy beam emission (10–18 MV), as shown in **Figure 5**, are dominated by two processes: positron creation, since annihilation peak at 511 keV is clearly visible, and neutron capture in hydrogen-rich material inside the door, due to the presence of a peak at 2224.6 keV, which is the neutron-binding energy in deuterium nucleus. The intensity of these processes could be correlated not only with neutron source strength of particular linac working at defined accelerating potential, but also with the amount of hydrogen-rich material used in door construction. The peak at 477.6 keV is due to the presence of boron (mostly in the form of borax–sodium tetraborate decahydrate) and is a consequence of $^{10}B(n,\alpha)^{7}Li$ reaction, where 477.6 keV is the deexcitation energy of lithium nucleus. The broadening of this peak has Doppler effect—origin, widely discussed in [56]. Since door construction is not unified and the usage of paraffin/polyethylene (as hydrogen-rich materials) or borax as neutron absorption agents depends on the construction concept, the intensity of the 477.6 keV line should not be directly connected with the therapeutic beam energy (see **Figure 5**: 15 MV vs. 18 MV cases) or even may not occur at all, whereas hydrogen capture of neutron is present for all linacs studied by us.

The gamma ray spectra are dominated by the abovementioned interactions; however, the minor contributions come from:

- (n,n'γ) and (n,γ) interactions in germanium crystal of HPGe spectrometer, which proves that neutrons contribute to the door-leakage radiation outside the treatment room;

Source	Origin	Photon energy [keV]
Germanium detector	$^{72}Ge(n,\gamma)^{73m}Ge$	66.7
	$^{74}Ge(n,\gamma)^{75m}Ge$	139.7
	$^{70}Ge(n,\gamma)^{71m}Ge$	174.9; 198.4; 499.9
	$^{73}Ge(n,\gamma)^{74}Ge^{*}$	595.9
	$^{74}Ge(n,n'\gamma)$	595.9; 608.3; 867.9; 1204.2
	$^{72}Ge(n,n'\gamma)$	689.6
Metal elements of door construction	$^{56}Fe(n,\gamma)$, $^{54}Fe(n,\gamma)$	352.4; 7631.1; 7645.5
	$^{56}Fe(n,\gamma)$	570.0; 810.6; 920.5; 1019.0; 1358.6; 1612.8; 1358.6; 1725.3; 1972.3; 2129.2; 2425.7; 2469.2; 2526.5; 2721.3; 3185.2; 3436.6; 3854.3; 4218.3; 4406.1; 5357.4; 5920.4; 6018.5; 7278.8; 7631.1; 7645.5; 9297.7
	$^{27}Al(n,\gamma)^{28}Al$	1779.0; 3033.9
	$^{12}C(n,\gamma)$	1261.8; 4945.3
Shielding elements of door construction	$^{1}H(n,\gamma)$	2224.6
	$^{10}B(n,\alpha)^{7}Li$	477.6
	$^{207}Pb(n,\gamma)$	7367.8
Concrete	$^{39}K(n,\gamma)$	770.3;
	$^{35}Cl(n,\gamma)$	516.7; 575.8; 1164.9; 1951.1; 1959.3; 2863.8; 2876.9; 3061.8; 3195.4; 4082.8; 4298.6; 4979.9; 5204.5; 6110.8; 6619.6; 6627.8; 7414.0; 7790.3
	$^{40}Ca(n,\gamma)$	707.7; 1942.7; 2001.3; 3610.2; 4418.5; 6419.6
	$^{23}Na(n,\gamma)$	834.7; 2517.8
	$^{28}Si(n,\gamma)$	1273.3; 2092.9; 3101.8; 3539.0; 3723.1; 4933.9; 6379.8
	$^{24}Mg(n,\gamma)$	3413.1; 3916.8
	$^{31}P(n,\gamma)$	3899.9

Table 1.
Characteristics of gamma ray spectra registered in the energy range of 10 keV–10 MeV near the door to linac therapy room during emission of 10–18 MV photon beams.

- (n,γ) reactions in concrete elements: ^{23}Na, ^{24}Mg, ^{28}Si, ^{31}P, ^{35}Cl, ^{39}K, ^{40}Ca;
- (n,γ) reactions in metals: ^{27}Al and ^{56}Fe.

Neutron capture nuclear reaction is accompanied by prompt gamma rays but also decay gamma radiation might be observed, when originated nucleus is radioactive. The first mentioned radiation type is observed only during emission of high-energy therapeutic beam, but observed energies are mostly above 2 MeV, whereas the second group of gammas has energies up to about 2 MeV and contributes to the increased background after the end of therapeutic beam emission, with a characteristic half-life.

Spectra registered in extended energy range prove that neutron capture process on light elements (mainly concrete) occurs intensively. The main differences between linac room shielding properties from occupational hazard point of view (for defined therapeutic beam) are due to the diverse construction of the door and specific material used, for example: borated polyethylene or paraffin alone will result in the presence of 477.6 keV line or not, which will affect the intensity of

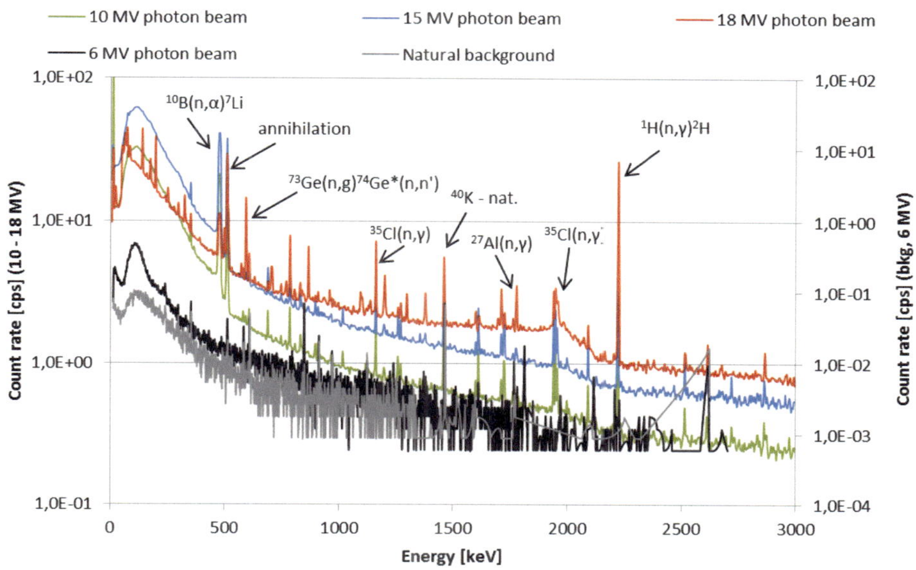

Figure 5.
The comparison of spectra registered behind the treatment room door during emission of linac photon beams. Natural background radiation is presented for reference. Only the most intense lines (mentioned in the text above) are marked for clarity of presentation. Detailed analysis is presented in **Table 1**.

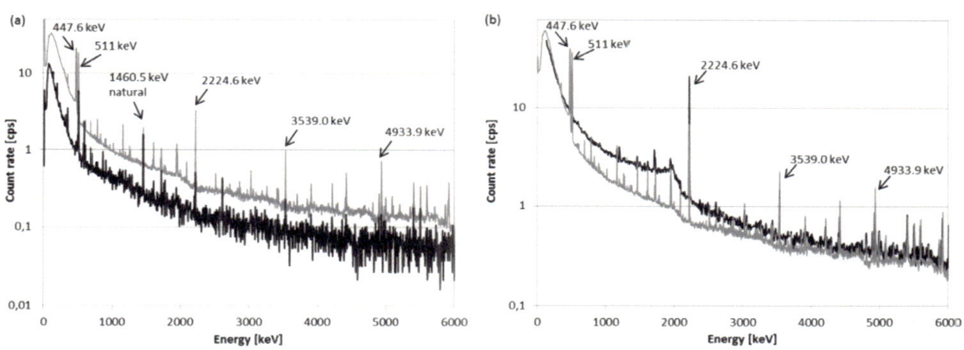

Figure 6.
Comparison of spectra registered for (a) 10 MV photon beams and (b) 15 MV photon beams, when thick (black line) and thin (gray line) door constructions are used. (See text above for the explanation of differences.)

2224.6 keV line. The comparison of spectra presented in **Figure 6** demonstrates the differences between thick and thin door constructions designed for (a) higher beam energy than currently used and (b) maximum beam currently in use.

3.2 Spectrometry—Based dose assessment

In the present study, the effective dose in front of linac room entrance door has been estimated on the base of photon flux density obtained according to the Eq. (1) and using conversion coefficients for AP geometry given in ICRP report 116 [54]. Specific values of these coefficients for energies registered on gamma ray spectra were calculated using Lagrange interpolation formula of third degree. The average total uncertainty of calculated effective dose values of 25% includes the accuracy of conversion coefficients as well as the uncertainty of photopeaks' area determination at the registered spectra.

Effective doses in studied location depend on the neutron source strength Q of particular linac as well as on the construction of the treatment room door. For 18-MV photon beam, more important is the first factor (linac construction), ranging the doses from 30.6 ± 7.7 to 56.2 ± 14.1 μSv/h. The second factor plays the crucial role for 10 MV photon beams, for which neutron generation is of the lowest intensity, ranging the doses from 1.8 ± 0.4 μSv/h (for flattening filter-free (FFF) beam), through 3.4 ± 0.8 μSv/h (for thick door construction) up to 10.5 ± 2.6 μSv/h (for thin door construction). However, also in this case, neutron production intensity in linac head plays significant role, what is concluded from the differences between FFF beam and conventional linac, since flattening filter takes part in neutron production [29, 30]. Effective dose rates measured during 15-MV beam emission using Geiger-Mueller radiometer (calibrated on ^{60}Co source) and the result obtained using spectrometry analysis presented here are 13.5 ± 3.0 μSv/h and 22.2 ± 5.5 μSv/h, respectively. This comparison shows that even 60% of dose could be omitted in the first case when excluding high-energy component of radiation leakage through the door due to prompt gamma rays accompanying the neutron capture process.

Production of neutron secondary radiation during emission of high-energy photon therapeutic beams is generally known and widely studied issue [29–46, 48]. Also, the phenomenon of high-energy X-rays and secondary neutron-induced radioactivity is well recognized [57]. However, the impact of photon radiation connected with neutron interaction in treatment room shielding materials on occupational safety is still difficult to assess experimentally in clinical conditions due to limited availability of high-resolution extended-energy range spectrometry systems, which often require special operating conditions (e.g., nitrogen cooling) and time-/labor-consuming data analysis. Nevertheless, recommendations concerning design of linac rooms [17] refer to publications devoted to this issue [58]. The use of a spectrometer (the usefulness of which has been demonstrated in presented study) is advised by IAEA [59] as a supplementary method for workplace monitoring, and its usage to characterize the energy spectrum of a given radiation type is recommended to support the performance of routinely used monitoring instruments.

4. Conclusion

The qualitative analysis performed by us has shown that the major component of gamma radiation field near the treatment room door comes from prompt photons emitted during neutron capture reaction and is common in door construction as well as in concrete materials. Comprehensive study of this issue requires extended energy range of spectrometric system, as demonstrated in presented investigations. High-energy gamma rays above 3 MeV (omitted in standard spectrometric measurements) contribute to the effective dose values from 26 to 58%, for low (10 MV FFF beam) and for high (18 MV beam) neutron source strength linacs, respectively.

Reactions intended for neutron capture in door construction: $^{10}B(n,\alpha)^{7}Li$ and $^{1}H(n,\gamma)^{2}H$ contribute to the effective dose of 0–17% and 4–19%, respectively. Borated inner layer of the door is not always used, whereas hydrogen-rich material is the commonly used neutron absorber.

Presented study proves the correctness of radiation protection guidelines to avoid the vicinity of treatment door during therapeutic beam emission and additionally provides the justification in terms of dose values and mechanisms of gamma ray production.

Acknowledgements

This work was possible due to one of the authors (KPG) involvement in scientific activity in University of Silesia in Katowice, Poland. Therefore, the authors express their gratitude for the opportunity to use in situ gamma spectrometric system.

Conflict of interest

The authors certify that they have no affiliations with or involvement in any organization or entity with any interest in the subject matter or materials discussed in this manuscript.

Author details

Kinga Polaczek-Grelik[1]*, Aneta Kawa-Iwanicka[1], Marek Rygielski[2] and Łukasz Michalecki[1]

1 Nu-Med Cancer Diagnosis and Treatment Centre, Katowice, Poland

2 Nu-Med Cancer Diagnosis and Treatment Centre, Tomaszów Mazowiecki, Poland

*Address all correspondence to: kinga.polaczek-grelik@nu-med.pl

References

[1] Glatzer M, Elicin O, Ramella S, Nestle U, Putora PM. Radio(chemo) therapy in locally advanced non small cell lung cancer. European Respiratory Review. 2016;**25**:65-70. DOI: 10.1183/16000617.0053-2015

[2] Joiner M, van der Kogel A, editors. Basic Clinical Radiobiology. 4th ed. London: Hodder Arnold; 2009. p. 391

[3] Sas-Korczynska B, Sokolowski A, Korzeniowski S. The influence of time of radio-chemotherapy and other therapeutic factors on treatment results in patients with limited disease small cell lung cancer. Lung Cancer. 2013;**79**:14-19. DOI: 10.1016/j.lungcan.2012.10.004

[4] Podgorsak EB. Radiation Physics for Medical Physicists. 1st ed. Berlin Heidelberg: Springer-Verlag; 2006. p. 437. DOI: 10.1007/978-3-642-008745-7

[5] Desobry GE, Boyer AL. Bremsstrahlung review: An analysis of the Schiff spectrum. Medical Physics. 1991;**18**(3):497-505. DOI: 10.1118/1.596653

[6] Rogers DWO, Faddegon BA, Ding GX, Ma C-M, BEAM WJ. A Monte Carlo code to simulate radiotherapy treatment unit. Medical Physics. 1995;**22**(5):503-524. DOI: 0094-2405/95/22(5)/503/22

[7] Sheikh-Bagheri D, Rogers DWO. Sensitivity of megavoltage photon beam Monte Carlo simulations to electron beam and other parameters. Medical Physics. 2002;**29**(3):379-390. DOI: 10.1118/1.1445413

[8] Sheikh-Bagheri D, Rogers DWO. Monte Carlo calculation of nine megavoltage photon beam spectra using the BEAM code. Medical Physics. 2002;**29**(3):391-402. DOI: 10.1118/1.1446109

[9] Ali ESM, Rogers DWO. An improved physics—Based approach for unfolding megavoltage bremsstrahlung spectra using transmission analysis. Medical Physics. 2012;**39**(3):1663-1675. DOI: 10.1118/1.3687164

[10] Ali ESM, Rogers DWO. Functional forms for photon spectra of clinical linacs. Physics in Medicine and Biology. 2012;**57**(1):31-50. DOI: 10.1088/0031-9155/57/1/31

[11] NCRP Report 49. Structural Shielding Design and Evaluation for Medical Use of X Rays and Gamma Rays of Energies Up to 10 MeV. 1976

[12] NCRP Report 147. Structural Shielding Design for Medical X-Ray Imaging Facilities. 2004

[13] NCRP Report 144. Radiation Protection for Particle Accelerator Facilities. 2003

[14] NCRP Report 151. Structural Shielding Design and Evaluation for Megavoltage X and Gamma-Ray Radiotherapy Facilities. 2005

[15] AAPM 19. Neutron Measurements Around High Energy X-Ray Radiotherapy Machines. 1986. p. 36

[16] AAPM 108. PET and PET/CT Shielding Requirements. 2003. p. 12. DOI: 10.1118/1.2135911

[17] IAEA Safety Reports Series, No 47. Radiation Protection in the Design of Radiotherapy Facilities. 2006. p. 129

[18] IPEM Report 75. Design of Radiotherapy Treatment Room Facilities. 1997. p. 304. DOI: 10.1088/978-0-7503-1440-4

[19] IPEM Reports. 2018. Available from: https://www.ipem.ac.uk/ScientificJournalsPublications/

IPEMReportSeries/AvailablePublications.aspx. [Accessed: 04-11-2018]

[20] ISO 16645. Radiological Protection—Medical Electron Accelerators—Requirements and Recommendations for Shielding Design and Evaluation. 2016. p. 84

[21] IEC 60976. Medical Electrical Equipment—Medical Electron Accelerators—Functional Performance Characteristics. 2007. p. 172

[22] IEC 60601-2-1. Medical Electrical Equipment—Part 2-1: Particular Requirements for the Basic Safety and Essential Performance of Electron Accelerators in the Range 1 MeV to 50 MeV. 2014. p. 134

[23] DIN EN 60601-2-1. Particular Requirements for the Basic Safety and Essential Performance of Electron Accelerators in the Range 1 MeV to 50 MeV. 2015

[24] DIN 6812:2013-06. Medical X-Ray Equipment up to 300 kV—Rules of Construction for Structural Radiation Protection. 2013. p. 43

[25] DIN 6847-2:2014-03. Medical Electron Accelerators—Part 2: Rules for Construction of Structural Radiation Protection. 2014. p. 48

[26] DIN 6875-20:2016-08. Special Radiotherapy Equipments—Part 20: Proton Therapy—Rules for Construction of Structural Radiation Protection. 2016

[27] IAEA Training Material on Radiation Protection in Radiotherapy [Internet]. 2018. Available from: https://www.iaea.org/resources/rpop [Accessed: 04-11-2018]

[28] National Nuclear Data Center. Sigma Evaluated Nuclear Data File (ENDF) Retrieval & Plotting [Internet]. Available from: https://www.nndc.bnl.gov/sigma/search.jsp [Accessed: 09-11-2018]

[29] Pena J, Franco L, Gomez F, Iglesias A, Pardo J, Pombar M. Monte Carlo study of siemens primus photoneutron production. Physics in Medicine and Biology. 2005;**50**:5921-5933. DOI: 10.1088/0031-9155/50/24/011

[30] Becker J, Brunckhorst E, Schmidt R. Photoneutron production of a Siemens primus linear accelerator study by Monte Carlo methods and a paired magnesium and boron coated magnesium ionization chamber system. Physics in Medicine and Biology. 2007;**52**:6375-6387. DOI: 10.1088/0031-9155/52/21/002

[31] Polaczek-Grelik K, Karaczyn B, Konefal A. Nuclear reactions in linear medical accelerators and their exposure consequences. Applied Radiation and Isotopes. 2012;**70**:2332-2339. DOI: 10.1016/j.apradiso.2012.06.021

[32] Janiszewska M, Polaczek-Grelik K, Raczkowski M, Szafron B, Konefal A, Zipper W. Secondary radiation dose during high-energy total body irradiation. Strahlentherapie und Onkologie. 2014;**190**(5):459-466. DOI: 10.1007/s00066-014-0635-z

[33] Gudowska I, Brahme A, Andreo P, Gudowski W, Kierkegaard J. Calculation of absorbed dose and biological effectiveness from photonuclear reactions in a bremsstrahlung beam of end point 50 MeV. Physics in Medicine and Biology. 1999;**44**(9):2099-2125

[34] Allen PD, Chaudhri MA. Charged photopatricle production in tissue during radiotherapy. Medical Physics. 1997;**24**(6):837-839

[35] Polaczek-Grelik K, Orlef A, Dybek M, Konefal A, Zipper W. Linear accelerator therapeutic dose—Induced radioactivity dependence. Applied Radiation and Isotopes.

2010;**68**(4-5):763-766. DOI: 10.1016/j.apradiso.2009.09.051

[36] Polaczek-Grelik K, Konefal A, Orlef A, Zipper W. Radioactivity induced in bones during radiotherapy treatment with the use of 20 MV accelerator beam. Polish Journal of Environmental Studies. 2006;**15**(1A):195-197

[37] McGinley PH. Air activation produced by high-energy medical accelerators. Medical Physics. 1983;**10**(6):796-800. DOI: 10.1118/1.595358

[38] Rawlinson JA, Islam MK, Galbraith DM. Dose to radiation therapists from activation at high-energy accelerators used for conventional and intensity-modulated radiation therapy. Medical Physics. 2002;**29**:598-608. DOI: 10.1118/1.1463063

[39] Konefal A, Orlef A, Dybek M, Maniakowski Z, Polaczek-Grelik K, Zipper W. Correlation between radioactivity induced inside the treatment room and the undesirable thermal and resonance neutron radiation produced by linacs. Physica Medica. 2008;**24**:212-218. DOI: 10.1016/j.ejmp.2008.01.014

[40] Polaczek-Grelik K, Karaczyn B, Grządziel M, Pieńkos M, Konefal A, Zipper W. The map of thermal and resonance neutron distribution inside the treatment room for accelerator therapy. In: Drzazga Z, Slosarek K, editors. Some aspects of medical physics – *In Vivo* and *In Vitro* Studies – Series of Monograph Polish Journal of Environmental Studies. Olsztyn, Poland: HARD; 2010. pp. 139-145

[41] Vega-Carrillo HR, Manzanares-Acuna E, Iniguez MP, Gallego E, Lorente A. Study of room-returned neutrons. Radiation Measurements. 2007;**42**:413-419. DOI: 10.1016/j.radmeas.2007.01.036

[42] Konefal A, Orlef A, Laciak M, Ciba A, Szewczuk M. Thermal and resonance neutrons generated by various electron and X ray therapeutic beams from medical linacs installed in Polish oncological centers. Reports of Practical Oncology and Radiotherapy. 2012;**17**(6):339-346. DOI: 10.1016/j.rpor.2012.06.004

[43] Martinez Ovalle SM. Neutron dose equivalent in tissue due to linacs of clinical use. In: Kataria T, editor. Frontiers in Radiation Oncology. Rijeka, Croatia: InTech; 2013. pp. 91-112. DOI: 10.5772/3065

[44] Howell RM, Kry SF, Burgett E, Hertel NE, Followill DS. Secondary neutron spectra from modern Varian, Siemens and Elekta linacs with multileaf collimators. Medical Physics. 2009;**36**(9):4027-4038. DOI: 10.1118/1.3159300

[45] Banuelos-Frias A, Borja-Hernandez CG, Guzman-Garcia KA, Valero-Luna C, Hernandez-Davila VM, Vega-Carrillo HR. Neutron spectra and H*(10) of photoneutrons inside the vault room of an 18 MV linac. Revista Mexicana de Fisica. 2012;**58**:192-194

[46] Kralik M, Turek K. Characterization of neutron fields around high-energy X-ray radiotherapy machines. Radiation Protection Dosimetry. 2004;**110**:503-507. DOI: 10.1093/rpd/nch274

[47] ICRP Publication 103. The 2007 recommendations of the international commission on radiological protection. Annals of the ICRP. 2007;**37**(2-4):334

[48] Mesbahi A, Ghisai H, Mahdavi SR. Photoneutron and capture gamma dose equivalent for different room and maze layouts in radiation therapy. Radiation Protection Dosimetry. 2010;**140**(3):242-249. DOI: 10.1093/rpd/ncp303

[49] ICRP Publication 26. Recommendations of the international commission on radiological protection. Annals of the ICRP. 1977;**1**(3):80

[50] ICRP Publication 60. The 1990 recommendations of the international commission on radiological protection. Annals of the ICRP. 1991;**21**(1-3):202

[51] International Commission on Radiation Units and Measurements. ICRU report 66, determination of operational dose equivalent quantities for neutrons. Journal of the ICRU. 2001;**1**(3):94

[52] International Commission on Radiation Units and Measurements. ICRU report 85a, fundamental quantities and unit for ionizing radiation. Journal of the ICRU. 2011;**11**(11):33

[53] ICRP Publication 74. Conversion coefficients for use in radiological protection against external radiation. Annals of the ICRP. 1996;**26**(3-4):205

[54] ICRP 116. Conversion coefficients for radiological protection quantities for external radiation exposures. Annals of the ICRP. 2010;**40**(2-5):1-257. DOI: 10.1016/j.icrp.2011.10.001

[55] ICRU Report 57. Conversion coefficients for use in radiological protection against external radiation. Journal of the ICRU. 1997

[56] Rawool-Sullivan M, Sullivan J. Understanding Doppler Broadening of Gamma Rays [Report]. Los Alamos National Laboratory. 2014. DOI: LA-UR-14-23048

[57] Thomadsen B, Ranvinder N, Bateman FB, Farr J, Glisson C, Islam M, et al. Potential hazard due to induced radioactivity secondary to radiotherapy: The report of task group 136 of the American Association of Physicists in Medicine. Health Physics. 2014;**107**(5):442-460. DOI: 10.1097/HP.0000000000000139

[58] Schmidt FAR. The Attenuation Properties of Concrete for Shielding of Neutrons of Energy Less than 15 MeV. Oak Ridge, Tennessee, USA: Oak Ridge National Laboratory: ORNL-RSIC-26 UC-34 – Physics. 1970. p. 160

[59] IAEA Safety Standards Series No. GSG-7. Occupational Radiation Protection—General Safety Guide. STI/PUB/1785. 2018. p. 360

Chapter 11

The Secondary Photoelectron Effect: Gamma Ray Ionisation Enhancement in Tissues from High Atomic Number Elements

Christopher Charles Busby

Abstract

The absorption of gamma rays is roughly proportional to the fifth power of the atomic number of an element. This immediately raises the issue of tissue ionisation enhancement effects from photoelectron production by elements of high atomic number incorporated into living tissue. The issue was raised in the 1950s in relation to calcium in the bone but has received little attention since then. New results, derived from mathematical modelling carried out at the University of Ulster, of photoemission from nanoparticles of gold and uranium are presented. These show that significant ionisation enhancement effects can occur when incorporated particles of high atomic number are exposed to natural background gamma radiation, effects which increase sharply at the lower energy end of the spectrum, around 150 keV. The effects must also occur for molecular species. The general problem is discussed, with reference to the literature, and approximate enhancement factors are derived for the effect. The implications for the evolutionary selection of elements by life are explored.

Keywords: gamma rays, ionisation, biological effectiveness, absorbed dose, radioprotection, photoelectron, secondary photoelectric effect, evolution

1. Introduction

Gamma is a high photon energy electromagnetic radiation which is absorbed by material with a number of different physical consequences. Its absorption results in the generation of fast electron tracks capable of breaking chemical bonds in living tissue with the generation of reactive ionic species and free radicals. These energetic fragments can then, themselves, migrate away from the ionisation track to react with other stable molecules and ions. The overall processes result in damage to living cells, either by direct interaction with a molecule or by indirect effects from the ionised or reactive species, and this can result either in cell death or in downstream genetic and genomic effects which are harmful to the health of the exposed individual. The mechanisms of genetic and genomic biological damage which follow from gamma ray exposure and X-ray exposures are described in the literature and are accepted by all the radiation risk agencies [1–3]. In this chapter, I will write *gamma ray*, but it is to be assumed that the processes I discuss apply to X-radiation also.

IntechOpen

It is generally accepted now that the biological effects of exposure are a consequence of either direct damage to cellular DNA or due to induction of instability in cellular DNA through a mechanism involving the detection of ionisation, expressed as an increased concentration of reactive oxygen species (ROS), generated by gamma ray interaction with water. Either way, the essential biological effective target for gamma ray (and indeed all ionising radiation) absorption is not primarily water but is the cellular DNA. Historically, the method developed for assessing exposure after 1950 involved defining quantities based on the absorption of energy per unit mass of material exposed to these high-energy photon radiations. Since the detection and quantification of gamma radiation (and X-rays) became most easily based on the ionisation of gases (Geiger Muller counters, proportional counters and ionisation chambers and, later, scintillation counters), it was a simple step to quantify absorption by living tissue in the same way. Thus, for ionising radiation, the quantity *absorbed dose* became the prime measure of risk. Since it became clear that for heavily ionising radiations, alpha and neutron radiations which have higher ionisation per unit track length, there must be allowance made, the later quantity, *equivalent dose*, was introduced whereby a weighting factor was added, based on the ionisation density or linear energy transfer of the radiation. However, for the purposes of this brief chapter, the concern is with *absorbed dose* and its calculation for the purposes of radiation protection.

Clearly, from the outline above, it is the ionisation density at the DNA which is the key factor defining radiation risk. But *absorbed dose* does not measure this. In the way it has come to be employed by the radiation risk agencies; it is a measure of mean ionisation density over significant masses of tissues and kilograms, modelled as water. The issue of anisotropy of ionisation density for *internal* radiation exposures to alpha and beta particles from incorporated radionuclides has been addressed elsewhere [4].

The calculation of absorbed dose assumes that the tissue in which the energy is dissipated is water or its tissue-equivalent substitute. Since all photon energy absorption from *external* exposure is converted ultimately to energetic electron tracks in tissue, either in the initial instance or as a result of the reabsorption of photons from other secondary sources (e.g. Compton, Bremsstrahlung), the averaging of these tracks over all tissue may seem reasonable as an approximation. But what this does (and this is the issue explored here) is it fails entirely to address or incorporate increases in absorption of photon radiations by elements of higher atomic number Z than water or tissue-equivalent material which is largely absorbed by the highest Z element in it, namely, oxygen ($Z = 8$). This would not matter much if any elements of higher Z were uniformly distributed in the tissue: in such a case, since gamma and X-ray absorption increases very quickly with atomic number, the overall absorption might be slightly increased, but where an incorporated elevated Z element is chemically bound to DNA, the transfer of energy into the DNA becomes very much greater than that which is assumed by conventional dosimetry. A similar enhancement of local dose occurs near high-Z nanoparticles incorporated into tissue. This is an interesting and important area of concern which has implications both for radiation safety and for the development of cancer therapy. Apart from some early work on the enhanced photoelectron density near the bone, it seems to have been entirely overlooked. The issue is also an important one for radiation protection in the nuclear industry and the military, especially in the case of uranium particle contamination, perhaps the reason why little research has been carried out on the subject. There are also other areas of interest, implications for medical prostheses and even for arguments about the development of living systems generally.

2. The absorption of gamma radiation by matter and the secondary photoelectron effect

Gamma radiation and matter interact mainly by three different mechanisms, Compton scattering, pair production and the photoelectric effect. The different contributions of these to absorption depend on the absorbing material, principally its atomic number Z and the quantum energy **E** of the incident photon, proportional to frequency $\mathbf{E} = \mathbf{h\upsilon}$.

In the photoelectric effect, incident photon energy causes the emission of an electron from the absorbing element. The electron has the energy of the absorbed photon minus the binding energy of the electron. For gamma radiation the binding energies are second order, and the emission electron carries almost all the initial gamma energy. Electrons may also lose energy in secondary processes occurring within the atom. For energies below 1 MeV, the photoelectric effect largely predominates. **Figure 1** illustrates the effects by incident gamma energy.

Thus, for energies below 1 MeV, the photoelectric effect predominates. The cross section for the photoelectric effect is approximately proportional to the atomic number Z to the power of five and to the incident photon energy to the power of −7/2 [5]. The sharp dependence of photoelectron generation on Z immediately raises interest in the resulting wide variation in absorption of gamma rays by high atomic number atoms and molecules in tissue. This concern is related to the range of the photoelectrons and their deposition of ionisation effects close to the atom. For low-energy photoelectrons generated by low-energy gamma and X-ray photons, the effects will be increasingly local to the atom, and if the atom is local to DNA, there will be an enhancement of radiobiological effectiveness of the absorbed energy. This may be termed the secondary photoelectron effect. The SPE will also occur in the vicinity of internal particles of high-Z elements and in the vicinity of metal prosthetic structures.

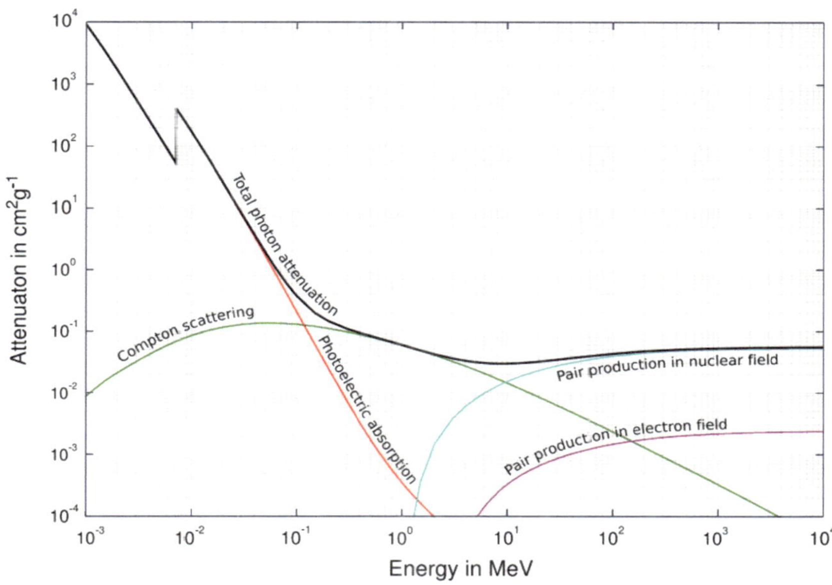

Figure 1.
Relative contributions of the main different types of energy conversion in materials following the absorption of gamma ray photons. The specific curves differ considerably for different elements, driven by the electronic structure of the element. Note that the attenuation coefficient is normally given in $cm^2\ g^{-1}$ and thus incorporates the density of the element.

E (keV)	Oxygen (8)	Water	Muscle (striated)
10	2.950E−4	2.515E−4	2.536E−4
50	4.992E−3	4.320E−3	4.356E−3
100	1.647E−2	1.431E−2	1.443E−2
150	3.325E−2	2.817E−2	2.841E−2
500	2.018E−1	1.766E−1	1.781E−1

Table 1.
Continuous slowing down range r_o in g cm^{-2} for electrons of different energies in oxygen, water and muscle tissue [6].

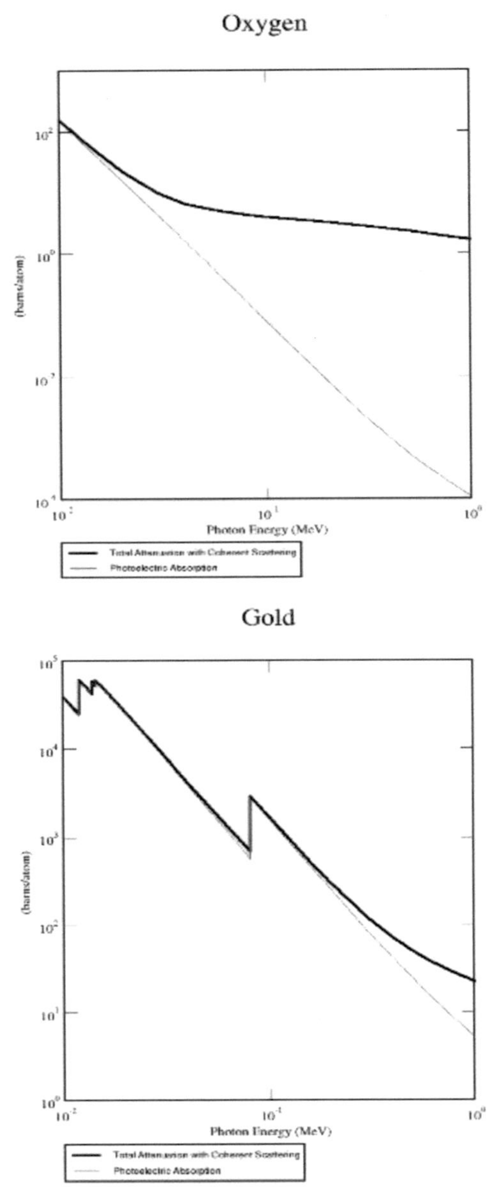

Figure 2.
Gamma ray absorption cross sections for oxygen (Z = 8) and gold (Z = 79).

DOI: http://dx.doi.org/10.5772/intechopen.86779

The starting point for examining this issue is the electron range in tissue by electron energy. This can be calculated on the basis of the continuous slowing down approximation (CSDA), and results for the lower energies for muscle tissue are given in **Table 1** [6]. In the lower-energy regions, the ranges of electrons in tissue are shown in **Figure 2**.

For electrons of energy <500 eV, the range in tissue is in the order of 1–10 nm [7]. This is of the order of the dimensions of the DNA molecule.

3. Absorption of photons by chemical elements

The photoionisation cross sections with photon energy of low Z (oxygen 8) and high Z (Gold 79) are shown in **Figure 2**. From **Table 1** we can see that oxygen may be used to approximate tissue absorption. In the low-energy region around 100 keV, it is clear from **Figure 2** that the absorption (and thus photoelectron production) of gold is several orders of magnitude greater than tissue. **Table 2** gives the photoionisation cross sections for a section of elements of interest [8].

If the absorption of gamma ray photons by chemical elements varies so widely, with such an increased cross section for the higher Z elements, it seems clear that the incorporation of high-Z elements in living tissue would be essentially harmful. There is evidence from evolution to support this idea, and this will be discussed below. Apart from contamination issues due to anthropogenic sources and the question of medical procedures, the problem arises because of the continuous irradiation of living creatures by natural background radiation (NBR). The gamma

Z	Element	10 keV	50 keV	100 keV	150 keV
1	Hydrogen	4.5E−3	1.8E−5	1.6E−6	4.1E−7
6	Carbon	4.1E+1	2.0E−1	2.0E−2	5.4E−3
8	Oxygen	1.5E+2	8.1E−1	8.2E−2	2.2E−2
11	Sodium	5.7E+2	3.6E00	3.7E−1	1.0E−1
15	Phosphorus	2.0E+3	1.5E+1	1.6E00	4.4E−1
16	Sulphur	2.6E+3	1.9E+1	2.2E00	5.9E−1
17	Chlorine	3.3E+3	2.5E+1	2.8E00	7.8E−1
19	Potassium	4.1E+3	3.3E+1	3.7E00	1.0E00
20	Calcium	6.2E+3	5.2E+1	5.9E00	1.7E00
26	Iron	1.6E+4	1.6E+2	1.9E+1	5.4E00
53	Iodine	3.4E+4	2.5E+3	3.6E+2	1.1E+2
74	Tungsten	2.8E+4	1.6E+3	1.3E+3	4.3E+2
78	Platinum	3.5E+4	2.0E+3	1.5E+3	5.1E+2
79	Gold	3.7E+4	2.1E+3	1.6E+3	5.4E+2
80	Mercury	3.9E+4	2.3E+3	1.7E+3	5.7E+2
82	Lead	4.3E+4	2.5E+3	1.8E+3	6.2E+2
92	Uranium	6.9E+4	4.0E+3	6.4E+2	9.4E+2

Table 2.
Photoionisation cross sections for a selection of elements of interest at different incident energies in the natural background low-energy region (barns) (Hartree-Fock approximation) [8].

spectrum of NBR increases rapidly to lower energies, roughly as the −7/2 power of the energy. From **Table 2**, it is clear that the absorption of photon energy in the NBR region (50 keV) from iodine is about 3000 times that from oxygen or water/tissue. It has been suggested that this may explain the radiosensitivity of the thyroid gland [9]. It should be noted in passing that the absorption coefficients at the energies tabulated do not generally reflect the overall absorption differences between the low-Z and high-Z elements over the whole-energy spectrum because of discontinuities in the absorption by the d- and f-orbital electrons in the heavier elements like gold and uranium. These discontinuities for gold are clear in **Figure 2**. For gold, the enhancement factor relative to water at the four energies tabulated (10, 50, 100 and 150 keV) are 246, 2592, 19,500 and 24,545. Similar variations in enhanced photon cross section are apparent for uranium which has 45,000 times the photoelectron cross section at 150 keV than the oxygen in water.

It is clear from this approach that the determining absorption of living tissue is defined not by water but by the higher Z elements present. This is starkly true for iron and iodine which must form centres for photon absorption and photoelectron production. It may therefore be plausible to argue that this is why that the main cancers associated with external radiation exposures are leukaemia and thyroid cancer.

4. The gamma energy spectrum of natural background radiation

Since secondary photoelectrons will be generated from all exposures to gamma radiation and since the local ionisation density near the absorbing atom, particle or metal prosthesis is the quantity of interest, it is clear that the energy spectrum of gamma NBR is an important component of any assessment. External gamma radiation degrades in energy as it passes through tissue as a result of the various processes which occur. Energy is lost by Compton scattering resulting in the production of a Compton photon of lower energy than the initial energy. Electrons generated by the photoelectric effect lose energy through collisions and the generation of Bremsstrahlung photons of low energy and so forth. Thus, the further the initial photon travels in tissue, the greater the number of low-energy photons there are in the medium. The natural background radiation spectrum in Burnham-on-Sea, Somerset, UK, is reproduced in **Figure 3**.

Note the sharp increase in the number of photons at low energy: the cut-off is a result of absorption by the shielding of the thallium-doped sodium iodide scintillation detector. The degradation of photon energy inside the human body can be examined by placing an insulated scintillation detector inside a water-filled container and comparing the spectrum with that obtained in air. The spectrum obtained in this way, which compares well with that employed by Pattison et al. (who attempted to model the photoelectron effects in uranium particles [10]), is shown in **Figure 4**. The cut-off at low energy 15 cm inside the water jacket is due to the absorption of the low-energy short-range photons. By subtraction it is possible to show that the number of photons of low energy increases inside the water sphere of 30 cm diameter (used to approximate the body). Thus, the dispersion curve shifts to lower energy. The enhancement of photon numbers by energy is shown in **Figure 5**.

What is clear from these results is that NBR delivers mainly low-energy photons. It turns out that 60% of in-air NBR photons have energy below 150 keV and the peak in photon numbers increases to low energy below 50 keV. Photoelectrons of this energy have a mean CSDA range (**Table 1**) which is comparable with the dimensions of a single cell or cell nucleus. Therefore, a high-Z atom in a cell will be continuously amplifying NBR in proportion to the photoionisation cross section

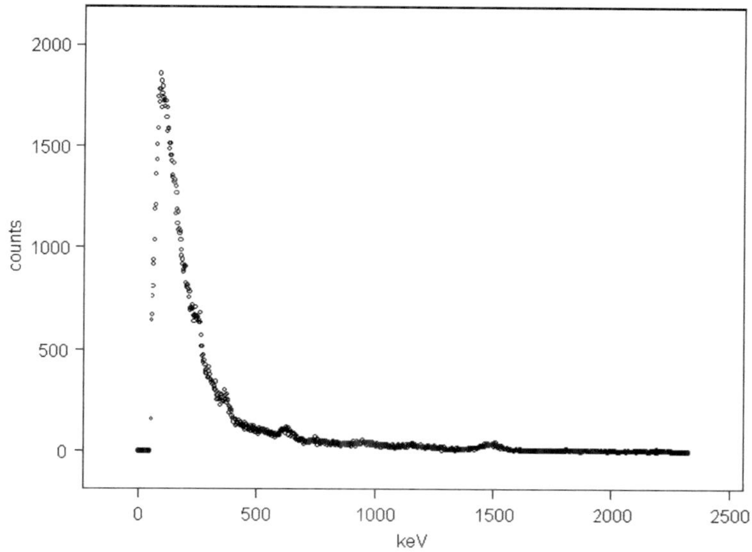

Figure 3.
Gamma ray spectrum obtained on beach at Burnham-on-Sea using a 2-in NaI (Tl) Scionix detector. Note rollover at about 60 keV.

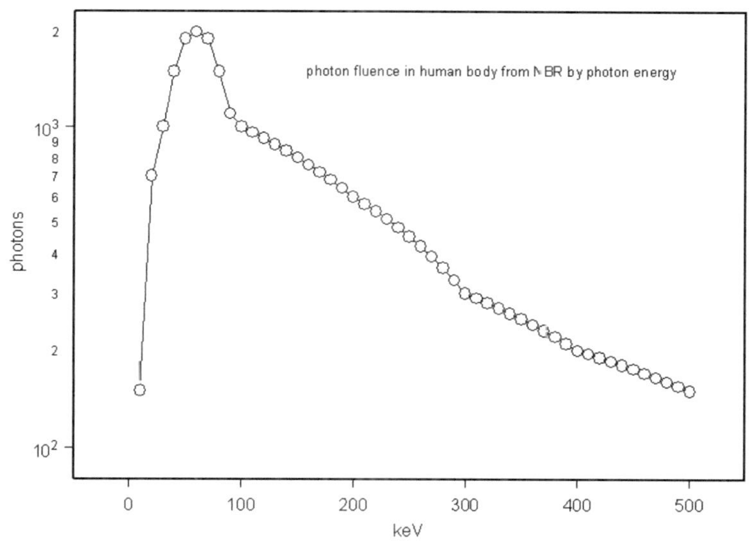

Figure 4.
Energy dispersion in the low-energy region 0–500 keV of the natural background gamma photons at 15 cm depth inside a human body. Based on Pattison et al., ***Figure 3*** *and unpublished work using a gamma probe packed with bags of water. Shielding effects on the primary in-air dispersion below 100 keV are uncertain, and the energy dispersion of photons inside the body is very uncertain.*

shown in **Table 2** and delivering enhanced ionisation to that cell or cell nucleus relative to that calculated using the concept of absorbed dose which is based on the assumption that the absorber is effectively water (i.e. oxygen). Further, the biological effectiveness of NBR, its damage to tissues, will be defined by the highest Z atoms in the tissue. This will also be true for other exposures, for X-rays, medical examinations and exposures to anthropogenic sources, indeed the entire range of exposures which are regulated by the law on the basis of the current risk models.

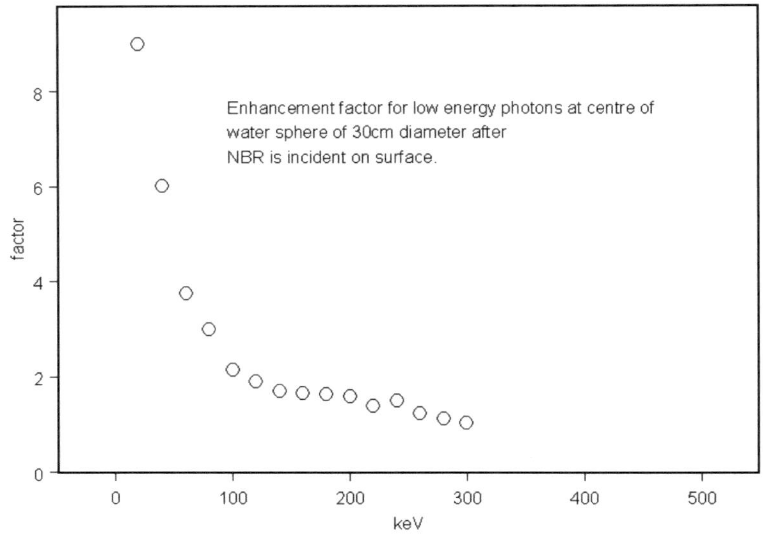

Figure 5.
Enhancement of photon energy at different energies on passage through 15 cm water. Internal photon fluence divided by external photon fluence. Unpublished measurements.

It will be the location in the body of a high-Z atom or particle relative to the target DNA which will be the determinator of biological risk. This is a phantom radioactivity: the atom is radioactive by virtue of its high atomic number and its amplification of NBR gamma radiation through photoelectron emission.

5. Internal particles

The radiobiological issue of photoelectron emission by internal high atomic number particles was raised in 2005 by Busby in connection with depleted uranium weapons which create respirable submicron particles on impact [11]. Research in Iraq, where DU weapons were deployed in 1991 and later in 2003, were shown to have caused high levels of congenital effects and cancer in a number of studies both of civilians in Iraq and of military veterans [12–14]. The concerns about the genotoxicity of DU particles led to research by a number of groups in the early 2000s. The laboratory researches demonstrated that both uranium and uranium particles were capable of causing measurable genetic effects, chromosome breakages and so forth [15–17]. In one study with mice, both embedded uranium and tungsten particles caused local cancer effects [18]. These findings have been reviewed in Busby [19, 20] and will not be rehearsed here. What will be presented here are some results from nanoparticle mathematical modelling studies carried out at the University of Ulster between 2009 and 2012 which looked at photoelectron production from water, gold and uranium spheres [21, 22].

5.1 The University of Ulster studies

Photoelectron emission from nanoparticles of water, gold and uranium was investigated by Elsaessar, Busby and Howard from 2009 to 2012. Preliminary results were presented at a conference [21], and the studies contributed to a PhD thesis [22]. The CERN FLUKA code was employed. The beam geometry is shown in **Figure 6**, and in **Figure 7** results are given for 10 nm particles of water, gold and uranium. Referring

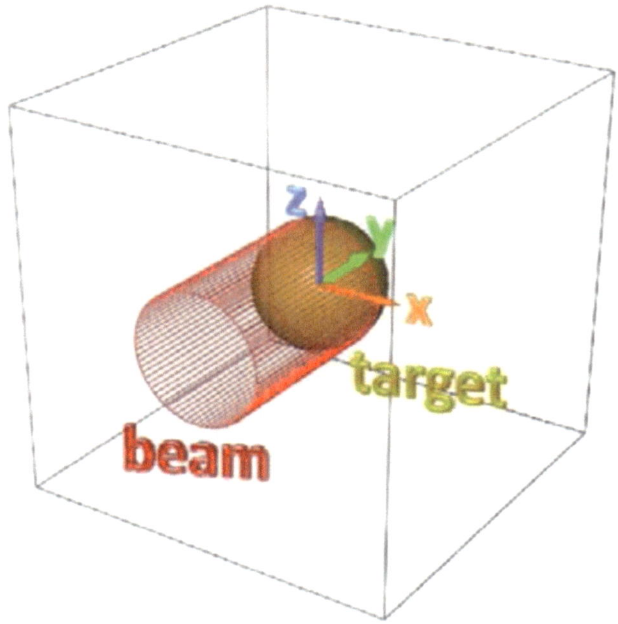

Figure 6.
Beam and target geometry for FLUKA calculations. A photon beam of cross-sectional diameter equal to that of a particle of water, gold (Z = 79) and uranium (Z = 92) [21, 22].

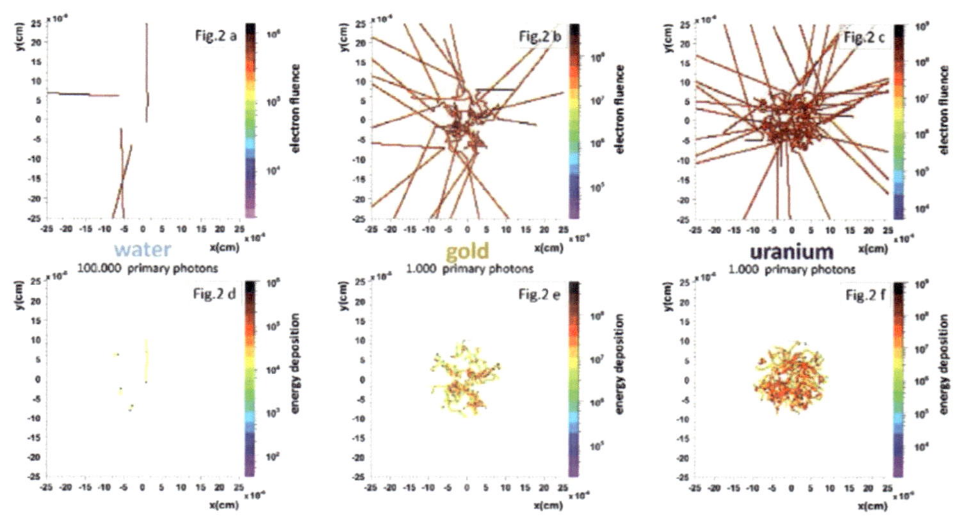

Figure 7.
*Secondary escaping photoelectron production (seen in two dimensions following incident 100 keV photon beam into 10 nm particles of water (Z = 7.5) [**Figure 2a** and **d**], gold (Z = 79) [**Figure 2b** and **e**] and uranium (Z = 92) [**Figure 2c** and **f**]). Below: corresponding energy deposition. Monte Carlo calculation with 1000 incident photons for gold and uranium but 100,000 for water [21, 22]. Individual figure numbers are from [21].*

to the numbering in **Figure 7**, which is from the conference presentation [21], the top row of **Figure 2a–c** shows photoelectron tracks induced by an incident photon beam of 150 keV involving 1000 photons in the cases of 10 nm diameter gold and uranium particles, whilst for the water particle, the number of photons was 10,000. Thus, it is clear that the photoelectron tracks of various energies (lengths) induced in the particles of the high atomic number elements gold and uranium are orders of magnitude greater than those in water. The emission of secondary photoelectron tracks from the three materials is roughly in agreement with a fourth or fifth power law. **Figure 2d–f** shows the energy deposition in the particles on a coloured scale given also in the picture. It is immediately clear from **Figure 7** how the internal particles of high-Z elements result in increased absorption of background radiation and its re-emission by photoelectrons and associated enhanced biological damage relative to the absorption by tissue (water). Due to self-absorption of the induced photoelectrons, the danger exists mainly from smaller particles. Results for different sizes of particles of gold and three different photon energies are shown in **Figure 8**. This shows the variation secondary photoelectron production with photon energy (100 keV, 250 keV, 500 keV and 1 MeV) in a gold target. Photon penetration depth decreases as energy decreases, but the number of electrons escaping the target increases.

To examine the deposition of photoelectron energy into the tissue surrounding the particles examined in the analysis presented in **Figures 7** and **8**, particles were modelled surrounded by water spheres, and the deposition of energy into the

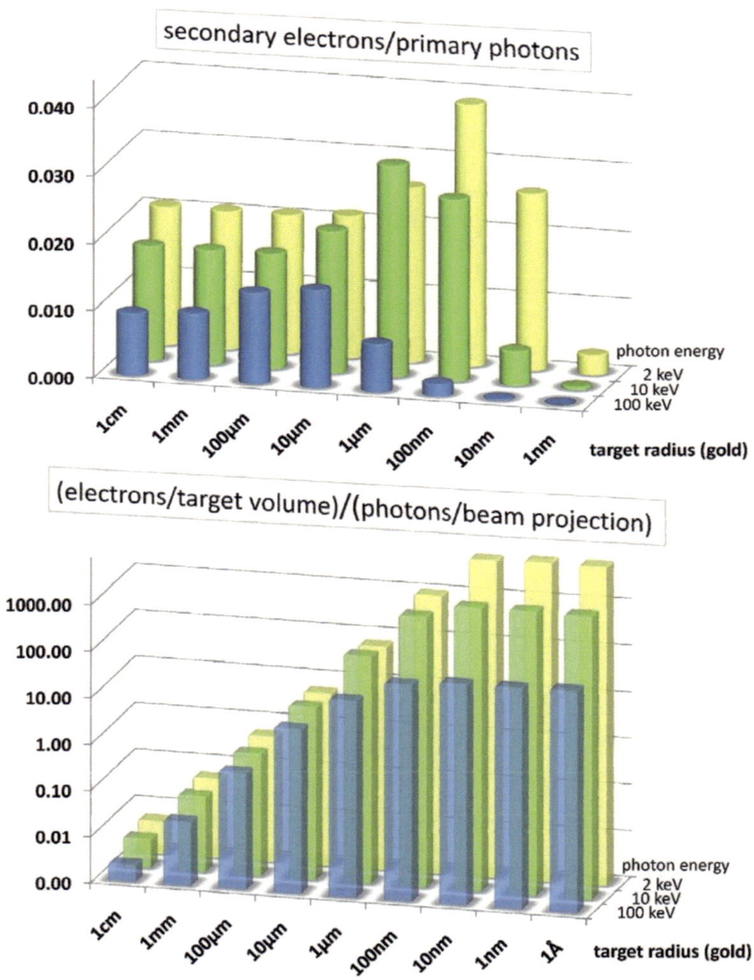

Figure 8.
Upper: secondary electrons/primary photons in gold particles of different diameters and photon energies 2, 10 and 100 keV. Lower: electrons per target volume/photons per beam projection for gold particles of different diameters and photon energies of 2, 10 and 100 keV [21, 22].

spheres was obtained. In **Figure 9**, results for different photon energies of 100, 250, 500 and 1000 keV are presented. As the photon energy was decreased, the penetration also decreased, as expected, but the photoelectron density in the local volume near the particle increased. This is not unexpected since the photoelectron range would be shorter with the low-energy photoelectrons.

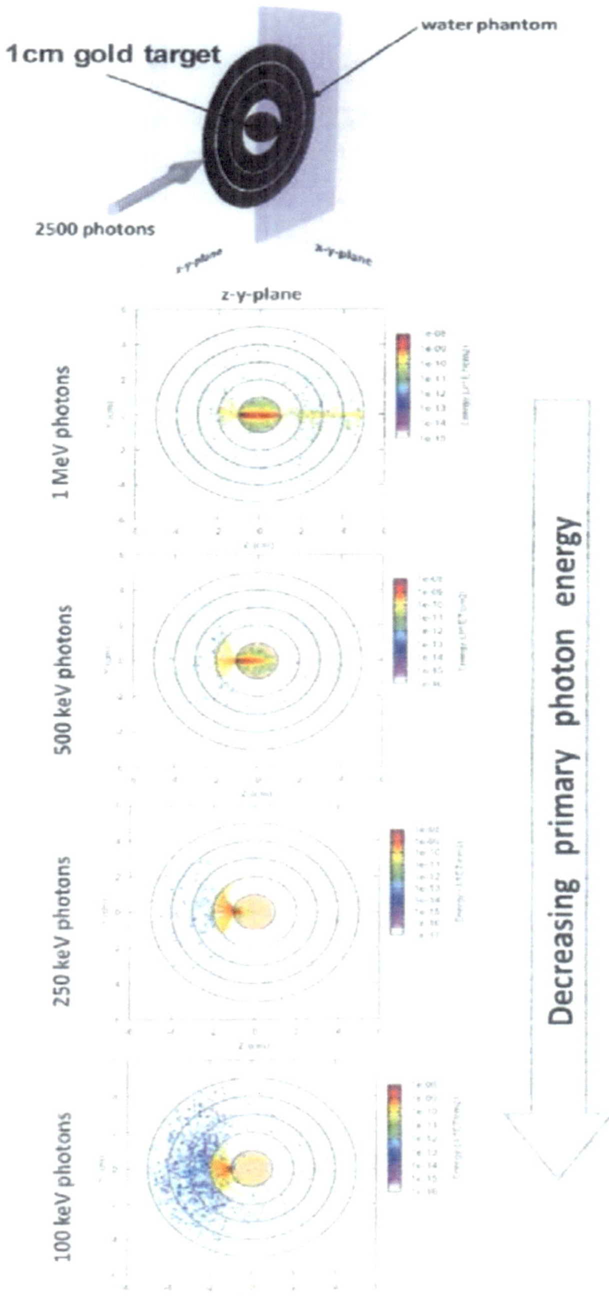

Figure 9.
The variation of secondary photoelectron production with photon energy (100 keV, 250 keV, 500 keV and 1 MeV) in a gold target. Photon penetration depth decreases as energy decreases, but the number of electrons escaping the target increases [21].

The Ulster results can be used to obtain enhancement factors for photoelectron production from 10 nm diameter gold and uranium particles relative to a water particle of the same size. This enhancement factor is compared with a fourth-power law comparison in **Table 3** [19].

The range of the photoelectrons increases as the photon energy increases, but the number of photons increases at low energy for natural background radiation as has been discussed above. The trade-off is shown in **Figure 10**. Dose enhancement (energy per unit mass) falls off rapidly with distance from the high-Z particle but is significant in the micron region. Results for a 400 nm uranium particle are given in **Figure 11** [19].

Figure 11 shows enhancement of dose close to a 400 nm uranium particle embedded in tissue and exposed to natural background radiation. For the method of obtaining this, see [19].

5.2 Other modelling studies of the secondary photoelectron effect

Because of the use in the battlefield of uranium weapons and the fact that there are other sources of uranium particles (which will be discussed below), there is considerable financial and military investment in showing that these photoelectron effects are not biologically important. The author was a member of the UK Ministry of Defence Depleted Uranium Oversight Board [23] from 2001 to 2005 and also the UK Committee Examining Radiation Risks from Internal Emitters (CERRIE) [24]. He also gave evidence to the Royal Society Committee on Depleted Uranium in 2001. In 2009 a paper describing the secondary photoelectron effect entitled "Phantom Radioactivity of Uranium" was sent by him to the chair of the Royal Society Committee which had published reports on the issue in 2001 and 2002. These reports argued that DU could have no adverse health effects as the absorbed doses from the particles were too low [25]. At the suggestion of the chair, Brian Spratt, the photoelectron paper was submitted to the *Journal of the Royal Society Interface* and sent for peer review. The three reviewers all advised that the idea was important and should be published. Despite this, the editor of the journal, William Bonfield, rejected the paper because of "lack of space". Nevertheless, the idea was next presented in a German conference [9] and was covered by the New Scientist in an article in 2009 [26]. Shortly after this a Monte Carlo study appeared in the same journal that had refused to publish the original idea, the *Journal of the Royal Society Interface*, by Pattison et al. arguing that there was no enhancement of dose by uranium particles [10]. A year later, another Monte Carlo study was published by Eakins et al. of the UK National Radiological Protection Board [27]. Both studies were badly flawed for various reasons which will be briefly summarised.

Pattison et al. carried out Monte Carlo modelling using a different code to that employed by Elsaessar, EGSnrc [10]. They modelled two sizes of cylindrical particles and hollow cylindrical particles of 10 μ diameter and length. Using input photons of 200 keV, they concluded that the enhancement of dose was significant and of the order of one to tenfold. Apart from the fact that the particles they modelled were too large to represent the respirable DU particles found in Iraq, and the input photons too energetic, the key to dismissing their approach was their finding that the dose enhancement was largest for the larger particles, the opposite result to that obtained at Ulster. This was because their method was to fix the spherical volume into which the photoelectrons were emitted and calculate energy per unit mass in the annular water shell. Clearly as the particle diameter approached the water shell diameter, the dose would become infinite, showing that the method was nonsensical, and it is hard to see how the paper passed the reviewers.

Calculation	Water Z = 7.5	Gold Z = 79	Uranium Z = 92
Elsaessar et al. [21]	1	12,900	29,200
Z^4	1	12,300	22,600

Ratio of gold and uranium photoelectron numbers to water photoelectron numbers. Also shown is the Z^4 predicted ratio [19].

Table 3.
Number of photoelectrons emitted following exposure of a 10 nm particle of water, gold and uranium to 100 keV photons.

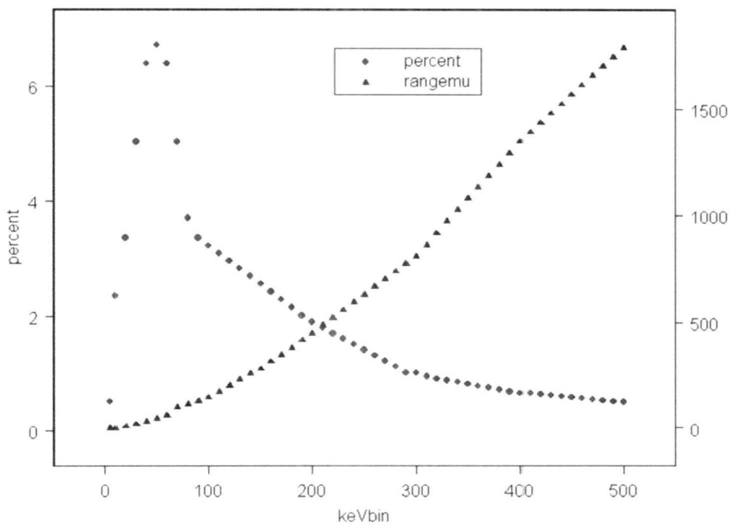

Figure 10.
*Percentage of all photoelectrons with energies equal to natural background radiation photons (blue diamonds) and range in tissue in microns (red triangles) (from results presented in **Figures 5** and **6**). Thirty percent of all NBR photons have energy <60 keV.*

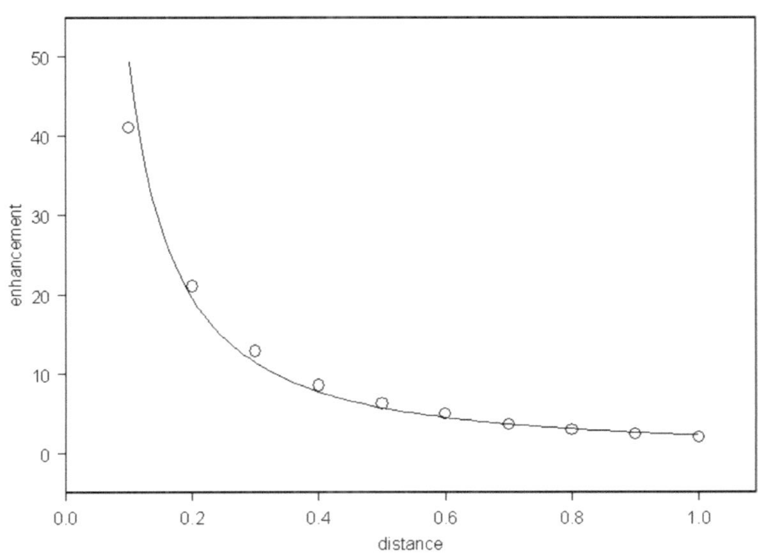

Figure 11.
Dose enhancement (energy per unit mass of tissue) by distance in nm from a 400 nm uranium particle.

Eakins et al. study was carried out by employees of the UK National Radiological Protection Board (NRPB) [27]. They used the computer code MCNP5 to model an arrangement consisting of concentric spheres with the particle at the centre and tissue shells surrounding the particle as had the Ulster modelling. However, like Pattison et al., Eakins et al. fixed the volume into which the photoelectron energy was converted into absorbed dose. The authors did, however, model a range of uranium particles, obtaining enhancements of 3-fold at 100 nm diameter and 20-fold for the 2.5 nm diameter particles. Like the Pattison et al. study, this was an absurd analysis since having a fixed volume for the dose absorption but increasing the particle size, the enhancement factor eventually becomes infinite.

6. High-Z elements and cancer radiotherapy

The augmentation of dose due to secondary photoelectron emission from high-Z elements is not a new concept; it is just that it has been ignored for the purposes of radioprotection. The idea of employing high-Z elements and their photoelectron emission to augment radiotherapy doses was advanced by Matsudeira et al. who measured the radio-enhancing effect of iodine on cell cultures [28]. Nath et al. incorporated iodine into cellular DNA with iododeoxyuridine in vitro and found a radiation enhancement of about 3-fold [29]. Herold et al. injected gold particles directly into a tumour followed by irradiation and found that the excised cells had reduced plating efficiency [30]. Mello et al. found that direct tumour injection with iodine contrast medium followed by 100 kVp X-rays completely suppressed growth of 80% of tumours in mice [31]. Norman et al. modified a CT scanner to deliver X-rays to spontaneous brain tumours in dogs after iodine injection and found a 53% longer survival [32]. Synchrotron radiation was used in combination with the tumour injected drug cisplatin to treat brain tumours in rats [33]. The issue of the mechanism of cisplatin is revisited below.

The photoelectron enhancement by high-Z nanoparticles was exploited in cancer radiotherapy by Hainfeld et al. who attempted to increase the dose delivered to tumours by injecting 1.9 nm gold nanoparticles into mice [34]. The authors also made the method the subject of a patent.

7. Radiobiological dose enhancement effects from molecular and atomic high-Z elements

It is curious that historically photoelectron emission by internal high-Z elements in living systems has received very little attention. The issue of enhanced doses near bones, due to the higher concentration of calcium in the bone, was addressed as long ago as 1949 by Spiers [35], and more recent work has also looked at photoelectron emission near the bone [36]. In 1988 Castillo reported burns and necrosis around reconstructive wires in mandibular cancer patients [37], and Regulla et al. employed a very sophisticated measuring apparatus to show a physical dose enhancement of about 100-fold and a biological enhancement into tissue of 50-fold within a range of 10μ from gold foil [38]. Despite work on enhancing radiotherapy which has been carried out, no authors appear to have related the question of photoelectron enhancement to health effects. One obvious question must be about the enhanced photoelectron doses near metal prosthetic structures containing zirconium (Z = 40). The element has a photoelectron cross section of 3.5E+3 at 150 keV compared with iron (Z = 26) at 5.4 and so would produce some 650 more photoelectrons.

The main question that has to be focused on is the enhancement of dose to the DNA from high-Z atoms or molecules which are attached to the DNA by chemical affinity. If a high-Z atom, ion or molecule were attached to the DNA, then it is easily predicted that this would cause enhanced genetic damage, measurable as downstream effects like cancer and congenital disease but also chromosome breakages and chromosome aberrations. The obvious candidate is uranium, which as the uranyl ion has been known to bind strongly to DNA since the 1960s when it began to be employed as a chromosome stain. The genotoxic effects of uranium exposure are by now well established both in human populations and in in vitro studies [12–20]. They cannot be explained by the intrinsic alpha activity, and indeed one experiment has revealed genetic effects in the absence of alpha decays [20]. The affinity of uranyl ion for DNA has been measured, and it is significant. So uranium (Z = 92) effects are one clear piece of evidence for the effects of secondary photoelectrons. But there is another one.

8. Cisplatin

There is further evidence from the anticancer agent cisplatin, cis-diamine-dichloro-platinum (II). Cisplatin has been a chemotherapeutic agent of choice since 1978 and is given to more than half of all cancer patients. Its mode of action has been variously described as "damaging nuclear DNA and arresting cell division". A recent review states: "Almost 30 years after its clinical benefits were first recognised, studies still continue in an effort to understand exactly how cisplatin works" [39].

Cisplatin also augments radiotherapy, that is to say, the combination of cisplatin and radiotherapy results in much higher cancer therapeutic effects than either agent on its own. This is, of course, a pointer to the mechanism [33, 39]. It is suggested here, based on what has been written above, that cisplatin, a simple diamine-dichloro-square planar complex of platinum (II), merely fixes the platinum atom (Z = 78) at the centre of the nuclear DNA where the secondary photoelectron doses are sufficient to fatally damage the DNA either from natural background radiation or in the case of the radiotherapy, from the induced photoelectrons. If this is the mechanism, then two suggestions are obvious: first, uranium as uranyl acetate, for example, also will act as a chemotherapeutic agent for cancer and will augment radiotherapy in the same way. Since it is suggested that it is the high-Z aspect of cisplatin that is the reason for its action, other high-Z molecular agents could be searched for or synthesised to act as DNA-seeking chemotherapeutic agents.

9. Evolution

The question of the spectrum of elements utilised by evolution of life on earth has been generally approached from the point of view of physical chemistry and more specifically redox equilibria [40]. There may be a separate or additional explanation for the reason why elements of high atomic number (e.g. mercury, bismuth, lead, uranium) although often commonly available on earth are not used by living creatures. As has been shown, chemical elements absorb gamma and X-rays of energy below about 250 keV approximately in proportion to the fourth power of their atomic number Z, and the energy is converted mainly to photoelectrons and local Auger recoil electrons resulting from internal rearrangements in the case of high-Z elements. For elements immobilised inside living tissue, this results in higher doses to components near high-Z atoms or nanoparticles than would be experienced by the same tissue in the absence of the contaminant. Thus, high-Z elements, inside the body, act as devices for focusing and enhancing the doses from natural background radiation and should be seen as phantom radioactivity sources.

If the phenomenon is significant, then it would seem reasonable that the contemporary spectrum of chemical elements employed by living systems will have been produced by evolutionary selection forces responding to such potentially critical damage.

It is a well-known fact that the effects of ionising radiation on living systems are mediated by genotoxicity. The damage can be seen as a consequence of both single- and double-strand breaks in DNA; the dose (D) response (E) can be written as [41]

$$E = aD + bD2 \tag{1}$$

But for the photoelectron effect being considered, dose (i.e. local dose at the DNA) can be written in terms of the atomic number Z or the elements:

$$D = \alpha Z4 \tag{2}$$

and thus

$$E = cZ4 + dZ8 \tag{3}$$

(a, b, c, α and d being arbitrary constants). For evolution it can be assumed that any stress S which prevents an individual from reproducing will represent an inhibitory effect of the survival probability of the species. S can be written in terms of the concentration C of the element in the individual and the radiation effect on the DNA from the element:

$$S = CE \tag{4}$$

$$S = C\,(cZ4 + dZ8) \tag{5}$$

Thus,

$$C = \text{Constant}/\left(cZ^4 + dZ^8\right) \tag{6}$$

If the log of the concentration of all elements found in living systems is plotted against the log of the atomic number Z, the theory predicts an approximately linear relation with slope of between −4 and −8 depending on the contributions of single- and double-strand breaks in DNA to the overall photoelectron and recoil genotoxicity. Of course, the proposed relation is for non-radioactive or weakly radioactive elements and assumes that only photoelectron and Auger effects contribute.

Figure 12 shows a log-log plot of concentration of elements vs. atomic number Z for standard man. Data were from the International Commission on Radiological Protection [42].

Results (**Figure 12**) for elements of Z > 5 seem to support the idea that the photoelectric conversion of natural background radiation has been a significant effect in evolution. The slope of the log correlation is −5.6, between −4 and −8 as predicted, suggesting that a significant component of the effect involved double-strand breaks of DNA and thus ionisation which is very local to the elements. Indeed, it is curious how very few of the elements available to life have been employed by

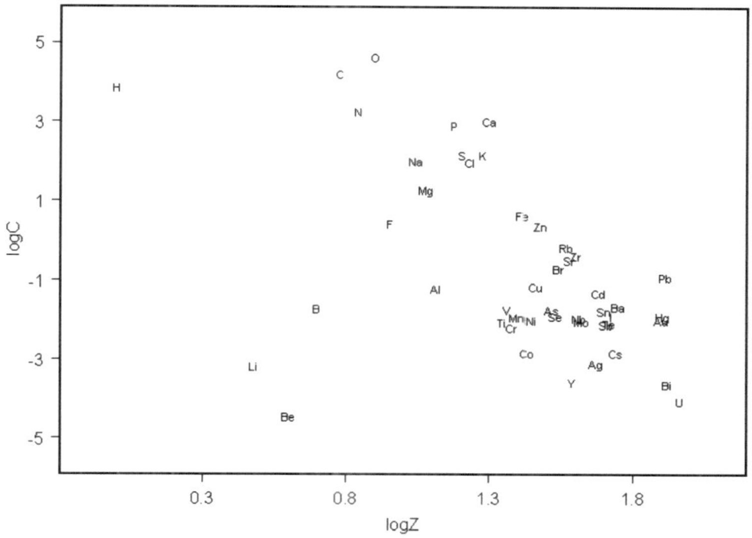

Figure 12.
Plot of log(C) vs. log(Z); investigating the relationship between concentration of elements in humans and the atomic number Z. Note H, Li, Be and B are significant outliers from a relation for which the slope of log(Z) is -5.6 *(*$R^2 = 0.514$*, F-statistic = 65.23 on 1 and 41 degrees of freedom;* $p < 10^{-10}$*).*

biological processes: evolutionary niches are generally found to be occupied but clearly not ones that involve utilising elements of high atomic number. This is not because these elements are scarce. The crustal concentrations of uranium are quite high; there is a significant quantity of uranium in seawater, yet the transfer coefficient for the gut (in mammals) ensures that the element is excluded quite efficiently. The same is true for many other high-Z elements that have been excluded from biological systems.

It is of interest that the elements lithium, beryllium and boron are significant outliers from the relation, and this needs addressing from within the general concept. One reasonable explanation is that all three elements are associated with neutron conversion effects, either the absorption of a neutron in a reaction that produces an alpha particle (boron, $^{10}B(n, \alpha)$; i.e. $^{10}B + n = {}^{7}Li + \alpha$; lithium, $^{6}Li(n, \alpha)$) or the absorption of an alpha particle in a reaction which produces a neutron (e.g. beryllium, $Be(\alpha, n)$; $^{4}He + {}^{9}Be = {}^{12}C + n$). Both alpha particles and neutrons are densely ionising and carry weightings of between 5 and 20 for radiobiological effectiveness in models which assess risk [41]. The thermal neutron cross sections of these three elements (in Barns, $^{10}Be = 3840$, $^{6}Li = 9400$ and $^{7}Be = 39,000$) are significantly higher than other higher Z elements ($^{238}U = 2.7$). The neutron cross section of hydrogen is modest (0.2), but the atomic concentration of the element in water ensures significant neutron absorption and the production of energetic protons by recoil. The natural background neutron fluence at ground level, produced by cosmic rays, has been measured at 46 $cm^{-2}\ h^{-1}$ equivalent to a dose of 10 nSv/h about 10% of the overall background dose [43]. Thus, the displacement of the "radiotoxicity relation" to the left by about one order of magnitude corresponds to the mean relative biological effectiveness of neutrons and alpha particles. It is therefore unsurprising that these elements are outliers in the general linear correlation of the log terms and this may be interpreted as a consequence of the existence of a natural background of these neutron radiations.

So, in general high-Z elements are not employed by life. Why then is there the utilisation by mammals of the element iodine (Z = 53)? The iodine-containing

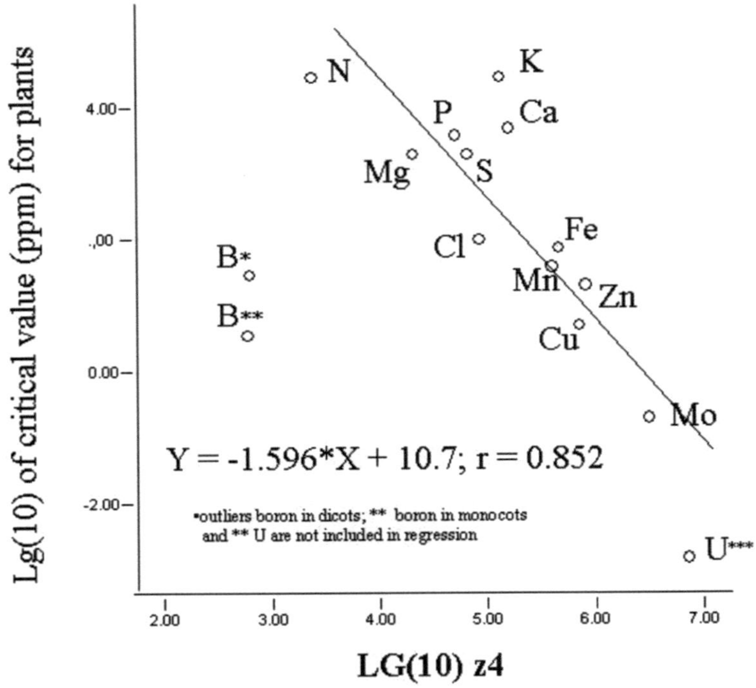

Figure 13.
Minimum concentrations of mineral elements essential for plants required for optimum growth as a function of the fourth power of the atomic number Z. The uranium data point is based upon detection of uranium in a wide range of plants [9].

systems (blood, thyroid) are those which are clinically most sensitive to radiation exposure (for reasons which are clear from the discussions above). It was suggested that the reason why iodine was employed is that the element is being exploited for its radiation detection quality and that the thyroid mediates an induced radiation damage address system through upregulation of genes associated with cellular surveillance and repair [9].

Finally, the relationships discussed here also obtain for plants. Plants are unable to move to avoid radiation exposure and might be expected to reflect responses to evolutionary stresses. The relationship between atomic number and the optimum concentration of elements for plants to thrive has been shown to conform to the same relationship [9]. The correlation is given in **Figure 13**.

10. Miscellaneous observations and suggested further research

The secondary photoelectron amplification of gamma radiation by different elements in living systems has importance in radiation dosimetry. For some inexplicable reason, elemental absorption has been entirely omitted from the calculations of absorbed dose published by radiation risk agencies like the International Commission on Radiological Protection (ICRP) which bases its recommendations of external dose limits on water- and tissue-equivalent phantoms. Furthermore, the phantom photoelectron radioactivity from this effect has considerable application to the element uranium which had been shown in a very large number of publications to have significant genotoxicity. This is particularly the case for internal uranium particles, generated from weapon use, from nuclear power station stacks, from global nuclear atmospheric testing, from nuclear fuel reprocessing and from uranium fuel manufacture. All the official risk agencies model uranium on the basis of its very low intrinsic alpha radioactivity and conclude that it cannot pose the risk that it clearly does.

The basis for the current radiation risk model is the lifespan study of the Japanese A-Bomb survivors, the LSS. One major confounding exposure to the LSS cohorts, upon which the current risk model depends, was the post detonation black rain, which consisted of uranium particles from the weapons [44, 45]. On the basis of the arguments and evidence submitted in this chapter, the uranium particle exposures of all the different dose groups that have been used to construct a linear dose response make any attempt to use these data to define radiation risk unsafe. The unusual cancer results which emerged as soon as 1970 resulted in the researchers deciding to discard the not-in-city unexposed groups that are anomalously healthy. This was an error since these were the only groups not exposed to the black rain, although no doubt, the residual contamination will have caused inhalation exposures after they entered the cities, some months and years after the detonation. The issue was raised by Busby 2017 [45]. Studies of the LSS groups based on truly unexposed control groups in neighbouring prefectures carried out in 2009 showed that the cancer rates in all groups, especially the low-dose groups in the LSS cohorts, were significantly high [46].

What is being suggested here is that the entire understanding of gamma ray interaction with living tissue needs to be rethought. Research must be carried out to quantify the extent to which certain elements with high gamma and X-ray absorption coefficients bind to DNA and the extent to which this causes genetic and genomic damage at background levels and during radiotherapy or other external radiation situations. It is astonishing that no one has questioned the method that has been developed to assess harm from external photon radiations, the simplistic physics-based dilution of energy into water phantoms. It is not as if there was no evidence that this might be an unsafe approach. The radiosensitivity of the iodine-rich thyroid gland should have supplied clues. The mechanism of the anticancer agent cisplatin should have supplied clues.

High atomic number particles have increased in the environment in the last 50 years or more. Platinum particles emerge from catalytic converters, thorium particles emerge from gas light filaments and uranium particles are released from nuclear power stacks, reprocessing plants and many other sources. The high-Z secondary photoelectron effect is used in cancer therapy. There is a whole field of development here where anticancer agents may be synthesised to bind to DNA and carry a high-Z warhead.

Finally, it is suggested that there is a simple experiment which will demonstrate and quantify this effect. It is to contaminate a system in which genetic damage may be measured with a uranyl salt, so that the DNA is contaminated with uranium, and then to irradiate the system with different doses of X-rays or gamma rays and then measure the genetic damage. To exclude alpha particle effects, the agent cisplatin could also be employed in a similar experiment.

11. Conclusions

Although the sharp dependence of the gamma- and X-ray-induced photoelectron yield of elements on atomic number has been known for more than 100 years, the implications for radiobiology have been hardly addressed. This chapter aims to open up this issue and call for more research attention. First, it can be concluded that high-Z elements, when inside living tissue, represent a focus for absorption of photon radiation and that the resulting ionisation density close to the element is much higher than what is calculated using conventional dosimetry such as that employed in current radiation protection, as in, e.g. [41]. This effect, the secondary photoelectron effect (SET), is most relevant to elements which also have affinity for DNA, the target for radiation-induced genotoxicity. The intrinsic radioactivity of such elements is not relevant, as can be seen by the genotoxicity and cancer therapy effect of the drug cisplatin. Results of Monte Carlo modelling carried out at the University of Ulster show that internalised high atomic number nanoparticles are likely to cause high local ionisation in living tissue. These effects are greatest for low-energy photons such as those in the natural background radiation spectrum. It is suggested that this may be one explanation for the anomalous genotoxicity of uranium particles found in many studies but hitherto dismissed as radiation effects on the basis of conventional dosimetry. Finally, an examination of the spectrum of elements employed by living systems reveals an interesting relationship which correlates the elemental composition adopted by life itself with the photoelectron cross section of the elements available to evolution. This relationship, which follows the photoelectron cross section and is highly statistically significant, suggests that living systems are exquisitely and critically sensitive to ionising radiation and have had to develop throughout evolution in such a way as to minimise the ionisation damage induced by background radiation. There are many important consequences of this approach, but the main ones are in the area of radiation risk assessment and in cancer therapy. Some approaches and experiments are suggested.

Author details

Christopher Charles Busby
Environmental Research SIA, Riga, Latvia

*Address all correspondence to: christo@greenaudit.org

References

[1] ICRP. The 2007 Recommendations of the International Commission on Radiological Protection. ICRP-Publication 103. Annals of the ICRP. 2007;**37**:19-27

[2] BEIR VII. The Health Effects of Exposure to Low Levels of Ionizing Radiation—BEIR VII. Washington: National Academy Press; 2007

[3] ECRR 2010. The 2010 recommendations of the European Committee on radiation risk. In: Busby C, Yablolov AV, Schmitz Feuerhake I, Bertell R, Scott Cato M, editors. The Health Effects of Ionizing Radiation at Low Doses and Low Dose Rates. Brussels: ECRR; Aberystwyth Green Audit; 2010

[4] Christopher B. Aspects of DNA damage from internal radionuclides. In: Chen C, editor. New Research Directions in DNA Repair. London: InTechOpen; ISBN: 978-953-51-1114-6; 2013. DOI: 10.5772/53942

[5] Huefner S. Photoelectron Spectroscopy: Principles and Applications. Berlin: Springer; 1996

[6] ICRU 35. Radiation Dosimetry: Electron Beams with energies between 1 and 50 MeV. Bethesda, MD, USA: ICRU; 1984

[7] Kenneth K. Introductory Nuclear Physics. New York: John Wiley; 1987

[8] Scofield JH. Theoretical Photoionization Cross Sections from 1 to 1500 keV. Lawrence Livermore National Laboratory Report; UCRL-51326; 1973

[9] Chris B, Ewald S. Advanced biochemical and biophysical aspects of uranium contamination. In: De Kok LJ, Schnug E, editors. Loads and Fate of Fertilizer Derived Uranium. Leiden, The Netherlands: Backhuys Publishers; 2008

[10] Pattison JE, Hugtenburg RP, Green S. Enhancement of natural background gamma-radiation dose around uranium micro-particles in the human body. Journal of the Royal Society Interface. 2010;7(45):603-611. DOI: 10.1098/rsif.2009.0300

[11] Busby CC. Depleted Uranium weapons, metal particles and radiation dose. European Journal of Biology and Bioelectromagnetics. 2005;**1**(1):82-93

[12] Busby C, Hamdan M, Ariabi A. Cancer, infant mortality and birth sex ratio in Fallujah Iraq 2005-2009. International Journal of Environmental Research and Public Health. 2010;7(7):2828-2837

[13] Muhammed AST, Christopher B, Hamdan M, Eleonore B-B. Uranium and other contaminants in hair from the parents of children with congenital anomalies in Fallujah, Iraq. Conflict and Health. 2011;**5**:1-15

[14] Araneta MR, Schlangen KM, Edmonds LD, Destiche DA, Merx RD, Hobbs CA, et al. Prevalence of birth defects among infants of Gulf War veterans in Arkansas, Arizona, California, Georgia, Hawaii, and Iowa, 1989-1993. Birth Defects Research. 2003;**67**(Part A):246-260

[15] Miller AC, Blakeley WF, Livengood D, Whittaker T, Xu J, Ejnik JW, et al. Transformation of human osteoblast cells to the tumorigenic phenotype by depleted uranium-uranyl chloride. Environmental Health Perspectives. 1998;**106**:465-471

[16] Miller AC, Stewart M, Brooks K, Shi L, Page N. Depleted uranium catalyzed oxidative DNA damage: Absence of significant alpha particle decay. Journal of Inorganic Biochemistry. 2002;**91**:246-252. https://www.newscientist.com/

article/mg19926723-800-how-war-debris-could-cause-cancer/

[17] Coryell V, Stearns D. Molecular analysis of hprt mutations generated in Chinese hamster ovary EM9 cells by uranyl acetate, by hydrogen peroxide and spontaneously. Molecular Carcinogenesis. 2006;**45**:60-72

[18] Miller AC, Brooks K, Smith J, Page N. Effect of the military relevant metals, depleted Uranium and heavy metal tungsten alloy on gene expression in human liver carcinoma cells (HepG2). Molecular and Cellular Biochemistry. 2004;**255**(1-2):247-256

[19] Christopher B. Anomalous health effects of Uranium exposure. Germany: Omniscriptum Publishing; 2017. ISBN 978-620-2-30250-0

[20] Christopher B. Editorial: Uranium epidemiology. Jacobs Journal of Epidemiology and Preventive Medicine. 2015;**1**(2):009

[21] Elsaessar A, Busby C, McKerr G, Howard CV. EMBO Conference: Nanoparticles. October 2007; Madrid; 2007

[22] Elsaessar A. Biophysical interaction of nanoparticles and radiation [PhD thesis]. Faculty of Life and Health Sciences, University of Ulster; 2011

[23] DUOB. Final Report of the Depleted Uranium Oversight Board. UK Ministry of Defence; 2005. Available from: www.bandepleteduranium.org/en/docs/27.pdf

[24] Busby CC et al. Report of the Committee Examining Radiation Risk from Internal Emitters (CERRIE). Chilton, UK: National Radiological Protection Board; 2004

[25] Royal Society. The Health Effect of Depleted Uranium Weapons. Vol. 1. London: Royal Society; 2001

[26] Tickell O. How war debris could cause cancer. New Scientist. 2008

[27] Eakins JS, Jansen JTM, Tanner RJ. A Monte Carlo analysis of possible cell dose enhancements effects by Uranium microparticles in photon fields. Radiation Protection Dosimetry. 2011;**143**(2-4): 177-180. DOI: 10.1093/rpd/ncq398

[28] Matsudeira H, Ueno AM, Furuno I. Iodine contrast medium sensitizes cultured mammalian cells to X-rays but not to gamma rays. Radiation Research. 1980;**84**:144-148

[29] Nath R, Bongiorni P, Rockwell S. Iododeoxyuridine radiosensitization by low and high energy photons for brachytherapy dose rates. Radiation Research. 1990;**124**:249-258

[30] Herold DM, Das IJ, Stobbe CC, Iyer RV, Chapman JD. Gold microspheres: A selective technique for producing biologically effective dose enhancement. International Journal of Radiation Biology. 2000;**76**:1357-1364

[31] Santos Mello R, Callisen H, Winter J, Kagan AR, Norman A. Radiation dose enhancement in tumours with Iodine. Medical Physics. 1983;**10**:75-78

[32] Norman A, Ingram M, Skillen RG, Freshwater DB, Iwamooto KS, Solberg T. X-ray phototherapy for canine brain masses. Radiation Oncology Investigations. 1997;**5**:8-14

[33] Biston M-C, Joubert A, Adam J-F, Ellaume H, Bohic S, Charvet AM, et al. Cure of Fisher rats bearing radioresistant F98 glioma treated with cis platinum and irradiated with monochromatic synchrotron X-rays. Cancer Research. 2004;**64**:2317-2323

[34] Hainfeld JF, Slatkin DN, Smilowitz HM. The use of gold nanoparticles to enhance radiotherapy in mice. Physics in Medicine and Biology. 2004;**49**:N309-N315

[35] Speirs FW. The influence of energy absorption and electron range on dosage in irradiated bone. The British Journal of Radiology. 1949;**22**:521-533

[36] Spiers FW, King SD. Photoelectron enhancement of the absorbed dose to human bone marrow: Experimental and theoretical studies. The British Journal of Radiology. 1985;**58**(688):345-356

[37] Castillo MH, Button TM, Doerr R, Homs MI, Pruett CW, Pearce JI. Effects of radiotherapy on mandibular reconstruction plates. The American Journal of Surgery. 1988;**156**:261-263

[38] Regulla DF, Hieber LB, Seidenbusch M. Physical and biological interface dose effects in tissue due to X-ray induced release of secondary radiation from metallic gold surfaces. Radiation Research. 1998;**150**:92-100

[39] Siddik ZH. Cisplatin: Mode of cytotoxic action and molecular basis of resistance. Oncogene. 2003;**22**:7265-7279

[40] Williams RJP, Frausto da Silva JJR. Evolution was chemically constrained. Journal of Theoretical Biology. 2003;**220**(3):323-343

[41] Beir V. Health Effects of Exposure to Low Levels of Ionizing Radiation. Washington: National Academy Press; 1990

[42] ICRP. Report of the Task Group on Reference Man. ICRP Publication 23. Oxford: Pergamon Press; 1975

[43] Weigel B, Alevra M, Matzke M, Schrewe J, Wittstock J. Spectrometry with the PTB multisphere spectrometer (NEMUS) at flight altitudes and at ground level. In: International Workshop on Neutron Field Spectrometry in Science Technology and Radiation Protection, 5-8 June 2000; Nuclear Instruments and Methods; A476; 2002. pp. 52-57

[44] Takada J, Hoshi M, Sawada S, Sakanone M. Uranium isotopes in Hiroshima black rain. Journal of Radiation Research. 1983;**24**:229-236

[45] Christopher B. Letter to the Editor on "The Hiroshima Nagasaki survivor studies. Discrepancies between results and general perception." by Bernard R. Jordan. Genetics. 2016;**204**(4):1627-1629

[46] Wanatabe T, Miyao M, Honda R, Yamada Y. Hiroshima survivors exposed to very low doses of A-bomb primary radiation showed a high risk of cancers. Environmental Health and Preventive Medicine. 2008;**13**:264-270

Section 4

Irradiation

Chapter 12

Modeling Plastic Deformation in Irradiated Materials

Nicholas Kamyshanchenko, Vladimir Krasil'nikov, Alexander Parkhomenko and Victor Robuk

Abstract

The classification of physical models of plastic deformation localization phenomena in the temperature range including the low-temperature radiation embrittlement effect is done. The new approach for the dislocation channeling mechanism description considering the collective behavior of dislocations and their interaction with radiation defects is proposed. The dislocation collective behavior model in materials irradiated, for example, by reactor radiation including neutron and accompanying gamma radiation is proposed on the basis of the evolution equation for dislocation density taking into account Burgers type nonlinearity. It is shown that the localized structures such as Danilov-Zuev's relaxation waves can be described. The possibility of localization effects (embrittlement) decreasing by the plastic deformation microlevel switching-on is demonstrated in relation with the thermal activated processes. The model describing dose dependence of uniform elongation of irradiated materials is constructed. This model is in good agreement with the experimental data of low-activated alloy based on chromium under (e, γ)—beam radiation.

Keywords: radiation, gamma-quants, embrittlement, deformation wave

1. Introduction

Since Paul Ulrich Villard's discovery of gamma rays, there are manifold phenomena occurring under the influence of high intensive flows of particles, including gamma radiation, in solid state physics. Gamma radiation is shortwave electromagnetic radiation with wavelength of $\lambda < 1$ nm. If an atomic nucleus is in an excited state, then its transition from the excited state into the normal state causes gamma-quant emission. Also gamma radiation can appear, for example, as a result of decay and annihilation of elementary particles and bremsstrahlung. Gamma radiation can be considered as a gamma-quant (photon) beam that demonstrates characteristic corpuscular properties due to its high energy. For example, gamma-quants with energy of from 10 keV to 10 MeV are produced in nuclear reactors.

Note, intensity and spectrum of gamma radiation uniquely characterize radioisotopes. Since decay of many nuclides, for example, ^{137}Cs, ^{60}Co, ^{22}Na, ^{152}Eu, is accompanied by gamma radiation, then knowledge of gamma radiation spectrum allows to identify composition and a number of radioactive isotopes in materials and also to define the degree of burnout of nuclear fuel in the reactors.

IntechOpen

A structure and properties of the substances exposed to radiation experience significant changes called radiation damages. The radiation damages created in solid crystalline bodies destroy the proper crystal structure due to a displacement of atom from the host crystal lattice and formation of radiation defect complexes (point defects, clusters, dislocation loops, pore lattices, and so on). Besides, nuclear reactions can originate under radiation; therefore transmutants (new elements) can arise.

There are basic physical phenomena in the solid bodies under radiation. This is radiation growth and radiation swelling, radiation-stimulated and radiation-induced processes, low-temperature and high-temperature radiation embrittlement, radiation creep, radiation doping and erosion of the surface, and others.

Radiation affects essentially on the operating ability of materials. For instance, the atomic defect excess created by radiation accelerates diffusion processes and produces selection centers of new phases in alloys that are easy getting old. If it accumulates significant amount of decay products, then they can segregate, for example, as gas bubbles.

Many of radiation phenomena appear directly in the conditions of operation of the nuclear reactors. It is considered that energy is directly realized by uranium fission in the reactor in the middle of the order of 195–200 MeV. Most of this energy (about 168 MeV) is realized as kinetic energy of fission fragments. A part of the energy (about 24 MeV) is realized as β-particles, γ-quants, and neutrino. The rest of the energy is carried away by fission neutrons. It is necessary to consider that gamma radiation accompanies a neutron flow at interaction of reactor radiation with radiation protection materials based by metal hydride compounds and affects radiation doze power behind protection.

In wide test temperature region, radiation embrittlement includes conditionally two temperature intervals:

LTRE: low-temperature radiation embrittlement—at the test temperatures up to 0.4 T_m (T_m is melting temperature).

HTRE: high-temperature radiation embrittlement—at the test temperatures higher than 0.5 T_m.

In the LTRE region, embrittlement can be accompanied by radiation hardening, that is, increasing a yield point of materials. In the HTRE region, embrittlement can be accompanied by hardening relaxation. But there is a common feature of both phenomena; exactly, they are accompanied by the localization effects of plastic flow. In the HTRE, localization evolves mainly along grain borders, and it leads to speeded up formation of wedge-shaped cracks. In the LTRE one, localization goes in a grain body. Upon that, the factor that leads directly to destruction of a deformed irradiated material is formation of localized high density dislocation charges that contain, according to a theory, the beginning of a crack [1].

Manifold complexes of defects and nuclear reaction products are formed in the materials irradiated in reactors by neutrons and γ-quants that are the base of LTRE mechanism models to be constructed. The helium LTRE theory is widely known. Helium is generated by (n,α)- or (γ,α)-reactions in the construction materials under radiation. For instance, helium is formed by the two-stage reaction of the thermal neutron: $Ni^{58} + n \rightarrow Ni^{59} + \gamma$ and $Ni^{59} + n \rightarrow Fe^{56} + \alpha$.

Helium is practically insoluble in metals and at elevated temperatures migrates to grain boundaries and other defects where gas bubbles are formed that essentially influence embrittlement of the materials.

From all that was said before, it follows the mechanisms of plastic flow of irradiated deformed materials are the LTRE mechanisms in fact. It is possible to propose the following classification of plastic flow localization phenomena in irradiated materials:

i. Stationary dissipative structures (dislocation channeling)

ii. Moving fronts of deformation localization (the Chernov-Luders band type)

iii. Macroscopic bands (the Danilov-Zuev relaxation wave type).

In the works [1–3], it was shown these effects were observed in the materials with any crystal lattice type. We consider briefly these phenomena with the synergetic point of view [4].

2. Stationary dissipative structures (dislocation channeling)

Electron microscope investigations of deformed irradiated materials showed that its structure has some features as long channels in order of several parts of micron width, without any radiation defects (see, for instance, [2]). In the work [5], it was proposed a new approach to describing mechanisms of dislocation channel formation on the basis of considering collective processes of interaction between dislocation ensembles and radiation defects. Upon that, it is supposed by the experimental facts that in the irradiated deformed material, ensemble dislocations move with velocities closed to 0.1 of the sound velocity, that is, in dynamic regime. On the basis of the general kinetic approach to evolving the ensemble of dislocations interacting with obstacles, the expression for part of dislocations q overcoming obstacles in dislocation channel regime is obtained:

$$q = \exp\left(-|\mathbf{v}|^{m+1}/(2A|a(m+1)|)\right), \tag{1}$$

where $\mathbf{v}$ is an initial velocity of the dislocations running across obstacles, a is dislocation acceleration, $m < -1$, and constant A is proportional to a radiation hardening power $A \sim \sigma_{irr}/\sigma_{init}$ (σ_{irr} is stress after radiation; σ_{init} is stress before radiation). At $|\mathbf{v}| \to \infty$ (or increasing a), this fraction goes to unity, that is, at the high velocities (energies), the dislocations bypass obstacles without stopping. Thus, in irradiated deformed materials, the effect of sharp increasing the part of dislocations overcoming obstacles in a dynamic regime can be observed. Upon that this effect can be gotten in lower deformation velocities in increasing a hardening power (concentration of the defects arising from radiation).

3. Moving fronts of deformation localization (the Chernov-Luders band type)

Recently, the synergetic approach is employed more and more to describe the evolution of plastic deformation in materials. As known, the balance equation for a local dislocation density $\rho(x,t)$ underlies the synergetic models. In the work [6], one of these is considered. There, the balance equation was written as

$$\frac{\partial\rho(\mathbf{x},t)}{\partial t} + div(\mathbf{v}\rho(\mathbf{x},t) - D\nabla\rho(\mathbf{x},t)) = J(\rho(\mathbf{x},t)) \tag{2}$$

where $\mathbf{v}$ is velocity vector of dislocation sliding, D is dislocation diffusion coefficient, and $J(\rho(\mathbf{x},t))$ is the dislocation density functional determined by interaction of dislocation with each other. The velocity of sliding dislocations $\mathbf{v}$ can be represented from three parts: $\mathbf{v} = \mathbf{v}_{\text{ext}} + m(f_{\text{int}} + f_{\text{cor}})$, where $\mathbf{v}_{\text{ext}}$ is velocity from external stress, m is dislocation mobility, and f_{int} origins from internal stress—$f_{\text{int}} = b\sigma_{\text{int}}$ supposing internal stress $\sigma = \alpha b G\sqrt{\rho}$, where b is Burgers vector quantity, α is a numerical coefficient, and G is shear modulus. f_{cor} is correlation force arising from mutual disposition of dislocations. We used the expression for it from [7]

$$f_{cor} = A_1 \frac{\partial \rho}{\partial x} \tag{3}$$

where $A_1 = \frac{Gb^2}{4\pi\rho_0}$ and ρ_0 is an average stationary dislocation density. Inserting expressions for forces in Eq. (2) and neglecting its right side ($J(\rho) \cong 0$ (argumentation in [6])), we obtain the basic equation of our model:

$$\frac{\partial \rho}{\partial t} + \frac{\partial}{\partial x}\left(V_{ext}\rho + mb^2 G\rho\sqrt{\rho} + \left(m\frac{Gb^2}{4\pi\rho_0} - D\right)\frac{\partial \rho}{\partial x}\right) = 0 \tag{4}$$

The solution of Eq. (4) is looked for as

$$\rho = \rho_0 + \rho_1(x, t) \tag{5}$$

where $\rho_1(x, t)$ is a dislocation density fluctuation near the average stationary dislocation density ρ_0. For the dislocation density fluctuation $\rho_1(x, t)$, we get the widely known Burgers equation:

$$\frac{\partial \rho_1}{\partial t} + \rho_1 \frac{\partial \rho_1}{\partial x} = \frac{1}{2Kb}\left(\frac{D}{m} - \frac{Gb^2}{4\pi}\right)\frac{\partial^2 \rho_1}{\partial x^2} \tag{6}$$

where K is determined by material constants. As well knowing the solution of Eq. (6) is a step, and $\rho(x, t)$ takes the form

$$\rho(x, t) = a\delta\left(1 + \tanh\frac{1}{2}\left(ax - a^2 t\delta\right)\right) \tag{7}$$

where $\delta = \frac{1}{2Kb}\left(\frac{Gb^2}{4\pi} - \frac{D}{m}\right)$ and a is constant determined by the bound condition, $\rho_1(x, t) \sim \tanh\frac{1}{2}\left(ax - a^2 t\delta\right) \to 0$, at $x - at\delta \to 0$. It corresponds to the edge of Chernov-Luders band or the area of a sharp stepwise transition from a certain value of dislocation density to another value. Irradiation increases the step height. This is shown qualitatively in **Figures 1** and **2**. There are three plots corresponding to Eq. (7) for three irradiation dose values: $p_1 < p_2 < p_3$.

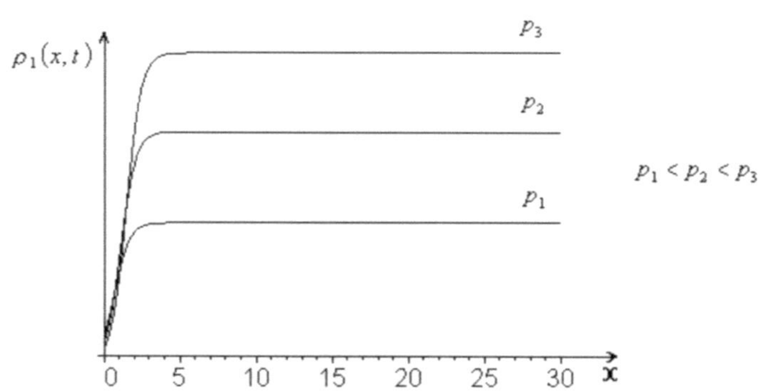

Figure 1.
The dependence of step height on irradiation dose: $p_1 < p_2 < p_3$. It corresponds to the edge of Chernov-Luders band.

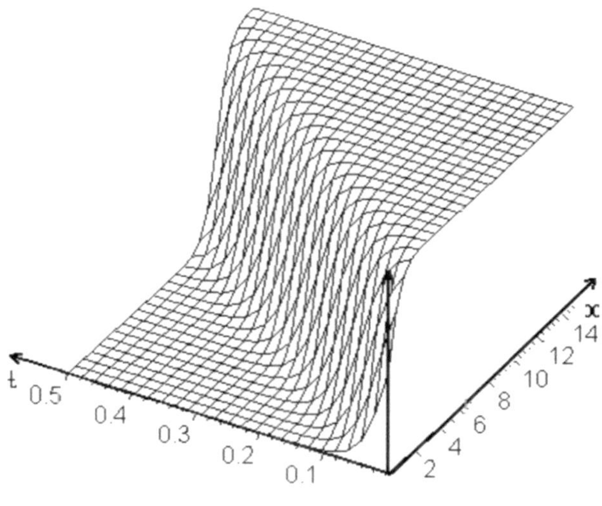

Figure 2.
Three-dimensional plot of step propagation of plastic deformation in sliding plane.

4. Macroscopic bands (the Danilov-Zuev relaxation wave type)

Now, in the dislocation density balance Eq. (2), we take into account correlation forces [7] originating from redistribution of energy between the interaction dislocations more exactly:

$$f_{cor} = A_1 \frac{\partial \rho}{\partial x} + A_2 \frac{\partial^3 \rho}{\partial x^3} \tag{8}$$

where $A_1 = \frac{Gb^2}{4\pi\rho_0}$, $A_2 = \frac{b\sigma_{ext}L}{8\pi\rho_0^2}$, and L is an average relaxation length of a dislocation ensemble [7]. The right side of Eq. (2) can be represented as $J(\rho) = k_1 - k_2\rho^2$ (see [6]) where k_1 characterizes a dislocation source and k_2 is responsible for interaction of the dislocations which can annihilate in particular.

Supposing $\rho_1 = \rho_0 u(x,t)$, one obtains the dimensionless equation for the relative density $u(x,t)$ of dislocations from Eq. (2):

$$\frac{\partial u}{\partial t} + \theta \frac{\partial u}{\partial x} + u \frac{\partial u}{\partial x} + \frac{\partial^2 u}{\partial x^2} + \frac{\partial^4 u}{\partial x^4} = -\chi_1 u - \chi_2 u^2, \tag{9}$$

where θ is the numerical coefficient and χ_1, χ_2 are responsible for the velocity of dislocation formation and their annihilation, respectively. In the work [8], it is shown that χ_1, χ_2 are extremely small for the parameter numerical values of a real standard metal. Due to this, the right side of Eq. (9) is considered to be equal to zero. Then Eq. (9) goes to Kuramoto-Sivashinsky's equation type:

$$\frac{\partial u}{\partial t} + \theta \frac{\partial u}{\partial x} + u \frac{\partial u}{\partial x} + \frac{\partial^2 u}{\partial x^2} + \frac{\partial^4 u}{\partial x^4} = 0, \tag{10}$$

solution [9] of which describes spatial quasi-periodical structures. For Eq. (10), the Cauchy problem was solved at the different initial conditions for the dimensionless function $u(x,t)$. The same level values of $u(x,0)$ were set at the initial time moment under a random distribution in x on the segment $[0, 30]$. In **Figure 3**, the plot I corresponds to the level $u(x,0) = 0.2$; the plots II and III do $u(x,0) = 0.5$ and

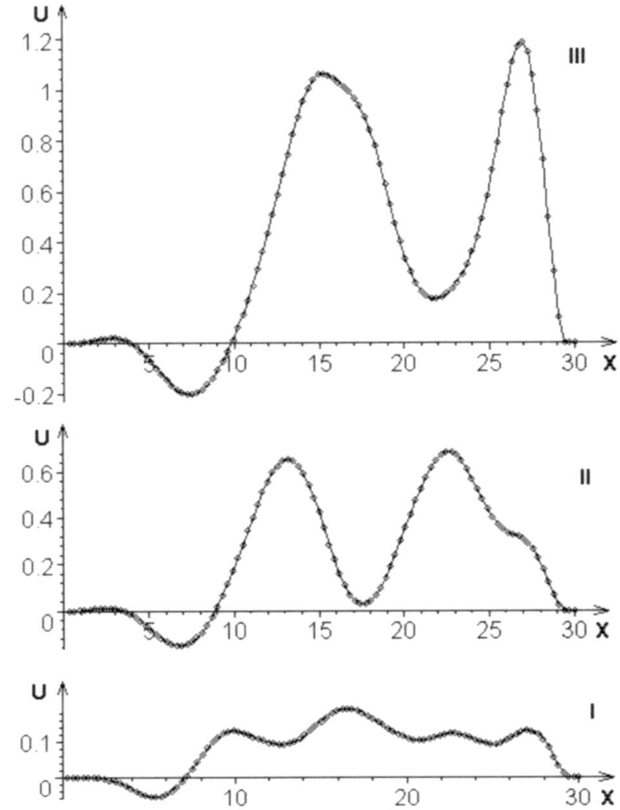

Figure 3.
The plots of the Eq. (10) solution constructed for the time interval $\Delta t = 7.3$. *The plot* ***I*** *corresponds to* $u(x, 0) = 0.2$; *the plots* ***II*** *and* ***III*** *do* $u(x, 0) = 0.5$ *и* $u(x, 0) = 1$, *respectively.*

$u(x, 0) = 1$, respectively. These plots show that the spatially inhomogeneous quasi-periodical dislocation structures are formed in a sample in a certain time interval Δt (here $\Delta t = 7.3$). The level of the $u(x, 0)$ initial density corresponds qualitatively to the certain irradiation dose level of the material. From **Figure 3**, one can see that the localized dislocation structures are formed faster and become more striking in increasing the irradiation dose. The similar wavy deformation distribution is experimentally got in a material sample [10] (**Figure 4** [11] shows the pattern of the deformation distribution along a sample during the initial straining stages).

The appreciable effect of the right side of Eq. (2) on a solution form $u(x, t)$ begins from the χ_1, χ_2 values of 10^{-3}. **Figure 5** shows relaxation of the random dislocation distribution $u(x, 0)$ during time on the x-interval equal to 45 of relative units.

The right side of Eq. (10) not equal to zero determines so-called "point" kinetics of dislocation interactions involving microlevels of plastic deformation. It may say the structures are formed when the point kinetics is absent.

Generalizing and following a concept of structure levels, it is possible to expect that switching-on micro- and meso-levels affects positively plastic deformation of irradiated and nonirradiated materials. So in materials deformed in the conditions of superplasticity, the effect of "running cell" is observed. When even extension exhausts its supply (it is small in these materials), the first neck stage must evolve, that is, deformation of a sample as a whole occurs (macro level N), the mechanism of grain border sliding (macro level N-1) is switched on and does not allow for long

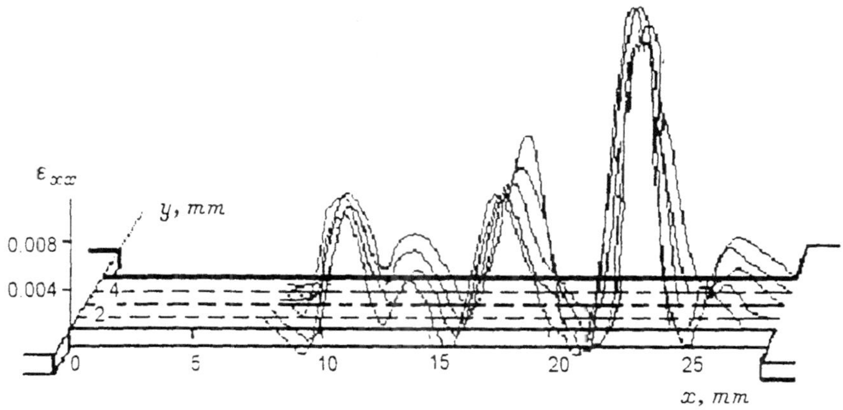

Figure 4.
*The pattern of the deformation distribution along a sample during the initial straining stages (**Figure 1** of the work [11]).*

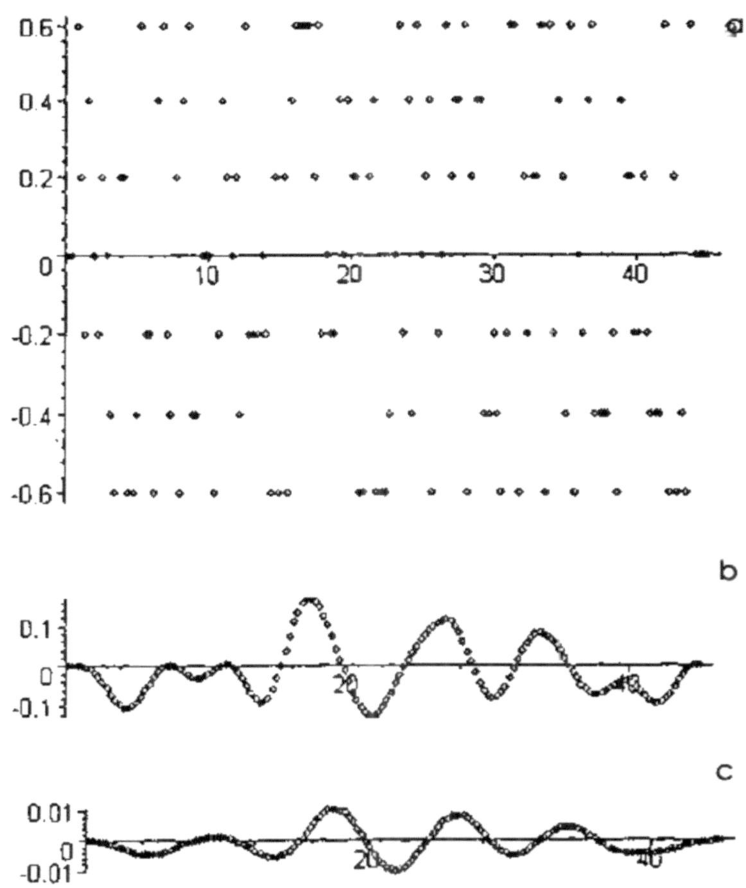

Figure 5.
The spatially temporary evolution of the dislocation density, according to Eq. (10) with the right side being equal to 10^{-3} in relative units. The account time is 0.3 (a), 1 (b), and 5 s (c).

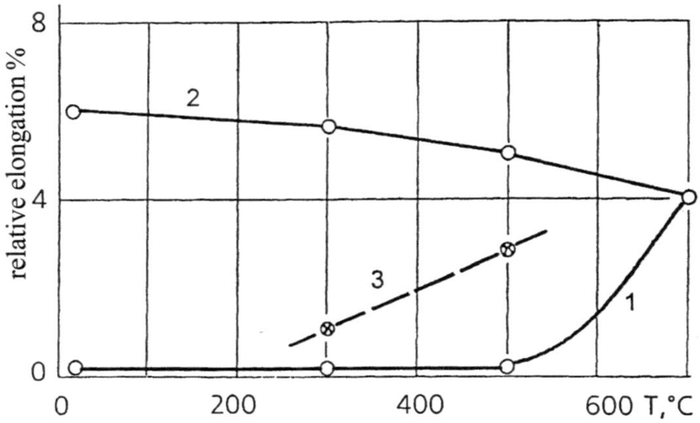

Figure 6.
The temperature dependence of relative extension of a low alloyed chromium irradiated with a fluence of approximately 0.1 displacements per atom (dpa) (1), the initial material (2), and the material irradiated and deformed by bends (3).

wave modes to be established. The neck only starting to arise relaxes due to the mechanism of grain border sliding.

The next example of the effect of switching-on (N-1) level is based on the investigation results of low alloyed chromium [12]. **Figure 6** shows this material when irradiated is absolutely brittle in the rather wide temperature range up to 500°C (plot 1). A part of amount of samples after irradiation was deformed by bend in 30–40°. As a result, new dislocations were created there, and plasticity became not equal to zero after radiation tests (plot 3). So in this case too, switching on the lower structure level (here microlevel) of plastic deformation leads to essential lowering material embrittlement. Thus, the synergetic law conception and applying the structure level conceptual design to study processes of material radiation embrittlement give us not only the investigation direction but the pointing of the ways to solve the specific problems for the material radiation embrittlement to be reduced.

5. Modeling dose dependence of uniform elongation of materials

Experiment shows it is possible for both the monotonous decrease [13] and non-monotonic behavior [14] of uniform elongation dependence ε_p to take place on radiation dose Φt as result of radiation. It does not take into consideration possible deformation hardening in traditional approaches to description of this phenomenon.

Here a new approach is represented allowing to explain peculiar properties of the $\varepsilon_p(\Phi t)$ dependence of irradiated materials considering deformation hardening [15].

The approach is based on the following equation system:

$$\frac{d\sigma}{d\varepsilon} = \sigma \tag{11}$$

$$\sigma = \alpha G b \sqrt{\rho(\varepsilon) + N(\Phi t) d} \tag{12}$$

$$\rho(\varepsilon) = \rho_\infty - (\rho_\infty - \rho_0) e^{-\beta\varepsilon} \tag{13}$$

$$\frac{d\sigma}{d\varepsilon} = A + B e^{-\gamma\varepsilon} \tag{14}$$

$$N = N(\Phi t) \tag{15}$$

where Eq. (11) is a known criterion of plastic stability loss of samples stretched by uniaxial stress with constant speed when a plastic flow localization occurs and a neck is formed. This equation defines a limit value of uniform elongation ε_p. $\frac{d\sigma}{d\varepsilon}$ is the coefficient of deformation hardening, σ plastic flow stress, and ε relative deformation. Eq. (12) determines the plastic flow stress dependence on dislocation density ρ as a function of deformation ε and the dislocation loop density $N(\Phi t)$ on dose (d—a size of a dislocation loop). Coefficient α is of the order of unity. Eq. (13) is an empiric formula of dislocation density dependence on relative deformation ε (see, e.g., [16]). Here ρ_0 is the value of dislocation density of undeformed material, and ρ_∞ is the saturation value of dislocation density at large values of ε. The values of ρ_0 and ρ_∞ are taken from the experimental data [16, 17].

Further, there are the experimental dependences of deformation hardening coefficient $d\sigma/d\varepsilon$ (hardening speed) on deformation ε at different dose values of material radiation in the proposed model (see **Figure 7**).

Computer processing of the curves depicted in **Figure 7** leads to Eq. (14) for the dependence of deformation hardening coefficient on relative plastic deformation ε: the values of parameters A, B, γ of Eq. (14) for corresponding doses are given in **Table 1**.

Eq. (15) is the dose dependence of radiation defect density N. In this model, N is considered as volume density of dislocation loops. Consider the options of the dose dependence Eq. (15) of dislocation loop density $N(\Phi t)$:

1. The monotonic dependence obtained in [18]

$$N(\Phi t) = N_0\left(1 - e^{-\frac{\xi \Phi t}{2}}\right)^{\frac{1}{2}}, \tag{16}$$

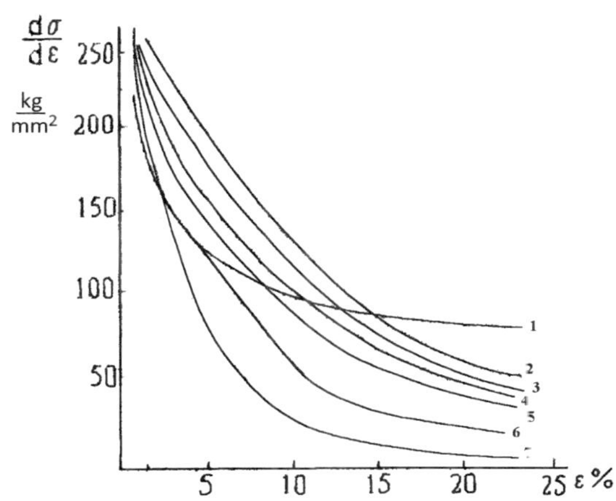

Figure 7.
The dependence of deformation hardening coefficient on relative plastic deformation ε: The curve 1 corresponds to unirradiated nickel; the curves 2, 3, 4, 5, 6, and 7 correspond to nickel irradiated by electrons with energy of 225 MeV (dose of radiation is 10^{-3}, 10^{-2}, 10^{-1}, 2×10^{-1}, 5×10^{-1}, and 5×10^{0} (dpa), respectively).

Dose dpa	**0**	**10^{-3}**	**10^{-2}**	**10^{-1}**	**2×10^{-1}**	**5×10^{-1}**
A (kg/mm^2)	816.59	222.84	59.33	257.88	171.96	430.47
B (kg/mm^2)	1454.48	1731.54	2653.8	2268.2	2399.68	2684.6
γ	0.24	0.09	0.09	0.11	0.13	0.27

Table 1.
Parameters of Eq. (4) for different doses.

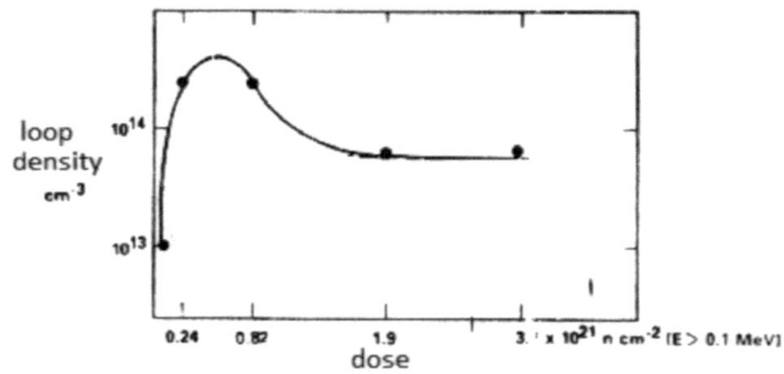

Figure 8.
The nonmonotonic dependence of dislocation loop density on radiation dose.

where ξ is a numerical coefficient and N_0 is the saturation value of loop density

2. The nonmonotonic dependence of a type presented in **Figure 8** [19].

The dependence of dislocation loop density on radiation dose can be described by the following analytic formula:

$$N(\Phi t) = N_0 e^{-(\Phi t - \Phi t_0)^2} \tag{17}$$

in the area of maximum. The nonmonotonic dependence of dislocation loop density on radiation dose is associated with the processes of growing a loop size. At the beginning of its evolution, the loops increase. Their density increases too. Upon reaching a certain radiation dose as a function of temperature and the speed of a displacement, the loops begin to interact with each other. As a result, the process of decreasing its density and loss of its defectiveness begins what is observed by [19].

Eq. (11) considering Eqs. (12)–(14) takes the form

$$\left(\frac{A + Be^{-\gamma\varepsilon}}{\alpha Gb}\right)^2 = \rho_0 + N(\Phi t) - (\rho_\infty - \rho_0)e^{-\beta\varepsilon} \tag{18}$$

Eq. (18) admits a numerical solution only. The numerical solution of Eq. (18) leads to the dependence of uniform elongation ε_p on dose for Eqs. (16) and (17) to be used. The respective curves are represented by **Figure 9**.

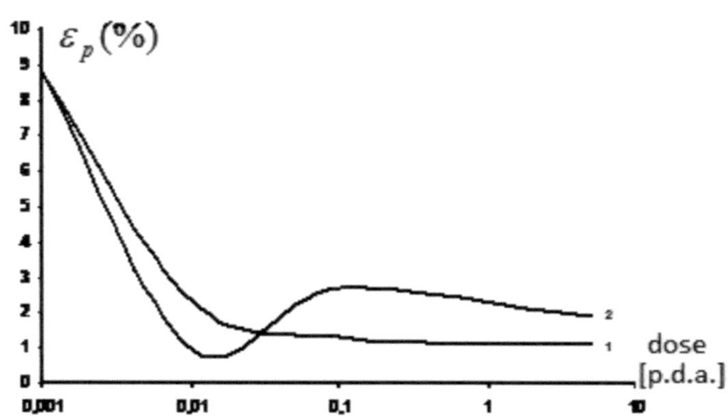

Figure 9.
The dependences of uniform elongation on radiation dose. The curve 1 corresponds to the monotonic dependence of the loop density (16) on dose. The curve 2 corresponds to the nonmonotonic dependence (17).

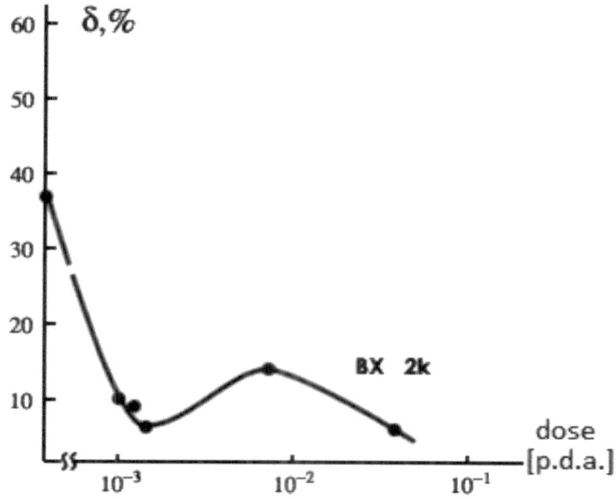

Figure 10.
The experimental dose dependence of elongation of low-activated alloy based on chromium VCh-2 K under (e, γ)—Beam radiation.

The experimental nonmonotonic dependence $\varepsilon_p(\Phi t)$ of low-activated alloy based on chromium VCh-2 K irradiated by (e, γ)—beams in a dose interval of $10^{-3} \div 10^{-1}$ (dpa)—is represented in **Figure 10**.

Comparison graphs of $\varepsilon_p(\Phi t)$ in **Figures 9** and **10** show alignment of modeling and experimental results.

6. Conclusions

1. Evolution of the ensemble of dislocations interacting with obstacles in irradiated materials is analyzes, and the expression for part of dislocations overcoming obstacles is obtained on the basis of the general kinetic approach.

2. Formation of the front Chernov-Luders band is due to the presence of a nonlinear term called the Burgers nonlinearity in the evolution equation for dislocation density.

3. The description of the process of formation of space–time self-organizing dislocation structures is offered. The qualitative agreement of a dislocation density distribution is shown along sample length with the experimentally detected deformation distribution called the Danilov-Zuev relaxation waves in irradiated materials.

4. The description of the dependences of radiation embrittlement of reactor materials on radiation dose is suggested. It is found out that monotonic and nonmonotonic dependences of uniform elongation of irradiated materials are determined by the form of the dislocation loop density dependence on radiation dose.

Author details

Nicholas Kamyshanchenko[1], Vladimir Krasil'nikov[1]*, Alexander Parkhomenko[2] and Victor Robuk[3]

1 Belgorod State University, Belgorod, Russian Federation

2 National Science Center Kharkov, Institute of Physics and Technology, Kharkov, Ukraine

3 Joint Institute for Nuclear Research, Dubna, Moscow Region, Russian Federation

*Address all correspondence to: kras@bsu.edu.ru

References

[1] AM Parshin IM, Neklyudov NV, Kamyshanchenko, et al. Physics of Radiation Phenomena and Radiation Material Science. Belgorod (Rus): BSU Publishing House; 1998

[2] Malygin GA. Self-organization of dislocations and localization of sliding in plastically deformable crystals. Fiz. Tv. Tela (Rus). 1995;**37**(1):3

[3] Kamyshanchenko NV, Krasil'nikov VV, Nekliudov IM, Parkhomenko AA. Kinetics of dislocation ensembles in deformable irradiated materials. Fiz. Tv. Tela (Rus). 1998;**40**(9):1631

[4] Likhachev VA, Panin VE, Zasimchuk YE, et al. Cooperative Deformation Processes and Deformation Localization. Kiyev: Publishing House Naukova Dumka; 1989

[5] Kamyshanchenko NV, Krasil'nikov VV, Nekliudov IM, Parkhomenko AA. Influence of irradiation on the dislocations kinetics with allowance for the dislocation velocity distribution. Journal of Nuclear Materials. 1999;**271, 272**:84

[6] Kamyshanchenko NV, Krasil'nikov VV, Sirota VV, Neklyudov IM, Parkhomenko AA. On the mechanism of development of plastic instability in irradiated materials. Izv. RAN. Metally (Rus). 2000;**4**:110

[7] Khannanov SK. Dislocation density fluctuations (DDF) in the plastic flow of crystals. Fiz. M. Met (Rus). 1994; **78**(2):31

[8] Kamyshanchenko NV, Krasil'nikov VV, Sirota VV, Neklyudov IM, Parkhomenko AA. On the mechanism of development of plastic instability in irradiated materials. Izv. RAN. Metally (Rus). 2001;**6**:53

[9] Alekseev AA, Kudryashov NA. Numerical simulation of self-organization process in dissipative-dispersion media with instability. Izv. USSR AN, Mekh. Zhidk (Rus). Gas. 1990;**4**:130

[10] Zuev LB, Danilov VI. A self-excited wave model of plastic deformation in solids. Phil. Mag. A. 1999;**79**(1):43

[11] Barannikova SA, Zuev LB, Danilov VI. Kinetics of periodic processes during plastic flow. Fiz. Tv. Tel. 1999;**41**(7): 1222

[12] Neklyudov IM. Twinning role in the radiation damage and plastic deformation of irradiation crystals. Functional Materials. 2000;**7**(1):77

[13] Parkhomenko AA. Electron Microscopy and Strength of Crystals. Vol. 9. Kyiev (Rus): Institute of Material Science Problems; 1998. p. 103

[14] Seeger A, Berner R, Wolf H. The deformation hardening mechanisms. Zs. Phys. 1959;**155**(2):247

[15] Kamyshanchenko NV, Krasil'nikov VV, Sirota VV, Neklyudov IM, Ozhigov LS, Parkhomenko AA, Voevodin VN. Mechanisms of localization of plastic deformation in irradiated materials. In: Proceedings of the XV International Conference on Physics of Radiation Phenomena and Radiation Material Science. Alushta (Rus): The Crimea; 2002. p. 70

[16] Dudarev EF, Kornienko LA, Bakach GP. The effect of the energy of the defect packing at the development of dislocation substructure, strain hardening and ductility of FCC solid solutions. Izv. Vuzov. Fizika (Rus). 1991;**3**:35

[17] Risbet A, Levy V. Ordre de cavites dans le magnetism et Ialuminium irradies aux neutron. Journal of Nuclear Materials. 1973;**50**(1):116

[18] Kozlov EV, Popova IA, et al. Gradient structures of nonequilibrium pearlite in deformable steel. Izv. Vuzov. Fizika (Rus). 1991;**3**:118

[19] Rybin VV. Structural and kinetic aspects of advanced plastic deformation physics. Izv. Vuzov. Fizika (Rus). 1991;**3**:7

Chapter 13

Influence of Gamma Radiation on Gas-Filled Surge Arresters

Luka Rubinjoni, Katarina Karadžić and Boris Lončar

Abstract

Current knowledge on the impact of γ radiation on gas-filled surge arresters (GFSA) is presented. Miniaturization of electronic components has led to their increased vulnerability to overvoltage. The combination of ionizing radiation and voltage surges is present in both space exploration and military applications. Some of the commonly used overvoltage protection components (transient suppressor diodes and metal-oxide varistors) perform poorly under ionizing radiation. GFSA demonstrate improved performance under γ irradiation. Performance of GFSA was tested under neutron + γ radiation, considering the effects of induced radiation. The effects of γ radiation were tested on commercially available GFSA components and on the purpose-built GFSA model. The model was used to measure the prebreakdown current and breakdown voltage of different electrode materials (aluminum, steel, brass) under varying gas pressures, under DC and pulse currents. The improvement of the performance of GFSA due to external γ radiation, combined with other improvements in the design (hollow cathode), can enable the use of GFSA without internal radiation sources in environments where γ radiation is present.

Keywords: overvoltage protection, gamma radiation, electromagnetic pulse, power surge, gas-filled surge arrestor

1. Introduction

The advances in complexity and continuous miniaturization of electronic components make them increasingly vulnerable to damage caused by overvoltage events. Modern integrated circuits may suffer temporary or permanent faults even from small overvoltage, due to the size of the internal components in the nanometer range [1, 2]. Overvoltage transients occur in both power and signal lines, due to natural events like lightning strike, normal operation of electric and electronic devices (commutation, powering the devices on or off), failure of equipment or network connections, electrostatic discharge, or the interaction of device/network conductors and electromagnetic fields. Proper design of circuitry protection is essential in preventing temporary or permanent malfunction caused by surge. The combined effects of fast electromagnetic pulse and ionizing radiation present a special challenge to the resilience of electronic components [3].

Depending on their principle of operation and response to voltage transients, overvoltage components can be linear and nonlinear. Linear overvoltage protection components are systems of coils and capacitors assembled into filters of different types. Nonlinear overvoltage protection components come into effect upon the voltage reaching a certain threshold. Commonly used are transient suppression diodes (TSD), metal-oxide varistors (MOV), and gas-filled surge arresters (GFSA) [4]. The advantages of GFSA compared to the other overvoltage components protection are (1) the ability to conduct high currents (up to 5000 A), (2) low intrinsic capacity (~1 pF), and (3) low costs [5]. The disadvantages of GFSA are (1) practical irreversibility of characteristics after the electric arc effect, (2) delayed response, and (3) unsuitability with respect to environmental protection (if GFSA have a radioactive filling) [6–8].

2. Gas-filled surge arresters (GFSA)

The basic design of gas-filled surge arresters is a simple configuration of two or three electrodes encased in a glass or ceramic enclosure filled with insulating gas (**Figure 1**) [9]. The common insulation medium of choice is noble gases [10]. The form of the electrodes is such that it provides a pseudo-homogeneous macro component of the electric field [11]. The strength of the electric field impacts the ratio of free electron generation and loss in the insulating gas. If the field is strong enough, a self-sustained avalanche process will lead to the electrical breakdown of the gas and discharge of the overvoltage [12, 13]. The performance of GFSA is best described by its pulse shape characteristic. The narrower the pulse characteristic is, and the smaller the slope is, the better protective characteristics of GFSA are [6, 7].

Solutions to improve the performance of GFSA in response time include the use of hollow cathodes, thus avoiding radioactive materials or implementing internal radioactive sources [14, 15]. The differences in regulations and standards and the lack of clear strategy for their clear storage and disposal make radioactive sources a challenging solution in the context of radiation protection and potential environmental contamination. Still, there are some natural (cosmic radiation, solar flares, coronal mass ejection) and man-made phenomena (nuclear explosions, especially at high altitude) where the power grid, electrical machinery, and electronic components can experience voltage transients under ionizing radiation.

In order to describe the electrical breakdown of the insulating gas, avalanche coefficients are used to specify the elementary processes in GFSA's gas. Avalanche coefficients that are most commonly used are α, the number of electronic ionization collisions per cm of a distance in the direction of the electric field; η, the number of electrons per cm of a distance in the direction of the field attached to the electrically negative atoms or molecules; and γ, the number of electrons generated

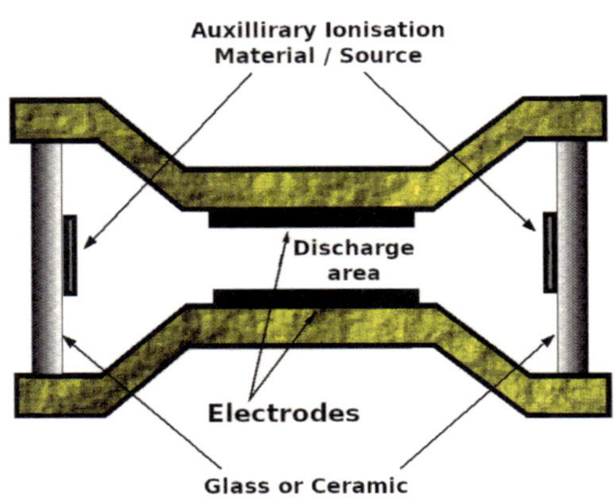

Figure 1.
Schematic of a gas-filled surge arrester [9].

from secondary processes per primary avalanches. The avalanche coefficients vary with pressure and electric field and are not constant for any particular gas or a gas mixture. This dependence has been experimentally determined for most insulating gases and is described as a range of values of the *pd* product, where p is the insulator gas pressure and d is the distance between the electrodes [1, 11, 16].

The electrical breakdown in gases can be dynamic or static, not depending on the precise mechanism of the electrical breakdown. In gases, static breakdown develops when the rate of voltage variation is significantly lower than the rate of elementary processes occurring during the electrical breakdown of the gas. The breakdown becomes dynamic in the moment these rates become comparable. Depending on the localization of the dominant secondary processes in the system, the static breakdown occurs through the Townsend mechanism or the streamer mechanism. Static breakdown voltage is a deterministic feature of the system, unlike dynamic breakdown that is a stochastic process occurring in a range of voltages [17].

Townsend mechanism is the main pathway of direct current (DC) breakdown, if the secondary processes on the electrodes (e.g., the ionic discharge, photoemission, metastable discharge, etc.) are more prevalent than the secondary processes in the gas (the ionization by positive ions, photo ionization, metastable ionization, etc.) [3]. The value of DC breakdown voltage by Townsend mechanism is described by the following equation:

$$\gamma \int_0^d e^{\int_0^d (\alpha-\eta)\mathrm{d}x} \alpha \mathrm{d}x = 1 \tag{1}$$

where α is the number of ionization collisions per cm along the electric field, η is the number of electrons per cm along the field associated to electrically negative atoms or molecules, and γ is the number of electrons generated from secondary processes per primary avalanche.

When the secondary processes in the gas are more prevalent than the secondary processes on the electrodes, DC breakdown will occur as streamers, and the breakdown voltage will be dictated by the streamer mechanism [18, 19]. The value of a DC voltage for streamer mechanism breakdown is described by the following equation:

$$\int_0^d [\alpha(x) - \eta(x)] \cdot \mathrm{d}x \geq k \tag{2}$$

To describe, determine, and compare the pulse shape characteristics of GFSA, an analytical algorithm has been developed. This approach allows a fast and accurate determination of the GFSA characteristics, without including the irreversible processes resulting from the repeated measurements of the "pulse breakdown voltage" random variable, using different shapes of rapid voltage changes. A pulse shape characteristic (volt-second) represents the breakdown voltage of the GFSA electrode configuration as a function of the applied voltage pulse duration. A decrease in the duration of the pulse results in the increase of the breakdown voltage. Determining the exact pulse characteristic experimentally would require a large number of tests with differently shaped voltage pulses. The application of the area law allows the determination of the pulse shape characteristic based on a single set of measurements using a single pulse voltage shape. The basic assumption in the area law is that plasma-spreading rate in the inter-electrode gap increases linearly due to the rise of the electric field:

$$V(x,t) = k \tag{3}$$

where k is the parameter that depends on electrical discharge mechanism and electrode polarization and E_s is the electric field value corresponding to the DC breakdown voltage U_s. The voltage pulse breakdown condition is given by $u(t) > U_s$, given that the DC breakdown voltage U_s represents the lowest possible value of the breakdown voltage for a specific electrode configuration. Without taking the charge spreading in the inter-electrode gap into account, the following stands:

$$E(x,t) = u(t) \cdot g(x) \tag{4}$$

where g(x) is a function that is to be determined from the specific electrode configuration.

If the complete breakdown occurs through the Townsend mechanism (i.e., k is assumed to be constant along the inter-electrode gap), then according to expressions (3) and (4)

$$\frac{1}{k}\int_{t_1}^{t_1+t_a} \frac{dx}{g(x)} = \int_{t_1}^{t_1+t_a} [u(t) - U_s]dt = P = \text{const} \tag{5}$$

where $x = x_k$ indicates the zone where breakdown through Townsend mechanism changes into breakdown by a streamer, in the moment $t = t_1 + t_a$. In order for a breakdown to occur, a constant surface has to be formed in the voltage-time plane between u(t) and U_s. Thus, measuring the surface P and the DC breakdown voltage U_s is sufficient to determine the pulse shape and pulse breakdown voltage, given that these characteristics of the system do not depend on the applied voltage [7].

A semiempirical method to determine the pulse shape characteristic is performed in the following way: a sequence of DC breakdown voltages is measured (with at least 20 measurements), followed by a sequence of pulse breakdown voltage measurements, by applying the pulse voltage of a stable and defined shape (at least 50 measurements). The corresponding distribution function is obtained by statistical analysis of the measured values. Distribution function allows the determination of the quantities of pulse shape characteristics U_x and U_y desired boundaries (most frequently one takes x = 0.1% and y = 99.9%). When the mean value of DC breakdown voltage U_s is known, and the quantities are determined, following system of equations can be solved:

$$\begin{aligned} u(t) &= U_s, t = t_1 \\ u(t) &= U_x, t = t_{ax} \\ u(t) &= U_y, t = t_{ay} \end{aligned} \tag{6}$$

Values t_1, t_{ax}, and t_{ay} allow the determination of surfaces P_x and P_y by applying the area law:

$$\begin{aligned} P_x &= \int_{t_1}^{t_1+t_{ax}} [u(t) - U_s]dt = \text{const} \\ P_y &= \int_{t_1}^{t_1+t_{ay}} [u(t) - U_s]dt = \text{const} \end{aligned} \tag{7}$$

Upon the determination of surfaces P_x and P_y, it is possible to calculate the xth and yth values of the "pulse breakdown voltage" random variable (by applying the area law) for any form of pulse voltage u(t). If the form of pulse voltage is taken as a parameter (in the time interval considered), it is possible to determine xth and yth pulse shape characteristics [7].

3. Radiation resistance of gas-filled surge arresters

The radiation resistance of gas-filled surge arresters (GFSA) is of great importance, especially if the devices are applied in operating regimes with constant, occasional, or potential exposure to ionizing radiation. Natural or artificial atmospheric electromagnetic pulses can cause varying levels of damage to electronic components. Thus, GFSA are very important overvoltage protection components for low-voltage applications in both the military industry and space exploration technologies. Testing of GFSA in comparison to other nonlinear surge arresters has indicated the feasibility of replacing commonly used semiconductor overvoltage components (transient suppressor diodes, metal-oxide varistor), whose protective characteristics significantly degrade when subjected to the radiation [4, 6].

4. Induced radiation effects on GFSA characteristics

In order to test the performance of GFSA under the influence of n + γ radiation, the following variables were determined in a n + γ field: (1) the random variable "pulse breakdown voltage," (2) the random variable "DC breakdown voltage," and (3) the volt-second characteristic. The n + γ source was californium isotope ^{252}Cf. This source was selected due to its neutron spectrum resemblance of a nuclear blast's neutron spectrum [20]. Since the nuclear cross section for capturing a neutron is large enough only for thermal and slow neutron capture and due to the structure of californium source fission spectrum, a relatively small part of neutrons takes part in the neutron activation of GFSA materials. GFSA was subjected to two neutron fluencies: 5.41×10^9 and 16.24×10^{11} n/cm^2. Along with the neutron component, emitted radiation also has a γ component. The latter influences the electric characteristics of GFSA only for the duration of the exposure to the radiation field. Also, the inelastic interaction cross section of the neutron component is larger than the corresponding γ component cross section [21]. This allows the observation of the effects of radiation resulting from the neutron fluency only. In the experiment, the type and pressure of gas varied in order to get a detailed insight into how the radiation influences the GFSA characteristics.

By measuring 1000 values, the influence on the "DC breakdown voltage" and "pulse breakdown voltage" random variables was tested. During the measurement series, discharge energy (current) was maintained constant. Results of the breakdown voltage obtained in the measurement series were divided into 10 groups of 50 successive values. Statistical tests were performed on each group of results by graphical visualization and chi-squared and Kolmogorov-Smirnov tests. Within each group of measurements, the measured values of breakdown voltage were tested with respect to the type of theoretical distributions (normal, exponential, double exponential, and Weibull's). U test was used to determine the groups of measurement series having the same random variable (with significance level $\alpha = 5\%$) [17, 22]. The area law was used to explore the effects on the pulse shape (volt-second) characteristic.

The experiments show that the standard deviation of the static breakdown voltage significantly decreased after the irradiation of the GFSA. The pulse voltage tests show that an irradiated GFSA reacts more readily and has somewhat narrower volt-second characteristic than unirradiated GFSA. Effectively, irradiation has improved GFSA's protective traits. GFSA DC breakdown voltage versus neutron fluency is presented in **Figure 2**. The GFSA volt-second characteristics before and after exposure to the radioactive source, respectively, is presented in **Figure 3A** and **B**.

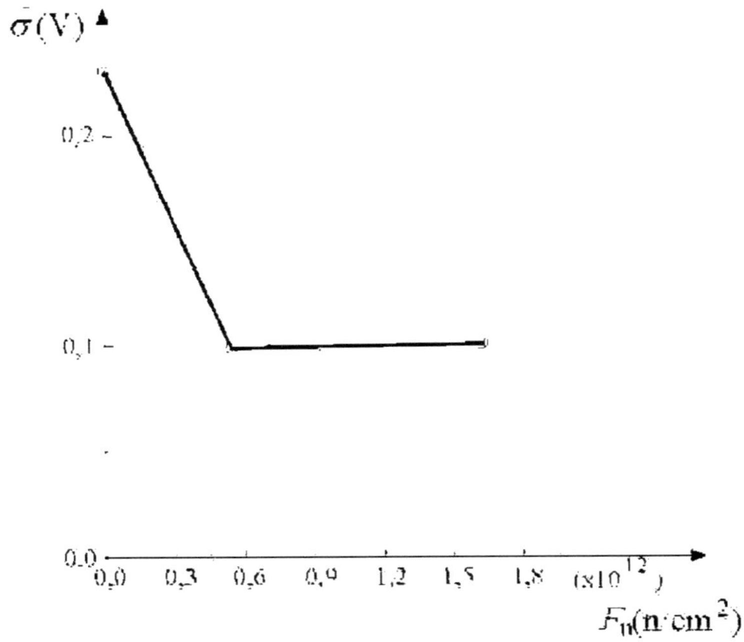

Figure 2.
The GFSA DC breakdown voltage versus neutron fluency characteristic [7].

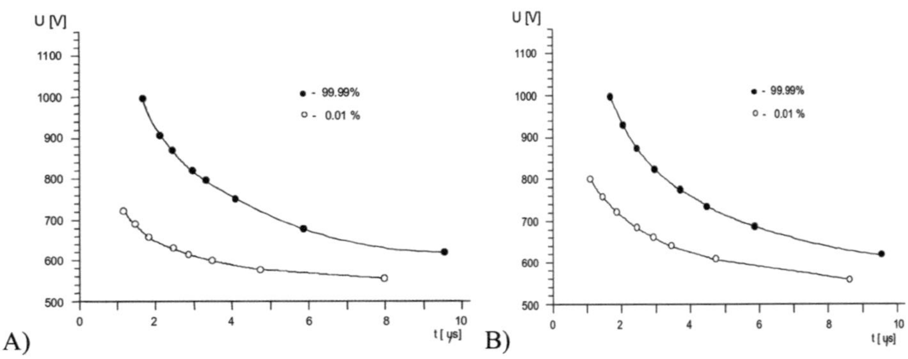

Figure 3.
The GFSA volt-second characteristic before (A) and after (B) radiation [7].

Elevated concentration of free electrons in the inter-electrode gap, resulting from the ionization of the insulating gas, has enabled a faster response of an irradiated GFSA. This ionization was induced by radiation of the GFSA material as a consequence of neutron activation.

GFSA activation analysis diagrams before and immediately after the exposure to the radioactive source, respectively, are presented in **Figure 4A** and **B**. The radioactive isotopes are identified and recorded close to the expected energy peaks. The activity of these isotopes consists of both γ and β component. This induced radioactivity ionizes the gas, leading to the reduction the stochastic dissipation of a pulse breakdown voltage random variable. The improvement to the pulse shape characteristic due to neutron radiation is short lasting, disappearing quickly as the half-lives of induced activities vary from several hours to mere minutes. A diagram of the activation

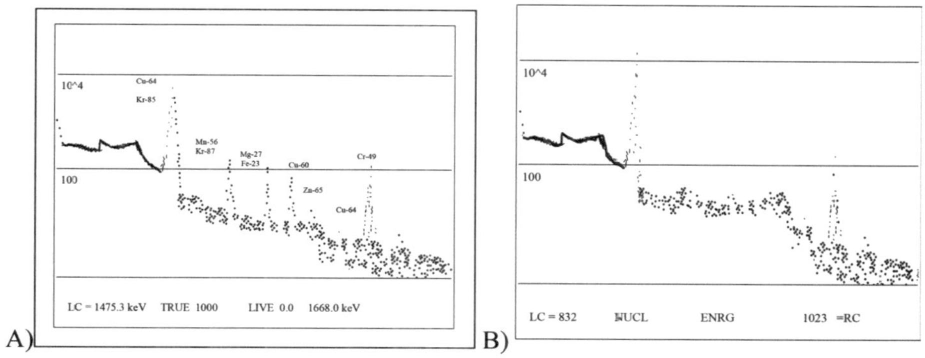

Figure 4.
Diagram of the GFSA activation analysis immediately (A) and 6 hours (B) after irradiation [7].

analysis of irradiated GFSA that is taken 6 hours after radiation clearly confirms this effect (**Figure 4B**). In that time frame, most of the active isotopes have degraded to trivial activities, and the GFSA characteristics have returned to unirradiated state.

5. Radiation resistance of commercial GFSA components

An external γ ray source was used to test commercially available GFSA components to analyze the effect of γ radiation. The examination was carried out on the following commercial components (1) SIEMENS (type A) gas surge arresters (nominal voltage 230 V), (2) CITEL BB (type B) bipolar ceramic gas surge arresters (DC spark overvoltage 230 V). The outer dimensions and shape of all components of the same type were the same. The effects of γ radiation on following GFSA characteristics were examined:

1. Prebreakdown current as function of applied voltage

2. Resistance as function of applied voltage

The scheme of the test cycle for investigating the radioactive resistance of GFSA by a DC voltage is depicted in **Figure 5**.

Examination of GFSA radioactive resistance was carried out in a gamma radiation field of ^{60}Co at the Institute of Nuclear Sciences "Vinča." The average energy of the applied gamma quantum was 1.25 MeV. The dose rate in air was 87.5, 875, and 1750 cGy/h, respectively. The distance between the radioactive source and the examined overvoltage components was 272, 86, and 60 cm, respectively. All tests were performed at room temperature, 20°C.

Test specimens, consisting of 50 commercial components of a single manufacturer, having identical characteristics, have been used in the experiment. During the formation of experimental groups consisting of 50 components each, the nominal characteristics of the tested components have been measured. When the measured values for a particular component exhibited significant discrepancy with respect to the declared values, they were excluded from further testing in accordance to the Sovene's criterion [2, 8].

The GFSA prebreakdown current as a function of applied voltage without radiation and with γ radiation is shown in **Figure 6A** and **B**, respectively. The diagrams demonstrate:

1. Before the breakdown in the absence of radiation, the current conducted by GFSA is constant and in the order of 0.1 nA. Upon the breakdown, the current rises very sharply (breakdown voltage being 212 V for type A components and 223 V for type B components). When voltage reaches the breakdown level, an abrupt increase of the current takes place, and current values reach μA level. Radiation causes increased numbers of electron-ion pairs in the area between electrodes, leading to the increase of prebreakdown current (consisting of the free electrons and ions reaching electrodes per unit of time). In the moment of breakdown (when one of the free electrons generated in this way becomes initial), an avalanche process generates a breakdown current, the magnitude of which is independent of the prebreakdown current. Ohm's law cannot be applied in this region, given that the observed two-electrode system saturated: all electron-ion pairs generated reach the electrodes in the observed time frame.

2. The influence of γ radiation on GFSA performance is significant. The prebreakdown current is 10 times larger than the current without radiation. The increase of the radiation dose rate causes an increase in the prebreakdown current. Breakdown occurred at lower voltages (205 V) in a γ radiation field. During the transition between nonconducting and conducting regime, the current increase is not as sharp as compared to the breakdown without irradiation.

Single electrons in atomic orbitals have low effective photoionization cross sections due to the small wavelength of high energy γ photons that cause ionization through the Compton effect [21].

The GFSA resistance versus applied voltage is shown in **Figure 7A** and **B**, without radiation and in Co γ field, respectively. From the volt-ampere curve, the volt-ohm characteristic can be easily determined. Formula-defining relationship between resistance and voltage is obtained by linear regression, using least-square minimal error method. The following conclusions are made:

1. GFSA resistance shows linear increase with the applied voltage in the prebreakdown regime. Increase is more prominent for type A commercial components than type B. Abrupt decrease of the resistance is observed as voltage reaches breakdown level.

2. Resistance also exhibits linear increase with the applied voltage in a γ radiation field but has one order of magnitude lower values than in operation without radiation. A slight decrease of the resistance is observed at voltage values near breakdown voltage; at breakdown voltage the decrease is pronounced.

The main conclusions of the experiment are:

1. In a γ radiation field, breakdown occurred at lower voltages (205 V). The values of breakdown voltage for two types of commercial GFSA have been determined in the presence of γ radiation and without radiation.

2. Prior to breakdown, the current had constant values, of the order of 0.1 nA. When voltage reached breakdown level, an abrupt increase of current was observed, current values reaching μA level.

3. In the prebreakdown regime, GFSA resistance increased linearly with the applied voltage. When the voltage reached breakdown level, an abrupt decrease of the resistance was observed.

4. γ radiation shows significant influence on GFSA performance. In prebreakdown regime, the current had one order of larger magnitude values than without radiation.

5. In a γ radiation field, resistance also showed a linear increase with the applied voltage, but one order of magnitude was lower than values without radiation.

6. All observed effects of γ radiation on GFSA commercial components had reversible character. Shortly after exposure (in a matter of hours), GFSA characteristics were the same as before irradiation.

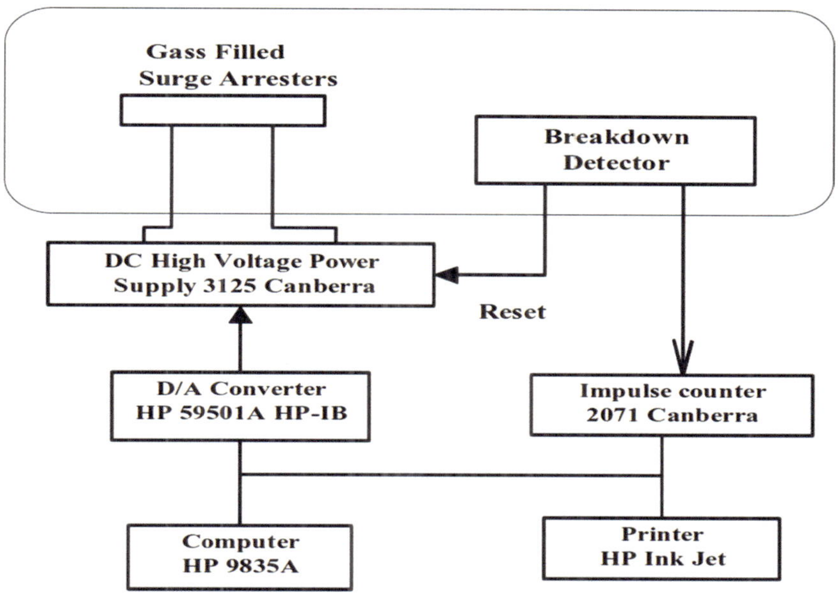

Figure 5.
Scheme of the test cycle for investigating the radioactive resistance of GFSA by a DC voltage.

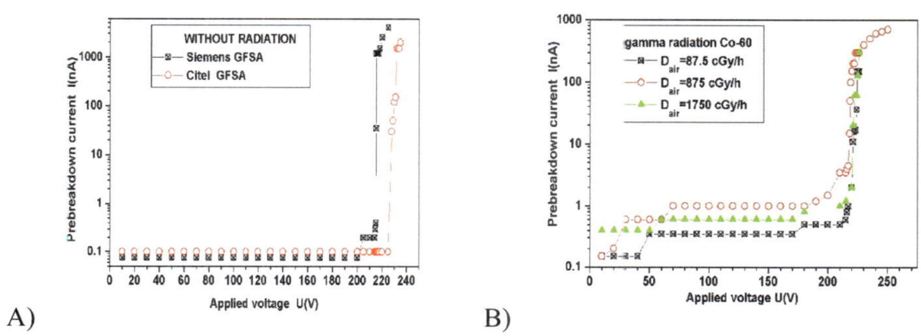

Figure 6.
GFSA prebreakdown current without radiation (A) and under γ radiation (B) [9].

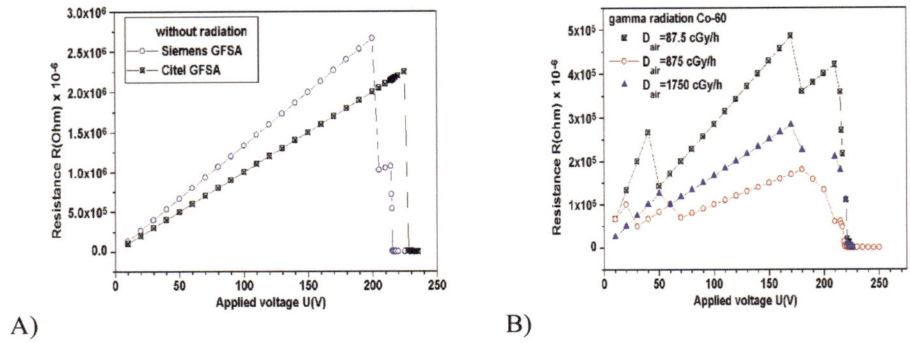

Figure 7.
GFSA resistance versus applied voltage without radiation (A) and under γ radiation (B) [9].

6. Testing the GFSA model in γ radiation field

A model of GFSA was constructed to allow the variation of relevant characteristics (chamber gas pressure and electrode materials), in order to test GFSA performance in various operational regimes. The experiment was performed as follows: (1) GFSA model was formed by choosing the appropriate material for the electrodes, placing the electrodes inside a gas-vacuum chamber, and setting the optimal electrode distance (**Figure 8A**); (2) the formed model (the gas tube) was connected to the gas-vacuum system with suitable valves, with a vacuum pump on one side and a steel gas supply cylinder on the other, and a pressure gauge; (3) the model was vacuumed by obtaining a stable pressure using valves leading to the vacuum pump and needle valves for grading the pressure; (4) the specific dose rate was set by appropriate positioning of the gas chamber relative to the source; (5) the electric circuit including the GFSA was closed; (6) the electrodes were conditioned by being kept in discharge state for a while in order to attain stable working conditions and insure the repeatability of measured results; (7) the value of prebreakdown current was measured, as the applied voltage was gradually increasing in each of the experimental dose rates; and (8) resetting the experimental system at a different working point (electrode material, gas pressure, dose rate) and performing a new measurement procedure. The scheme of the experimental setup is presented in **Figure 8B**. Measuring equipment consisted of (1) gas-vacuum chamber, (2) pressure gauge Speedivac, (3) steel cylinder with pressurized Ar gas, (4) vacuum pump Edwards 5, (5) DC high voltage source, CANBERRA, (6) AVOmeter Iskra MI 7006, (7) digital multimeter LDM—852 A, (8) variable resistance MA 2110, and (9) coaxial cables and connectors.

Examination of the GFSA was carried out in a gamma radiation field of ^{60}Co. The average energy of the applied gamma quanta was 1.25 MeV. The absorbed dose rate in air was, respectively, 96, 960, and 1920 cGy/h. The distance between the examined overvoltage components and the radioactive source was, respectively, 272, 86, and 60 cm. The distance between electrodes was 0.5 cm. All tests were carried out at pressures of 4666.27 Pa (35 Torr) and 2666.45 Pa (20 Torr), at room temperature of 20°C. The electrodes were made either of aluminum, steel, or brass.

Investigation of the dependence of the prebreakdown current on the applied voltage was performed under various experimental conditions: in the absence of radiation and in a gamma radiation field (for aluminum, steel, and brass electrodes), under two gas pressures. The results for DC current are presented in **Figures 9–11**, for aluminum, steel, and brass, respectively. In captions shown in these figures, current I_1 corresponds to the measurements without the radiation, current I_2 corresponds to the γ radiation absorbed dose rate of 0.96 Gy/h, current I_3

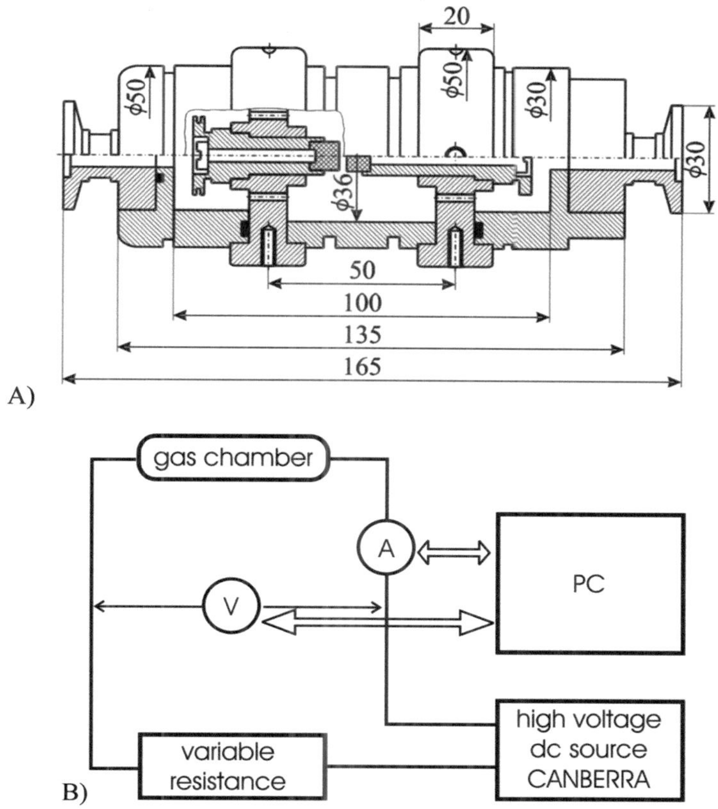

Figure 8.
GFSA model (dimensions in mm, A) [23] and the scheme of the test cycle (B).

corresponds to the γ radiation absorbed dose rate of 9.6 Gy/h, and current I_4 corresponds to the γ radiation absorbed dose rate of 19.2 Gy/h.

Gamma radiation has a strong influence on the prebreakdown current in GFSA, as can be deduced from the presented graphs. Prebreakdown current is constant and independent of the applied voltage up to the value of the breakdown voltage in the absence of radiation. When ^{60}Co source is present, a steady rise of the prebreakdown current is observed, increasing with the increase of applied voltage. For all three-electrode materials, the rise of the prebreakdown current is more pronounced as the γ radiation dose increases, effective under both of the tested pressures. Breakdown voltage increased under higher gas pressure regime. The highest breakdown voltages were obtained using brass electrodes (up to 450 V), and the lowest were obtained using steel electrodes (from 320 to 350 V, depending on the radiation dose). Under lower pressure, for both steel and brass electrodes, higher radiation doses resulted in lower breakdown voltages, with the highest breakdown voltage measured when no radiation was applied. For aluminum, the highest breakdown voltage was obtained under highest radiation dose. The best performing GFSA for DC current was the brass electrode under 20 Torr pressure.

Pulse shape (volt-second) characteristic is shown in **Figures 12–14**, respectively, for aluminum, steel, and brass electrodes. Experimental data indicates the following: γ radiation leads to a decrease in standard deviation and the narrowing of pulse shape characteristics of the arresters, which leads to an increase in the response speed. Because of that, we can conclude that γ radiation improves the performance of GFSA. This phenomenon is most prominent in aluminum

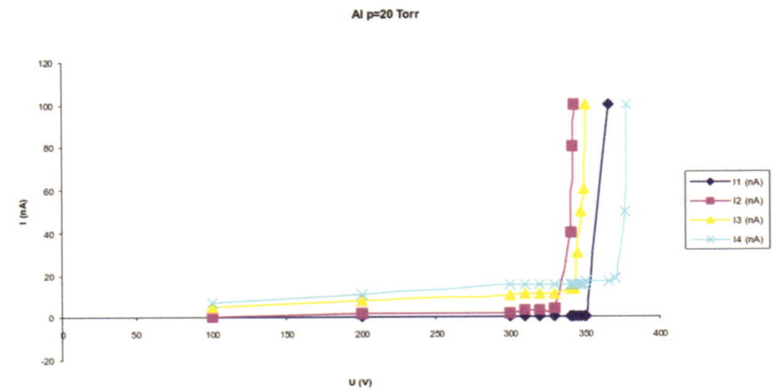

A)

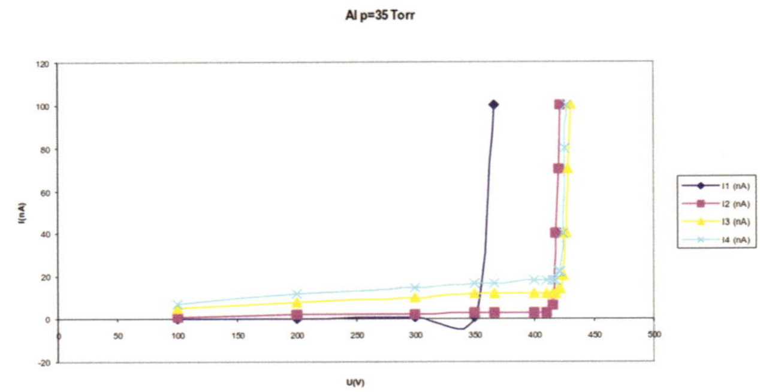

B)

Figure 9.
Prebreakdown current versus applied voltage in γ radiation field with aluminum electrodes under pressure of 20 Torr (A) and 35 Torr (B) [9].

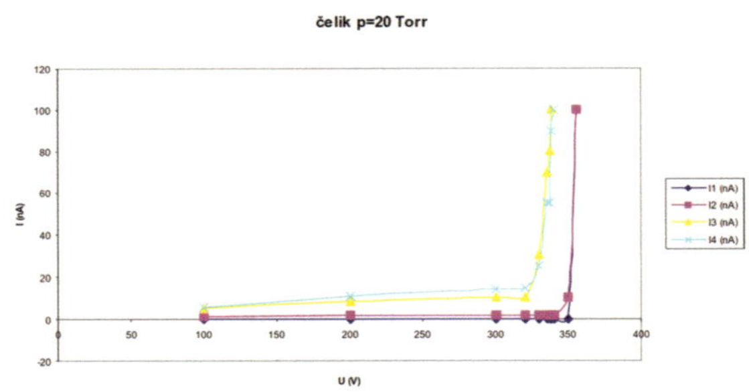

A)

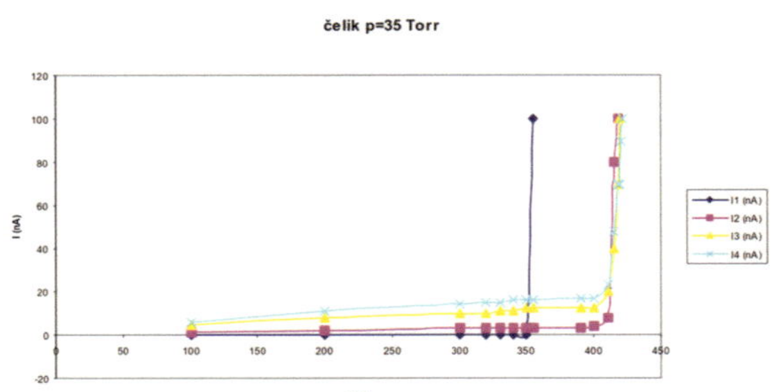

B)

Figure 10.
Prebreakdown current versus applied voltage in γ radiation field with steel electrodes under pressure of 20 Torr (A) and 35 Torr (B) [9].

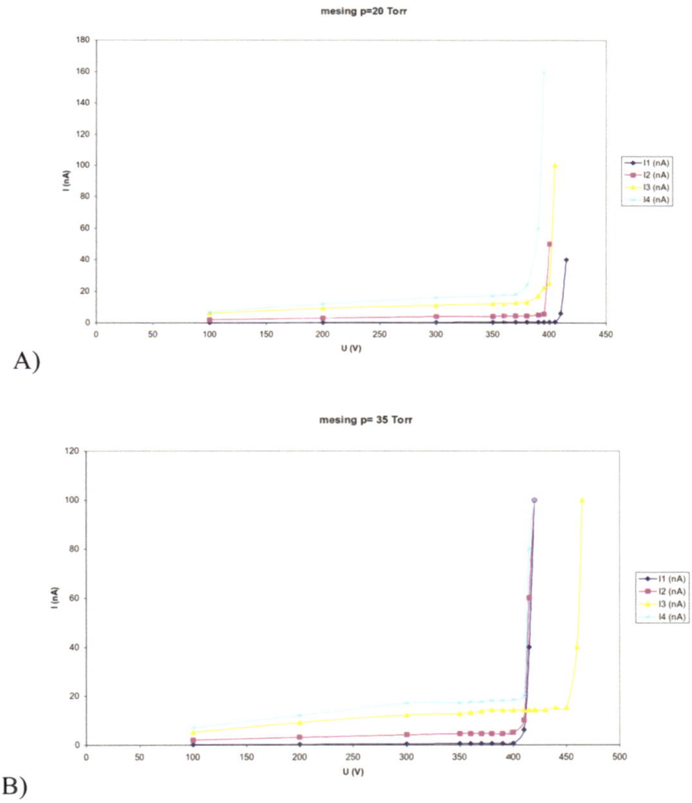

Figure 11.
Prebreakdown current versus applied voltage in γ radiation field with brass electrodes under pressure of 20 Torr (A) and 35 Torr (B) [9].

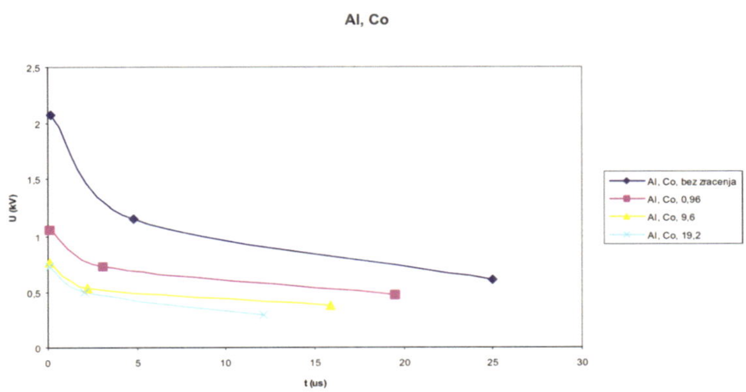

Figure 12.
Pulse shape (volt-second) characteristic for aluminum electrodes in γ radiation field [9].

electrodes and least prominent in steel electrodes. Breakdown voltage deviation was least under the shortest pulses (1.2/50 μs) and highest under the longest ones (100/700 ms). All the observed changes lasted only during γ radiation and reversed as soon as the radiation exposure ceased. The best pulse shape characteristic was obtained for aluminum electrodes, which had its performance improved under γ radiation [9].

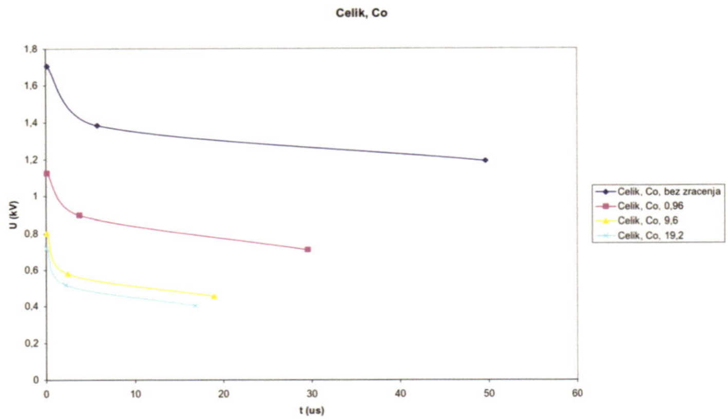

Figure 13.
Pulse shape (volt-second) characteristic for steel electrodes in γ radiation field [9].

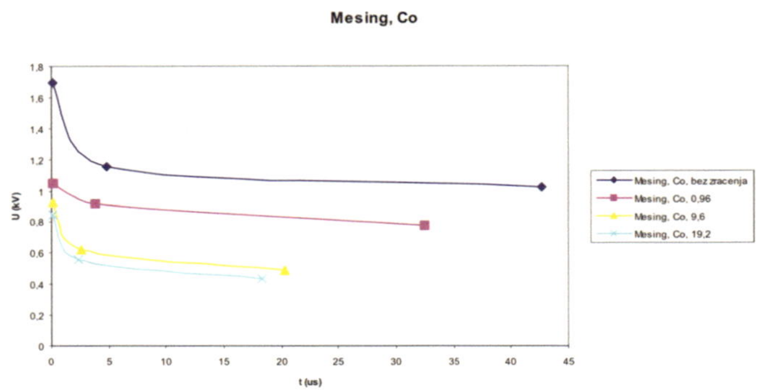

Figure 14.
Pulse shape (volt-second) characteristic for brass electrodes in γ radiation field [9].

7. Conclusion

In this chapter, the influence of γ radiation on gas-filled surge arrester operation is discussed. An experimental model has been developed that allows easy modification of elements of the system and tests under different operational regimes. The experimental setup has also been used to test commercial GFSA components. An analytical method to describe GFSA pulse shape characteristics using area law has been established. These theoretical and empirical tools were used to measure and analyze the performance of different GFSA components exposed to combine n + γ and pure γ radiation.

The experiments demonstrated that γ radiation improves the performance of GFSA. This effect was observed both in commercial components and the experimental model. The prebreakdown current had increased when GFSA were exposed to γ radiation. Beneficial effect of γ radiation on pulse shape characteristics was determined: due to the reduction of standard deviation, response time of GFSA was improved. These effects were consistent under different insulating gas pressure regimes. Among the metals tested as electrode materials using the model, brass was the best performing one. The effects of γ radiation were lasting only as long as the components were exposed. The performance of GFSA under γ radiation makes them suitable for overvoltage protection of electronic circuitry constantly or occasionally exposed to that type of radiation.

Acknowledgements

The authors thank the Ministry of Education, Science and Technological Advancement, Republic of Serbia, for supporting the research through projects no. 171007 and 43009.

Conflict of interest

The authors declare no conflict of interest.

Author details

Luka Rubinjoni[1], Katarina Karadžić[2] and Boris Lončar[2*]

1 Faculty of Technology and Metallurgy, Innovation Center, University of Belgrade, Serbia

2 Faculty of Technology and Metallurgy, University of Belgrade, Serbia

*Address all correspondence to: bloncar@tmf.bg.ac.rs

References

[1] Loncar B, Osmokrovic P, Stankovic S. Temperature stability of components for overvoltage protection of low-voltage systems. IEEE Transactions on Plasma Science. 2002;**30**(5):1881-1885

[2] Osmokrovic P, Loncar B, Stankovic S, Vasic A. Aging of the over-voltage protection elements caused by over-voltages. Microelectronics Reliability. 2002;**42**(12):1959-1966

[3] Osmokrovic P, Loncar B, Sasic R. Influence of the electrode parameters on pulse shape characteristic of gas-filled surge arresters at small pressure and inter-electrode gap values. IEEE Transactions on Plasma Science. 2005;**33**(5):1729-1735

[4] Osmokrovic P, Stojanovic M, Loncar B, Kartalovic N, Krivokapic I. Radioactive resistance of elements for over-voltage protection of low-voltage systems. Nuclear Instruments and Methods in Physics Research Section B: Beam Interactions with Materials and Atoms. 1998;**140**(1-2):143-151

[5] Markov Z. The comparison of modern over-voltage protection components. Electrotechnics. 1987;**10**(1):961-963

[6] Loncar B, Osmokrovic P, Stankovic S, Filipovic D. Investigation the optimal method for improvement the protective characteristics of gas filled surge arresters-W/O the built in radioactive sources. In: Pulsed Power Plasma Science 2001 "PPPS-2001", vol. 1. Digest of Technical Papers; 17 June 2001. IEEE. pp. 475-478

[7] Loncar B, Osmokrovic P, Stankovic S. Radioactive reliability of gas filled surge arresters. IEEE Transactions on Nuclear Science. 2003;**50**(5):1725-1731

[8] Loncar B, Osmokrovic P, Stankovic S, Sasic R. Influence of electrode material on gas-filled surge arresters characteristics in γ and X radiation field. Journal of Optoelectronics and Advanced Materials. 2006;**8**(2):863-866

[9] Lončar B. Radijaciona otpornost memorijskih I prenaponskih zaštitnih komponenata. Beograd: Zadužbina Andrejević; 2006. 82 p

[10] Loncar B, Osmokrovic P, Vasic A, Stankovic S. Influence of gamma and X radiation on gas-filled surge arrester characteristics. IEEE Transactions on Plasma Science. 2006;**34**(4):1561-1565

[11] Osmokrovic P, Krivokapic I, Matijasevic D, Kartalovic N. Stability of the gas filled surge arresters characteristics under service conditions. IEEE Transactions on Power Delivery. 1996;**11**(1):260-266

[12] Pejovic MM, Ristic GS, Karamarkovic JP. Electrical breakdown in low pressure gases. Journal of Physics D: Applied Physics. 2002;**35**(10):R91

[13] Osmokrovic P. Mechanism of electrical breakdown of gases at very low pressure and interelectrode gap values. IEEE Transactions on Plasma Science. 1993;**21**(6):645-653

[14] Loncar B, Novakovic D, Matijasevic D, Osmokrovic P. The influence of radioactive filling on efficient of gas filled surge arresters of small values of pd product. In: Proceedings of XLIV Conference on Society for Electronics, Telecommunications, Computers, Automatic Control and Nuclear Engineering. Belgrade, Serbia; 2000. Vol. 4. pp. 129-132

[15] Matijasevic D, Loncar B, Krivokapic I, Novakovic D. Improvement of the protective characteristics of gas filled surge arresters using hollow cathode method. In: Proceedings of XLII Conference on Yugoslav Society for

Electronics, Telecommunications, Computers, Automatic Control and Nuclear Engineering. Belgrade, Serbia; 1998. Vol. 4. pp. 229-232

[16] Loncar B, Osmokrovic P, Stankovic S, Vasic A. Static and dynamic radioactive resistance of gas filled surge arresters. In: Proceedings of 14th IEEE Pulsed Power Conference; 15-18 June 2003; Dallas, United States: IEEE; 2003

[17] Djogo G, Osmokrovic P. Statistical properties of electrical breakdown in vacuum. IEEE Transactions on Electrical Insulation. 1989;**24**(6):949-953

[18] Osmokrovic P, Krivokapic I, Krstic S. Mechanism of electrical breakdown left of Paschen minimum. IEEE Transactions on Dielectrics and Electrical Insulation. 1994;**1**(1):77-81

[19] ISO I, OIML B. Guide to the Expression of Uncertainty in Measurement. Geneva, Switzerland; 1995

[20] International Organization for Standardization. Neutron Reference Radiations for Calibrating Neutron-Measuring Devices Used for Radiation Protection Purposes and For Determining Their Response as a Function of Neutron Energy. International Organization for Standardization; 1989

[21] Messenger G, Ash M. The Effects of Radiation on Electronic Systems. New York: Van Nostrand Reinhold; 1992

[22] Rizk FA, Eteiba MB. Impulse breakdown voltage-time curves of SF6 and SF6-N2 coaxial-cylinder gaps. IEEE Transactions on Power Apparatus and Systems. 1982;(12):4460-4471

[23] Stanković K, Brajović D, Alimpijević M, Lončar B. Long-term deconditioning of gas-filled surge arresters. Radiation Effects and Defects in Solids. 2016;**171**(7-8):678-691

Chapter 14

Gamma Irradiation Causes Variation and Stability of Artemisinin Content in *Artemisia annua* Plants

Thongchai Koobkokkruad, Praderm Wanichananan, Chalermpol Kirdmanee and Wanchai De-Eknamkul

Abstract

Artemisinin is an anti-malarial sesquiterpene lactone isolated from *Artemisia annua* L., a traditional Chinese herb of the family Asteraceae. The plant contains relatively low artemisinin content, ranging from 0.01 to 0.8% of the plant dry weight, depending on the geographical origin, seasonal, and somatic variations. Ionizing radiation has been recognized as a powerful technique for plant improvement, especially in crop plants. This technique creates genetic variability in plants, which can be screened for desirable characteristics. Very little is known about the effect of gamma irradiation on the potential increase of artemisinin production in *A. annua*. In this study, 130 shoot tips excised from the population of in vitro *A. annua* plantlets (with an average leaf artemisinin content of 0.18 ± 0.09%) were exposed to 5 Gy ^{60}Co gamma irradiation and subsequently transferred to a suitable medium for in vitro development of plantlets. The resulting 90 stable survived after four passages appeared to have a wide variation of artemisinin content, ranging from 0.02 to 0.68% of dry weight. All the viable plantlets were then transferred from the in vitro cultures to ex vitro conditions both in a greenhouse and an open field. A significant correlation was observed between artemisinin content among individual pairs of the vitro plantlets and ex vitro mature plants, with the correlation coefficient (R^2) values of 0.915 for the greenhouse plants and 0.797 for the open field plants. Among these, the highest artemisinin-containing plant appeared to accumulate 0.84% artemisinin of dry weight in the open field, which is almost five times higher than the original plants. These results suggest that gamma irradiation with 5-Gy dose can produce viable variants of *A. annua* that can maintain the biosynthetic capability of artemisinin throughout the in vitro-ex vitro transfer and development of the first generation of mature plants.

Keywords: artemisinin, *Artemisia annua* L., gamma irradiation, in vitro plantlets, ex vitro plants

1. Introduction

Artemisinin is a natural sesquiterpene lactone containing an unusual peroxide bridge (**Figure 1**) [1]. It is present mainly in the leaves of *Artemisia annua L.* (family Asteraceae) by storing in the glandular trichomes, which are tiny specialized hair-like epidermal cells found on the epidermis of leaves [2]. This traditional Chinese herb is a wild growing species with relatively low artemisinin content, ranging from 0.01 to over 1% of the plant dry weight, depending on the geographical origin, seasonal, and somatic variations [3, 4] and density of glandular trichomes in the leaves and aerial parts [5]. At present, the only commercial source of artemisinin is by extraction from field-grown leaves and flowering tops of the plant although many attempts to obtain higher artemisinin yield have been made from using simple breeding programs to complicated biotechnological approaches (for review, see [6]). Total synthesis of the compound has been reported [7, 8], but many chemical steps are required and the yields are low. In vitro cultures of *A. annua*, such as cell suspension and callus [9], shoot [10, 11] and hairy root cultures [12–15], have also been established for studying their potentials of producing artemisinin, but in vitro culture for artemisinin production has yet to prove commercially feasible. Therefore, the whole plant of *A. annua* is still the most economic source of artemisinin, and the development of high-producing plants of *A. annua* seems to be the main direction to obtain large quantities of relatively cheap artemisinin.

Ionizing radiation has been recognized as a powerful technique for plant improvement, especially in crop plants [16–18] and medicinal plants (for review, see [19]). This technique creates genetic variability in plants, which can be screened for desirable characteristics. So far, very little is known about the effect of gamma irradiation on the potential of artemisinin biosynthesis in *A. annua*, which involves several steps in its pathway [6]. Previously, we have reported a method for establishing in vitro plantlet variants of *A. annua* using low-dose gamma irradiation (less than 10 Gray) [20]. The survived plantlet variants maintained under the in vitro conditions for more than 6 months appeared to have stable content of artemisinin. However, it remained unknown whether the changes in artemisinin biosynthesis in in vitro plantlets would be maintained when the plants are grown ex vitro in a greenhouse or an open field. This question prompted us to investigate the process of acclimatization of the plantlets, followed by evaluation of the artemisinin content in the resulting whole plants.

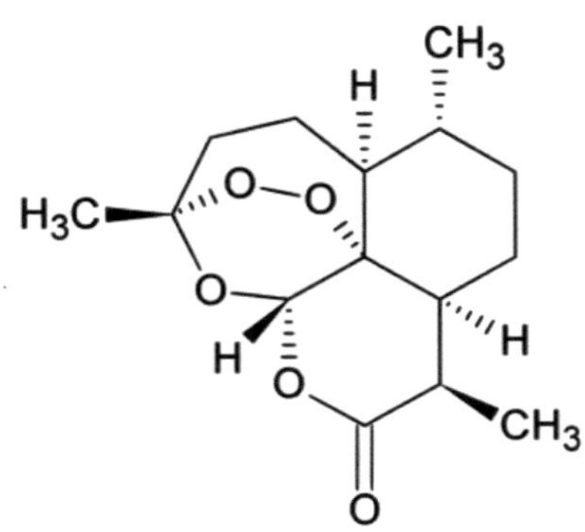

Figure 1.
The structure of artemisinin [1].

2. Application of the gamma irradiation technique for potential increase of artemisinin accumulation in *A. annua*

2.1 Effects of gamma irradiation on the morphology and survival of *A. annua* plantlets

In this study, mature seeds of *A. annua* were first surface sterilized and germinated on the MS medium, supplemented with 3% sucrose and solidified with 0.8% agar. The cultures were incubated for 2 months at 25°C with an exposure to 16 h light (ca. 3000 lux) and 8 h dark cycle. The obtained plantlets were then subcultured for five times before their shoot tips were excised and treated with gamma radiation. Practically, 1000 shoot tips (ca. 5 mm) excised from the in vitro plantlets were placed onto the same MS medium and irradiated with gamma rays generated by Cobalt-60 at the dose rate of 8.56 Gy min^{-1} (using the facilities at the Office of Atomic Energy for Peace (OAEP), Bangkok, Thailand). With this dose rate, the amount of irradiation energy absorbed by the shoot tips from 1 to 10 Gray (Gy) was conducted using the irradiation times from 7 to 70 s. After the irradiation, the exposed shoots were transferred to the fresh hormone-free MS medium. The shoots with subsequent active growth were subcultured every six weeks for four times on the same hormone-free MS medium. All the cultures were grown under the same conditions. After the forth passage, each vigorous shoot was cultured on the hormone-free MS medium in a 230-ml glass bottle. After culturing for six weeks, survival percentage and regrowth ability were recorded.

Figure 2 shows the morphology of *A. annua* plantlets derived from shoot tips gamma irradiated with a low dose range from 1 to 10 Gy. The plantlets depicted are representative of populations irradiated with the indicated gamma ray doses that survived four subsequent passages over a period of more than 6 months. The lowest dose of 3 Gy appeared to promote the growth of the plantlets, whereas doses above 5 Gy led to significant growth and morphological abnormalities. As shown in **Figure 2**, the 8-Gy dose gave rise to plantlets with pale green, fully expanded leaves, and the 10-Gy dose resulted in dwarf plantlets with no root differentiation.

In terms of survival rate, the results showed that there was a continual reduction in the survival percentage of the in vitro plantlets with increase in gamma

Figure 2.
Effects of low dose of gamma irradiation on the morphology of A. annua plantlets.

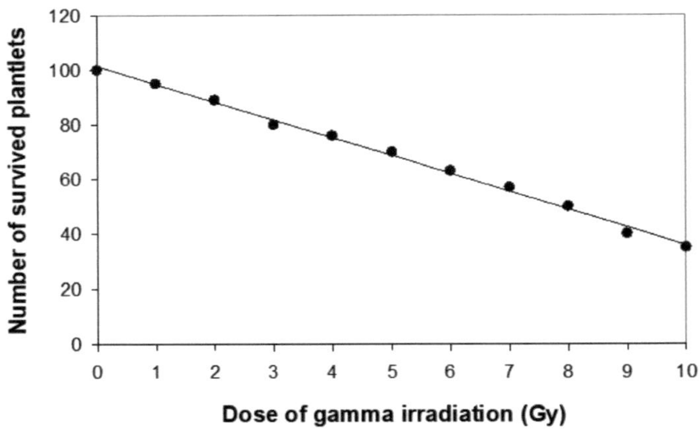

Figure 3.
The effect of various doses of gamma irradiation on the survival of in vitro plantlets of A. annua.

irradiation dosage from 1 to 10 Gy (**Figure 3**). The lethal dose of gamma rays that causes 50% survival reduction (LD_{50}) was 8 Gy. Again, the doses lower than this LD_{50} value showed essentially normal morphology of the survived plantlets, whereas the higher doses seemed to cause significant abnormalities as shown by dwarf plantlets with pale leaves (**Figure 2**). This LD_{50} value was obtained from the in vitro plantlets survived for at least 6 months (four subsequent subcultures) after the irradiation. It is, therefore, likely that they are genetically stable variants.

2.2 Effects of gamma irradiation on artemisinin accumulation in *A. annua* in vitro plantlets

To perform a rapid analysis of artemisinin in a large population of the irradiated plantlets, we used our own developed simple and sensitive TLC-densitometric method for artemisinin analysis which was reported previously [21]. Practically, fresh leaves obtained from various in vitro plantlets were collected, dried at 60°C, and ground to fine powder in a grinder. Each powder sample (100 mg) was extracted under reflux in 10 ml hexane (70°C) for 1 h. The extract was then filtered, and a 10 μl aliquot was spotted onto a pre-coated silica gel TLC plate. Up to 15 samples could be applied onto each plate which was developed using the solvent system of hexane:ethyl acetate:acetone, 16:1:1. The plate was dried and exposed for 2 h with saturated ammonia vapor (in a closed TLC tank) for complete derivatization of artemisinin. The TLC plate was then taken from the tank, air dried, and observed in a light box under the wavelength of 366 nm which could be seen variation of artemisinin band intensity among various extract samples (**Figure 4A**). The plate was then scanned by a TLC densitometer under the wavelength of 320 nm to obtain corresponding TLC-densitometric chromatograms. The area under artemisinin peak of each sample was then converted to artemisinin content based on a calibration curve that showed linearity from 0.06 to 12 μg ml^{-1} of artemisinin.

Using this TLC-based technique, the surviving plantlets (obtained after 8 Gy treatment and four subsequent subcultures) were analyzed for their ability to accumulate artemisinin. The results showed a wide variation of artemisinin content, ranging from 0.03 to 0.70% dry weight (**Figure 5**). The control plantlets derived from the shoot tips not exposed to gamma rays showed their artemisinin levels of as low as 0.18 ± 0.09% of dry weight. In terms of content distribution, almost 80% of the irradiated plantlets showed artemisinin content less than 0.3%,

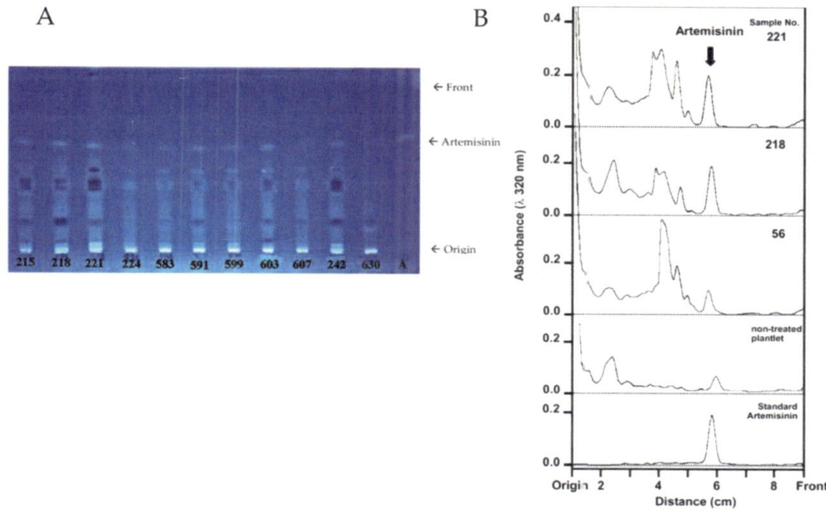

Figure 4.
(A) Typical TLC-patterns of some A. annua crude extracts prepared from the population of gamma-irradiated plantlets. Variation of artemisinin band intensity among the samples can be observed under 360 nm after exposed with ammonia at 100°C for 2 h. (B) TLC-densitometric chromatograms obtained by scanning at 320 nm of some samples of the TLC plate shown in (A).

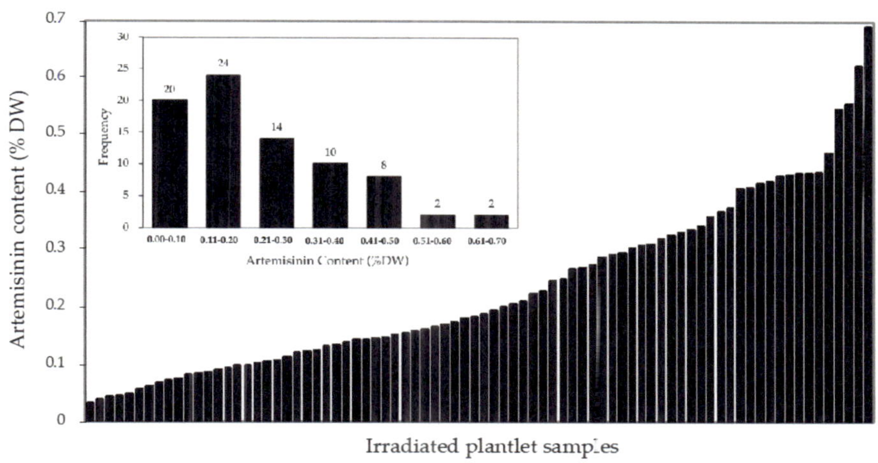

Figure 5.
Variation of artemisinin content in various plantlets of A. annua that their shoot tips had been exposed to a dose of 8 Gy of gamma rays.

and approximately 5% of the population showed higher than 0.5% of dry weight (**Figure 5** inset). Thus, it was clear that there were variations in the potential of artemisinin biosynthesis among the irradiated plantlet population.

It should be noted that more than 50% of the plantlet population accumulates artemisinin in the content higher than the original untreated plants (0.18 + 0.09% of dry weight). This is probably due to the use of low-artemisinin containing plants as a starting material, which allow higher artemisinin-containing variants be obtained more easily upon the irradiation. The observed quantitative and some extent of qualitative variations indicate that the secondary metabolism in the irradiated plantlets of *A. annua* is affected considerably by the treatment of the low-dose gamma irradiation.

2.3 Correlation between artemisinin accumulation and enzyme activity of amorpha-4,11-diene synthase in irradiated plantlets

The observed variation of artemisinin accumulation in the irradiated plantlets raised a question on the possible site genes of mutation, especially of the genes involved in the biosynthetic pathway of artemisinin. Since amorpha-4,11-diene synthase (ADS) has been known as one of the key enzymes of the pathway in *A. annua* [22], it was selected as a target for studying the relationship between its enzyme activity and artemisinin accumulation. In this study, a crude enzyme extract of each plantlet sample was prepared by quick freezing the fresh leaves (5 g) in liquid nitrogen and ground in a pre-cooled mortar. The resulting fine powder was added with 15 ml cold extraction buffer containing 5 mM 3-(N-morpholino) propanesulfonic acid (Mops) buffer, pH 7.0, 10% (v/v) glycerol, 1 mM ascorbic acid, 10 mM $MgCl_2$, and 2 mM dithiothreitol (DTT). After 15 min of stirring, the suspension was passed through four layers of cheesecloth, and the filtrate was centrifuged 100,000 × *g* at 4°C. The supernatant was then desalted by passing through a PD-10 column. The filtrate was used as crude enzyme extract and kept at −80°C before being used for determining the enzyme activity of ADS.

The enzyme activity of ADS was then determined by modifying the radioisotopic method described previously [22]. The reaction mixture contained [1-^{3}H(N)] farnesyl diphosphate (100,000 dpm), 5 mM Mops buffer, pH 7.0, 10%(v/v) glycerol, 10 mM ascorbic acid, 10 mM $MgCl_2$, 2 mM DTT, and 10 mM Na_2MoO_4 in a total volume of 70 μl. After 30 min of incubation at 30°C, the reaction mixture was extracted with 1 ml hexane, taking the hexane layer to evaporate followed by spotting onto a TLC plate (aluminum sheet, silica gel 60 F254, 0.25 mm thickness). The resulting TLC plate was then developed in a solvent system of hexane:ethylacetate:acetic acid (25:7:1) and was scanned to obtain radio-chromatograms by a TLC-radioscanner. The area under amorpha-4,11-diene peak of each radio-chromatogram was then used for calculating the synthase activity. **Figure 6** shows typical TLC-radiochromatograms of the reaction mixtures catalyzed by enzyme preparations obtained from some irradiated plantlets. It can be seen that the radioactive peaks of amorpha-4,11-diene (Rf value of 0.55) could be clearly detected with different peak sizes from different samples, suggesting that the TLC-radio assay worked well for determining the enzyme activity of ADS.

Subsequently, 18 plantlet samples with different artemisinin contents were assayed for their ADS activity. Again, it was found that the selected samples showed high variation in the enzyme activity, ranging from 0.02 to 0.18 pkat mg^{-1} protein. When the results of the enzyme activity and artemisinin content were plotted together (**Figure 7**), it was found that the 18 plantlets showed their poor value of the correlation coefficient at R = 0.300 (**Figure 7A**). However, by excluding only two outliners with extremely high ADS activities, the correlation coefficient appeared to be much better, with R = 0.717 (**Figure 7B**). Among these, 11 of the 16 samples showed quite high value of the correlation coefficient, R = 0.922 (**Figure 7C**). This suggested that the gamma irradiation did affect the gene of ADS, and thus the biosynthetic capability of artemisinin in the mutant plants.

2.4 Variations in artemisinin content in plantlets irradiated with the 5-Gy dose

As mentioned earlier (Section 2.1), the doses above 5 Gy led to significant growth and morphological abnormalities, with the 8-Gy dose giving rise to

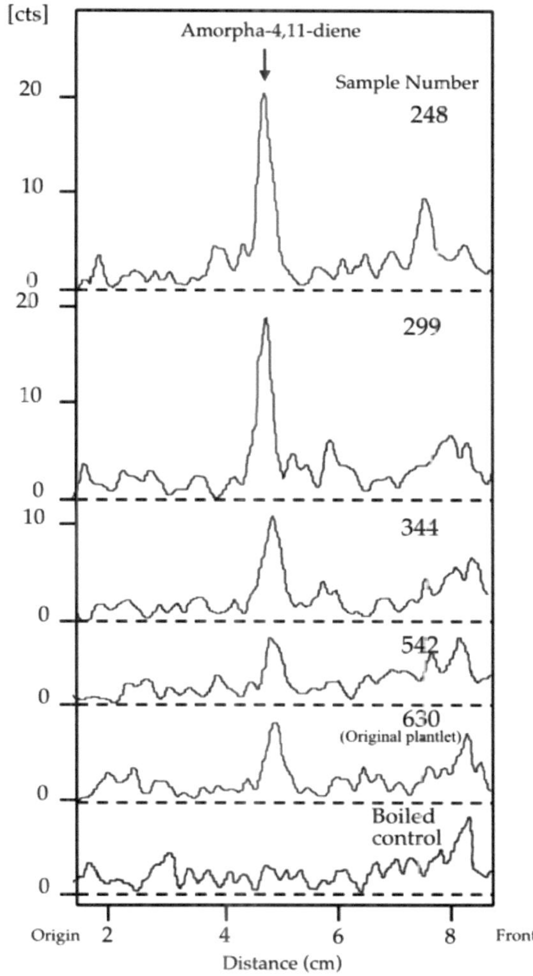

Figure 6.
Typical TLC-radiochromatograms obtained from a radio-based enzyme assay of ADS of some irradiated plantlet samples (indicated by the sample number) of A. annua. The reaction mixture of each ADS enzyme-assay sample was extracted with hexane followed by separation and detection of ADS product by a TLC-radioscanner.

pale-green and fully expanded leave plantlets, and the 10-Gy dose showing dwarf and no differentiate root plantlets with no root differentiation. Therefore, the 5-Gy dose was chosen for mass irradiation of shoot tips to obtain a population of *A. annua* plantlet variants to characterize the stability of changes in artemisinin biosynthesis in mature plants.

The results showed that among 130 shoot tips irradiated with a 5-Gy dose, 90 plantlets (69% survival) were obtained using established in vitro culture conditions. These plantlets were then evaluated for their ability of artemisinin accumulation. The results revealed again wide variation in artemisinin content, ranging from 0.02 to 0.68% dry weight (**Figure 8**). Among these plantlets, 6 individuals had an artemisinin content greater than 0.5% dry weight, 39 plantlets had an artemisinin content in the range of 0.21–0.50%, and 45 plantlets had an artemisinin content below 0.02%. As control plantlets derived from shoot tips not exposed to gamma rays exhibited an average artemisinin content of 0.18 + 0.09%, treatment with the

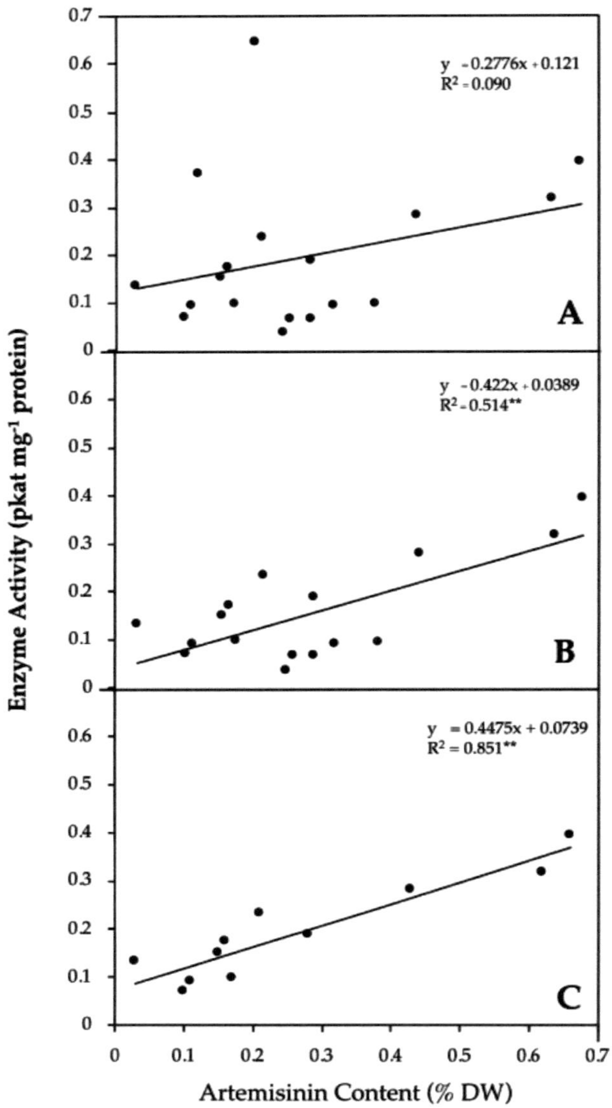

Figure 7.
Relationship between artemisinin content and ADS activity of various plantlets cultured in vitro. (A) The content-activity plot among 18 plantlet samples shows a poor value of the correlation coefficient (R) of 0.300, (B) the plot among 16 samples shows a moderate value of the correlation coefficient, R = 0.717 (P = 0.001), and (C) the plot among 11 samples shows a good value of the correlation coefficient, R = 0.922 (P = 0.001).

5-Gy dose resulted in an artemisinin content above that of the control for approximately half of the irradiated plantlets (**Figure 8**).

2.5 Ex vitro acclimatization of *A. annua* irradiated plantlets

In this study, actively growing shoot tips (length, ca. 5 mm) were excised from in vitro plantlets and stripped of their leaves. The resulting shoot tips (with 130 tips) were then inserted vertically 2 mm in depth into the MS medium containing 3% sucrose and 0.8% agar. Induction of variation in *A. annua*

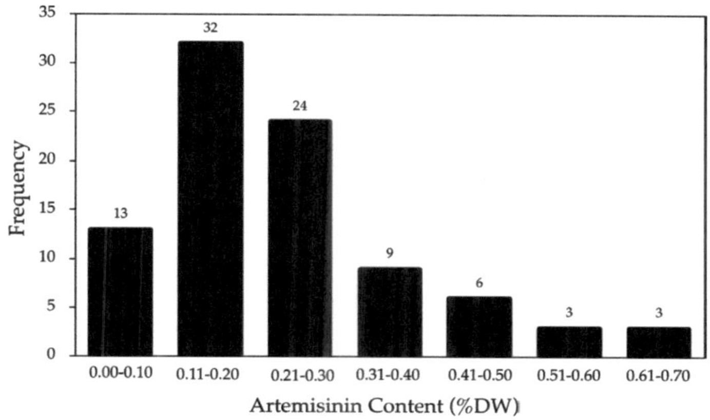

Figure 8.
Variations in artemisinin content in A. annua plantlets derived from shoot tips irradiated with 5 Gy of gamma rays.

plantlets using gamma irradiation (5 Gy) was then performed as described above (Section 2.1) and was transferred to the hormone-free MS medium for plantlet development. From each developed plantlet, selected shoots with two buds were excised, so that the upper bud would be regenerated to from shoots, and the lower bud regenerated to form roots. After four subculture passages, the leaves from healthy plantlets were harvested for artemisinin determination, and the shoots were used for a second round of plant regeneration. After 6 weeks, the in vitro plantlets were cleaned to remove the agar medium and were transferred into acclimatization conditions in 230-ml glass bottles. Each bottle contained 50 ml vermiculite and 50 ml MS medium without sucrose. The transferred plantlets were maintained at 25°C, with a photoperiod of 16 h of light at 40 μmol/m^2 s photosynthetic photon flux density provided by cool-white fluorescence lamps. After 8 weeks, a piece of membrane filter with a pore size of 0.5 μm was placed over a hole (area, 0.8 cm^2) in each plastic cap to increase air exchange in the culture vessel [23]. When the shoot of each plantlet grew sufficiently to fill the container, the plastic cap was loosened for one week and removed a week later to allow the shoot to continue growing out of the top of the bottle. The surviving plants (60 out of 90 plants) were then transferred from the vermiculite-based bottles to 10-cm pots containing sterile soil to grow for an additional 15 days under the same temperature and light conditions. Among the resulting 40 plants survived from the process of acclimatization, a group of 20 plants was transferred to a greenhouse at Chulalongkorn University, Bangkok, Thailand, and another group of 20 plants was transferred to an open field (30–38°C day air temperature and 20–29°C night air temperature) in Kanchanaburi Province, Thailand. These ex vitro plants were grown for a period of 6 months to reach maturity. At this point, leaves from the surviving plants (13 from the greenhouse group and 10 from the open field group) were harvested by random cuts of three different branches from each plant and prepared for determination of artemisinin content.

The development from the step of in vitro irradiated plantlets to ex vitro *A. annua* plants during the acclimatization process is shown in **Figure 9.** In general, both plant height and number of leaves increased continuously for plants grown in agar-MS (**Figure 9a**) and vermiculite-MS media (**Figure 9b** and **c**). In pots containing sterile soil, the plant height increased rapidly over 4 weeks, with the development

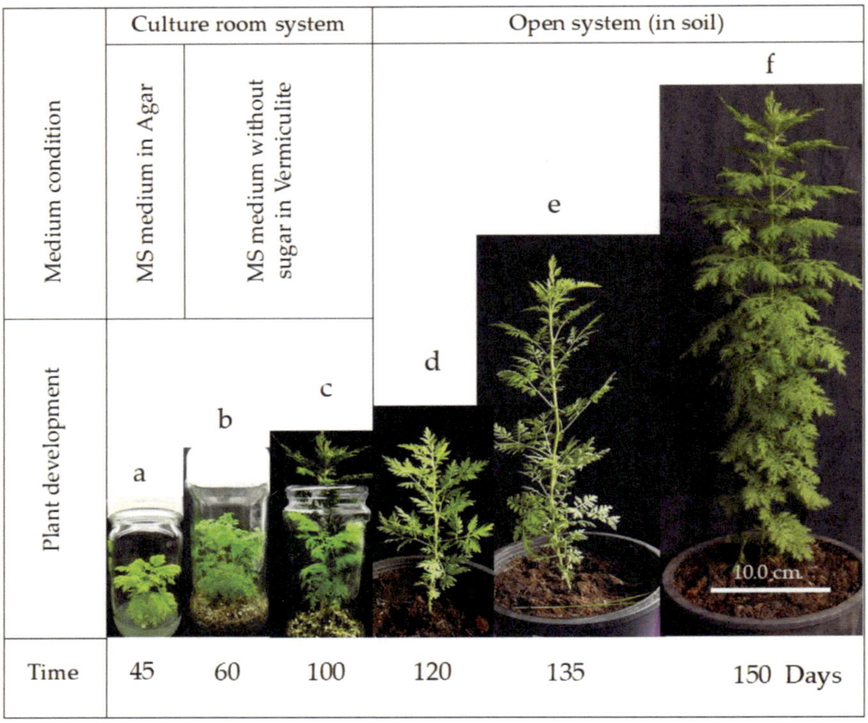

Figure 9.
Development of A. annua plants from in vitro plantlets to ex vitro plants. Six-week-old plantlets (a) were transferred to acclimatization conditions in 16-oz glass bottles containing 50 ml vermiculite and 50 ml MS medium without sucrose. After 8 weeks, (b) a piece of membrane filter (pore size: 0.5 μm) was placed over a hole (0.8 cm^2) in each plastic cap to increase air exchange within the culture vessel. When the shoot of each plantlet completely filled the bottle, the plastic cap was loosened for one week and removed a week later to allow the shoot to grow above the top of the bottle (c). The surviving plants were then transferred from the vermiculite-based bottles to four-inch pots containing sterile soil for an additional 15 days of growth (d) prior to transfer to the greenhouse or an open field (135 days for (e) and 150 days for (f)).

of leaves similar to that of normal mature *A. annua* plants (**Figure 9d–f**). With respect to percent survival, the acclimatization protocol resulted in survival of 60 of the total 90 plantlets, corresponding to a survival rate of 66.7%.

2.6 One-to-one correlation of artemisinin content between in vitro plantlets and ex vitro plants

Quantitative analysis revealed that the 13 surviving plants from the in vitro-ex vitro transfer to the greenhouse had their artemisinin content ranging from 0.12 to 0.42% dry weight. This range was slightly narrower than that of the in vitro plantlets, which exhibited artemisinin contents ranging from 0.06 to 0.66%. Interestingly, comparison of in vitro plantlets and ex vitro plants grown in the greenhouse on a one-to-one basis revealed that there was a significant individual correlation between the artemisinin content of paired in vitro plantlets and ex vitro plants grown in the greenhouse, with a very good correlation coefficient (R^2) value of 0.915 (**Figure 10a**). For plants transferred to the open field, the 10 in vitro-ex vitro pairs exhibited artemisinin contents for in vitro plantlets ranging from 0.25 to 0.69% and for ex vitro plants ranging from 0.31 to 0.84% (**Figure 10b**). A one-to-one comparison of the artemisinin content of plants grown in the open field with their corresponding plantlets also revealed a good R^2 value of 0.797 (**Figure 10b**).

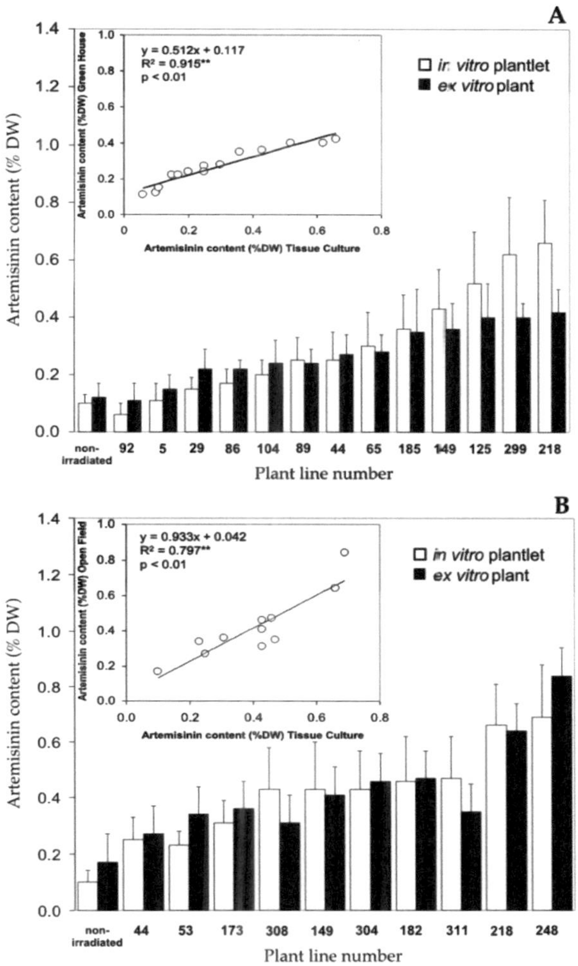

Figure 10.
One-to-one correlation of artemisinin content between in vitro gamma-irradiated plantlets and the ex vitro plants grown in a greenhouse showing a very good correlation coefficient value (R^2) of 0.915 (A, inset) and in an open field showing a relatively good R^2 value of 0.797 (B, inset).

3. Discussion

A. annua tissues appear to be sensitive to gamma rays, as a low-dose range (1–10 Gy) of radiation can generate viable plant variants. Within this range, variations in both plant growth and artemisinin content were observed in the resulting population of in vitro plantlets. Doses greater than or equal to 8 Gy, previously reported as the 50% lethal dose (LD_{50}) [20], clearly caused growth inhibition and leaf abnormalities in the plantlets, while the lower dose of 3 Gy resulted in growth promotion. The 5-Gy dose, on the other hand, has no significant effect on the plantlet morphology. Previous reports have indicated that low doses of gamma irradiation affect seed germination and seeding growth [24–27]. In red pepper (*Capsicum annuum*), three irradiated groups exposed to 2, 4, and 8 Gy exhibited enhanced seedling development [24]. In Arabidopsis, ultrastructural changes in cellular organelles have been observed after gamma irradiation, and seedlings treated with 0–5 Gy developed normally, while vertical growth in plants exposed to 50 Gy was significantly inhibited [27].

In terms of formation of secondary products, little information is available regarding the use of gamma irradiation for yield improvement. To our knowledge, the only related report characterized the effect of low-dose gamma irradiation (2–16 Gy) on the increased production of shikonin derivatives in callus cultures of *Lithospermum erythrorhizon* [28]. In agreement with these results, we also found a significant effect on artemisinin content caused by a similar low-dose range of gamma irradiation. However, in the present study, we observed these effects in mature plants, rather than in disorganized tissues. Our results show that treatment with either 8 Gy [19] or 5 Gy can create a population of plantlets with a high range of artemisinin contents. Due to the minimal associated morphological effects, the 5-Gy dose was chosen to produce *A. annua* variants that were presumably affected primarily at the biochemical and physiological levels.

The specific genes affected by the low-dose gamma radiation were observed to be at least on the gene of ADS of the biosynthetic pathway of artemisinin in *A. annua* plantlets. More than half of the variant population appeared to have a high correlation coefficient value (R = 0.922) between artemisinin content and ADS enzyme activity. The reason why ADS gene is particularly sensitive to the irradiation is still not clear. However, it might be that the low doses of 5–8 Gy of the gamma irradiation are just mild enough to affect this ADS gene. In principle, some lesions of ADS gene caused by the irradiation are likely to be repaired through the action of intracellular DNA repair process, while ADS gene of some other samples might remain unpaired or misrepaired, giving rise to permanent changes in the affected ADS gene. This would lead to a cellular response including a wide range of the enzymatic systems, as observed in this case with the variable ADS enzyme activities in *A. annua*. In the literature, there has been a report supporting our results. That is the case of low-dose irradiated callus cultures of *Lithospermum erythrorhizon* in which the enzyme activity of p-hydroxybenzoic acid geranyltransferase involving in the shikonin biosynthesis is boosted after the gamma irradiation [28]. Thus, it was suggested that the creation of plant variation through gamma irradiation has significant effect on ADS gene which is likely to be related to the enhancement of the artemisinin content in *A. annua*.

For the ex vitro acclimatization of the plants, we have previously characterized the conditions and supporting material important for photoautotrophic growth of *Eucalyptus camaldulensis* plantlets, both in vitro and ex vitro [23]. Adoption of this protocol resulted in a survival rate of 67% for *A. annua* plants after the in vitro-ex vitro transfer and a survival rate of 38% after 6 months of ex vitro growth to obtain mature plants. Clearly, the stresses generated in weakened irradiated plants during the process of acclimation lead to significant mortality. However, whether the mortality is more prevalent among in vitro individuals with low or high artemisinin content remains unclear.

For the 23 surviving mature plants, we observed an individual correlation in artemisinin content between the in vitro plantlets and the ex vitro mature plants. This one-to-one correlation was strongly positive for plants grown in the greenhouse, with $R^2 = 0.915$, and relatively positive for field-grown plants, with $R^2 = 0.797$. These results suggest that the capability for artemisinin biosynthesis in each in vitro plantlet is maintained throughout the in vitro-ex vitro transfer and the subsequent development into a mature plant. With respect to the greenhouse plants, although the correlation coefficient value was quite high, the high-yield plants did exhibit a reduction in artemisinin content. This observation is likely due to the high biomass weight per leaf for high-content leaves, which clearly appear thicker than low-content leaves found in greenhouse conditions.

The differences in biomass associated with artemisinin content are not so obvious among the established in vitro plantlets, resulting in a decrease in the

DOI: http://dx.doi.org/10.5772/intechopen.82385

degree of the one-to-one correlation observed strictly for high-yield plants. For the field-grown plants, the relatively positive value of the correlation coefficient (R^2 = 0.797) may be due to two outliers present among the 10 samples (nos. 308 and 311) that deviate from the rest of the population. A pairwise comparison of the remainder of the samples would result in a higher R^2 value, which is reflective of a good correlation in artemisinin content between the in vitro plantlets and ex vitro field-grown plants. In addition, the lower correlation could also be attributed to the less controlled conditions of the open field compared with those of the greenhouse.

To be certain that changes in artemisinin biosynthesis in the ex vitro plants are genetically stable, it is necessary to test the next generation. However, many of the established mature plants could not produce seeds. Therefore, stability tests assessing the next generation through seed germination are not possible. Alternatively, this analysis can be performed through a second round of in vitro-ex vitro transfer. No attempt was made to use this method in the present study due to the high mortality associated with this long process that would have resulted in an insufficient number of pairs of plants for a one-to-one analysis.

4. Conclusion

Based on these results, we conclude that the technique of gamma irradiation can produce viable variants of *A. annua* that are capable of maintaining changes in gene expression associated with the artemisinin biosynthetic pathway (such as ADS) throughout the in vitro-ex vitro transfer process and, at minimum, through the first generation of mature plant development. Relatively low doses of gamma irradiation (ca. 3–8 Gy) can be effective for yield enhancement of artemisinin in *A. annua*. A mechanistic understanding of the increased biosynthesis of artemisinin in response to gamma irradiation is important for the development of a production-scale operation.

Acknowledgements

This work was supported by Chulalongkorn University's Ratchadaphiseksomphot Endowment Fund (to Natural Biotechnology Research Unit) and Thailand National Center for Genetic Engineering and Biotechnology, National Science and Technology Development Agency (NASDA).

Conflict of interest statement

The authors declare that the research was conducted in the absence of any commercial or financial relationships that could be construed as a potential conflict of interest.

Author details

Thongchai Koobkokkruad[1,3], Praderm Wanichananan[2], Chalermpol Kirdmanee[2] and Wanchai De-Eknamkul[1,4]*

1 Biomedicinal Chemistry Program, Department of Biochemistry, Faculty of Pharmaceutical Sciences, Chulalongkorn University, Bangkok, Thailand

2 Plant Physio-Biochemistry Laboratory, National Center for Genetic Engineering and Biotechnology, National Science and Technology Development Agency, Pathum Thani, Thailand

3 Nano-Cosmeceutical Laboratory, National Nanotechnology Center, National Science and Technology Development Agency, Thailand Science Park, Pathum Thani, Thailand

4 Natural Product Biotechnology Research Unit, Department of Pharmacognosy and Pharmaceutical Botany, Faculty of Pharmaceutical Sciences, Chulalongkorn University, Bangkok, Thailand

*Address all correspondence to: dwanchai@chula.ac.th

References

[1] Liu JM et al. Structure and reaction of qinghaosu [in Chinese]. Acta Chimica Sinica. 1979;**37**:129-143

[2] Duke MV, Paul RN, Elsohly HN, Sturtz G, Duke SO. Localization of artemisinin and artemisitene in foliar tissues of glanded and glandless biotypes of *Artemisia annua* L. International Journal of Plant Sciences. 1994;**155**:365-372

[3] Wallaart TE, Pras N, Quax WJ. Seasonal variations of artemisinin and its biosynthetic precursors in tetraploid *Artemisia annua* plants compared with the wild-type. Planta Medica. 1999;**65**:723-728

[4] Ferreira JFS, Simon JE, Janick J. Developmental studies of *Artemisia annua*: Flowering and artemisinin production under greenhouse and field conditions. Planta Medica. 1995;**61**:167-170

[5] Kjær A, Grevsen K, Jensen M. Effect of external stress on density and size of glandular trichomes in full-grown *Artemisia annua*, the source of anti-malarial artemisinin. AoB Plants. 2012;**2012**:pls018. DOI: 10.1093/aobpla/pls018

[6] Ikram NKBK, Simonsen HT. A review of biotechnological artemisinin production in plants. Frontiers in Plant Science. 2017;**8**:1966. DOI: 10.3389/fpls.2017.01966

[7] Schimid G, Hofheinz W. Total synthesis of qinghaosu. Journal of the American Chemical Society. 1983;**105**:624-625

[8] Avery MA, Chong WKM, Jennings WC. Stereoselective total synthesis of (+)-artemisinin, the antimalarial constituent of *Artemisia annua* L. Journal of the American Chemical Society. 1992;**114**:974-979

[9] Paniego NB, Giulietti AM. Artemisia annua L.: Dedifferentiated and differentiated cultures. Plant Cell, Tissue and Organ Culture. 1994;**36**:163-168

[10] Chun-Zhao K, Chen G, Yu-Chun W. Fan: Comparison of various bioreactors on growth and artemisinin biosynthesis of *Artemisia annua* L. shoot cultures. Process Biochemistry. 2003;**39**:45-49

[11] Ferreira JFS, Janick J. Roots as an enhancing factor for the production of artemisinin in shoot cultures of *Artemisia annua*. Plant Cell, Tissue and Organ Culture. 1996;**44**:211-217

[12] Putalun W, Luealon W, De-Eknamkul W, Tanaka H, Shoyama Y. Improvement of artemisinin production by chitosan in hairy root cultures of *Artemisia annua* L. Biotechnology Letters. 2007;**29**:1143-1146

[13] Weathers PJ, DeJesus-Gonzales L, Kim YJ, Souret FF, Towler F. Alteration of biomass and artemisinin production in *Artemisia annua* hiary roots by media sterilization method and sugars. Plant Cell Reports. 2004;**23**:414-418

[14] Smith TC, Weathers PJ, Cheetham RD. Effects of gibberellic acid on hairy root cultures of *Artemisia annua*: Growth and artemisinin production. In Vitro Cellular & Developmental Biology: Plant. 1997;**33**:75-79

[15] Jaziri M, Shimomura K, Yoshimatsu K, Fauconnier M. Establishment of normal and transformed root cultures of *Artemisia annua* L. for artemisinin production. Journal of Plant Physiology. 1995;**145**:175-177

[16] Al-Safadi B, Ayyoubi Z, Jawdat D. The effect of gamma irradiation on potato microtuber production in vitro.

Plant Cell, Tissue and Organ Culture. 2000;**61**:183-187

[17] Sen CB. Enhancement of regeneration potential and variability by gamma-irradiation in cultured cells of Scilla indica. Biologia Plantarum. 2001;**44**:189-193

[18] Dong SK, In SL, Cheol SJ, Sang JL, Hi SS, Young IL, et al. AEC resistant rice mutants induced by gamma-ray irradiation may include both elevated lysine production and increased activity of stress related enzymes. Plant Science. 2004;**167**:305-316

[19] Vardhan PV, Shukla LI. Gamma irradiation of medicinally important plants and the enhancement of secondary metabolite production. International Journal of Radiation Biology. 2017;**93**:967-979

[20] Koobkokkruad T, Chochai A, Kerdmanee C, De-Eknamkul W. Effects of low-dose gamma irradiation on artemisinin content and amorpha-4,11-diene synthase activity in *Artemisia annua* L. International Journal of Radiation Biology. 2008;**84**:878-884

[21] Koobkokkruad T, Chochai A, Kerdmanee C, De-Eknamkul W. TLC-densitometric analysis of artemisinin for rapid screening of high producing plantlets of Artemisia annua. Phytochemical Analysis. 2007;**18**:229-234

[22] Bouwmeester HJ, Wallaart TE, Janssen MHA, Loo BV, Jansen BJM, Posthumus MA, et al. Amorpha-4,11-diene synthase catalyses the first probable step in artemisinin biosynthesis. Phytochemistry. 1999;**52**:843-854

[23] Kirdmanee C, Kitiya Y, Kozai T. Effects of CO_2 enrichment and supporting material in-vitro on photoautotrophic growth of eucalyptus plantlets in-vitro and ex-vitro. In Vitro Cellular & Developmental Biology: Plant. 1995;**31**:144-149

[24] Kim DS, Lee IS, Jang CS, Lee SJ, Song HS, Lee YI, et al. AEC resistant rice mutants induced by gamma-ray irradiation may include both elevated lysine production and increased activity of stress related enzymes. Plant Science. 2004;**167**:305-316

[25] Kim JH, Baek MH, Chung BY, Wi SG, Kim JS. Alterations in the photosynthetic pigments and antioxidant machineries of red pepper (*Capsicum annuum* L.) seedlings from gamma-irradiated seeds. Journal of Plant Biology. 2004;**47**:314-321

[26] Ková E, Keresztes Á. Effect of gamma and UV-B/C radiation on plant cells. Micron. 2002;**33**:199-210

[27] Wi SG, Chung BY, Kim JH, Baek MH, Yang DH, Lee JW, et al. Ultrastructural changes of cell organelles in Arabidopsis stems after gamma irradiation. Journal of Plant Biology. 2005;**48**:195-200

[28] Chung BY, Lee Y-B, Baek M-H, Kim J-H, Wi SG, Kim J-S. Effects of low-dose gamma-irradiation on production of shikonin derivatives in callus cultures of *Lithospermum erythrorhizon* S. Radiation Physics and Chemistry. 2006;**75**:1018-1023